METHODS IN MOLECULAR BIOLOGY

Series Editor
John M. Walker
School of Life and Medical Sciences
University of Hertfordshire
Hatfield, Hertfordshire, UK

For further volumes:
http://www.springer.com/series/7651

For over 35 years, biological scientists have come to rely on the research protocols and methodologies in the critically acclaimed *Methods in Molecular Biology* series. The series was the first to introduce the step-by-step protocols approach that has become the standard in all biomedical protocol publishing. Each protocol is provided in readily-reproducible step-by-step fashion, opening with an introductory overview, a list of the materials and reagents needed to complete the experiment, and followed by a detailed procedure that is supported with a helpful notes section offering tips and tricks of the trade as well as troubleshooting advice. These hallmark features were introduced by series editor Dr. John Walker and constitute the key ingredient in each and every volume of the *Methods in Molecular Biology* series. Tested and trusted, comprehensive and reliable, all protocols from the series are indexed in PubMed.

Fe-S Proteins

Methods and Protocols

Edited by

Patricia C. Dos Santos

Department of Chemistry, Wake Forest University, Winston Salem, NC, USA

Humana Press

Editor
Patricia C. Dos Santos
Department of Chemistry
Wake Forest University
Winston Salem, NC, USA

ISSN 1064-3745 ISSN 1940-6029 (electronic)
Methods in Molecular Biology
ISBN 978-1-0716-1604-8 ISBN 978-1-0716-1605-5 (eBook)
https://doi.org/10.1007/978-1-0716-1605-5

This Humana imprint is published by the registered company Springer Science+Business Media, LLC part of Springer Nature.
The registered company address is: 1 New York Plaza, New York, NY 10004, U.S.A.

Preface

Proteins containing inorganic complexes of Fe and S (Fe-S proteins) participate in a wide range of biochemical processes. Fe-S proteins are found in all domains of life and are critical for life on Earth. The reactions enabled by these metalloproteins are mechanistically diverse, ranging from simple one-electron transfer to thermodynamically unfavorable reactions like dinitrogen reduction and more complex organic rearrangements. Functional assignment of Fe-S proteins is challenging given the oxygen sensitivity of their inorganic counterparts, the requirement for specific assembly pathways for the synthesis of Fe-S clusters, and the involvement of metallochaperones in the insertion of the metal cofactor. This volume provides a description of current methodologies used to investigate the formation, insertion, and function of metalloclusters associated with proteins.

This project was only possible with the valuable contributions of experts in this field of study. Collectively, they provide a description of current and established methods covering various aspects of Fe-S cluster metabolism. Chapters 1–6 describe genetic manipulations to study genes involved in Fe-S cluster synthesis and function and strategies to optimize expression and isolation of Fe-S proteins. Chapters 7–11 report on biochemical approaches to probe steps involved in the synthesis of Fe-S clusters. Lastly, Chapters 12–17 explore biophysical methods to study metalloclusters and reactions promoted by metalloenzymes. Following the format used for the *Methods in Molecular Biology*™ series, each method's chapter includes an introduction to the topic and technique, lists of materials and reagents, a description of established laboratory protocols, and troubleshooting tips. These methods aim to benefit readers with a new interest in Fe-S proteins as well as those who seek to expand the range of approaches used to study metalloenzymes.

I am incredibly thankful to the authors who shared their expertise. Together, they provided a complete description of established methodologies used by their groups that continue to expand the frontiers of an exciting field study of Fe-S proteins.

Winston Salem, NC, USA *Patricia C. Dos Santos*

Contents

Contributors

MAAME A. ADDO • *Department of Chemistry, Wake Forest University, Winston-Salem, NC, USA*

SALVATORE ADINOLFI • *Department of Pharmaceutical Technology, University of Turin, Turin, Italy*

EDWIN ANTONY • *Department of Biological Sciences, Marquette University, Milwaukee, WI, USA; Department of Biochemistry and Molecular Biology, Saint Louis University School of Medicine, St. Louis, MO, USA*

ALBERTO BARBIROLI • *Section of Chemistry and Biomolecular Sciences, DeFENS, University of Milan, Milan, Italy*

AMIE K. BOAL • *Department of Biochemistry and Molecular Biology, The Pennsylvania State University, University Park, PA, USA; Department of Chemistry, The Pennsylvania State University, University Park, PA, USA*

FRANCESCO BONOMI • *Section of Chemistry and Biomolecular Sciences, DeFENS, University of Milan, Milan, Italy*

SQUIRE J. BOOKER • *Department of Biochemistry and Molecular Biology, The Pennsylvania State University, University Park, PA, USA; Department of Chemistry, The Pennsylvania State University, University Park, PA, USA; The Howard Hughes Medical Institute, The Pennsylvania State University, University Park, PA, USA*

JEFFREY M. BOYD • *Department of Biochemistry and Microbiology, Rutgers, the State University of New Jersey, New Brunswick, NJ, USA*

VALERIE J. CARABETTA • *Department of Biomedical Sciences, Cooper Medical School of Rowan University, Camden, NJ, USA*

TERRELL D. CARTER • *Department of Chemistry and Biochemistry, University of South Carolina, Columbia, SC, USA*

ELLIOT I. CORLESS • *Department of Biological Sciences, Marquette University, Milwaukee, WI, USA*

JASON C. CRACK • *Centre for Molecular and Structural Biochemistry, School of Chemistry, University of East Anglia, Norwich, UK*

BENOIT D'AUTRÉAUX • *Université Paris-Saclay, CEA, CNRS, Institute for Integrative Biology of the Cell (I2BC), Gif-sur-Yvette, France*

KINJAL DAVE • *Department of Chemistry, Boston University, Boston, MA, USA*

PATRICIA C. DOS SANTOS • *Department of Chemistry, Wake Forest University, Winston-Salem, NC, USA*

IAN J. DRAKE • *Corporate Strategic Research, ExxonMobil Research & Engineering Company, Annandale, NJ, USA*

YOHANN DUVERGER • *Laboratoire de Chimie Bactérienne, Institut de Microbiologie de la Méditerranée, CNRS UMR 7283, Aix-Marseille University, Marseille, France*

ASHLEY M. EDWARDS • *Department of Chemistry, Wake Forest University, Winston-Salem, NC, USA*

OLIVER EINSLE • *Institut für Biochemie, Albert-Ludwigs-Universität Freiburg, Freiburg, Germany*

KARLA ESQUILIN-LEBRON • *Department of Biochemistry and Microbiology, Rutgers, the State University of New Jersey, New Brunswick, NJ, USA*

Sylvain Gervason • *Université Paris-Saclay, CEA, CNRS, Institute for Integrative Biology of the Cell (I2BC), Gif-sur-Yvette, France*
Jakob Gies-Elterlein • *Institut für Biochemie, Albert-Ludwigs-Universität Freiburg, Freiburg, Germany*
Stefania Iametti • *Section of Chemistry and Biomolecular Sciences, DeFENS, University of Milan, Milan, Italy*
Vivian Robert Jeyachandran • *Department of Biochemistry and Molecular Biology, The Pennsylvania State University, University Park, PA, USA*
Michelle Langton • *Department of Biochemistry, Brandeis University, Waltham, MA, USA*
Nick E. Le Brun • *Centre for Molecular and Structural Biochemistry, School of Chemistry, University of East Anglia, Norwich, UK*
Yuchen Liu • *Corporate Strategic Research, ExxonMobil Research & Engineering Company, Annandale, NJ, USA*
Mauro Marengo • *Department of Pharmaceutical Technology, University of Turin, Turin, Italy*
Lydie Martin • *Univ. Grenoble Alpes, CEA, CNRS, IBS, Metalloproteins Unit, Grenoble, France*
Erin L. McCarthy • *Department of Biochemistry and Molecular Biology, The Pennsylvania State University, University Park, PA, USA*
Christa N. Molé • *Department of Chemistry, Boston University, Boston, MA, USA*
Yvain Nicolet • *Univ. Grenoble Alpes, CEA, CNRS, IBS, Metalloproteins Unit, Grenoble, France*
Simonetta Oliaro-Bosso • *Department of Pharmaceutical Technology, University of Turin, Turin, Italy*
F. Wayne Outten • *Department of Chemistry and Biochemistry, University of South Carolina, Columbia, SC, USA*
Maria-Eirini Pandelia • *Department of Biochemistry, Brandeis University, Waltham, MA, USA*
Katharina Parison • *Institut für Biochemie, Albert-Ludwigs-Universität Freiburg, Freiburg, Germany*
Annalisa Pastore • *UK Dementia Research Institute at the Maurice Wohl Institute of King's College London, London, UK*
Jay V. Pendyala • *Department of Biochemistry and Molecular Biology, The Pennsylvania State University, University Park, PA, USA*
Deborah L. Perlstein • *Department of Chemistry, Boston University, Boston, MA, USA*
Rita Puglisi • *UK Dementia Research Institute at the Maurice Wohl Institute of King's College London, London, UK*
Béatrice Py • *Laboratoire de Chimie Bactérienne, Institut de Microbiologie de la Méditerranée, CNRS UMR 7283, Aix-Marseille University, Marseille, France*
Christina A. Roberts • *Astrix Technology Group, Red Bank, NJ, USA*
Batoul Srour • *Université Paris-Saclay, CEA, CNRS, Institute for Integrative Biology of the Cell (I2BC), Gif-sur-Yvette, France*
Troy A. Stich • *Department of Chemistry, Wake Forest University, Winston-Salem, NC, USA*
Christian Trncik • *Institut für Biochemie, Albert-Ludwigs-Universität Freiburg, Freiburg, Germany*
Chie Ueda • *Department of Biochemistry, Brandeis University, Waltham, MA, USA*

XAVIER VERNÈDE • *Univ. Grenoble Alpes, CEA, CNRS, IBS, Metalloproteins Unit, Grenoble, France*
EHUD ZELZION • *Department of Biochemistry and Microbiology, Rutgers, the State University of New Jersey, New Brunswick, NJ, USA*
CUIPING ZHAO • *Department of Biochemistry and Microbiology, Rutgers University, New Brunswick, NJ, USA*

Part I

Methods for Genetic Manipulation and Fe-S Protein Expression

Chapter 1

Molecular Biology and Genetic Tools to Investigate Functional Redundancy Among Fe-S Cluster Carriers in *E. coli*

Yohann Duverger and Béatrice Py

Abstract

Iron-sulfur (Fe-S) clusters are among the oldest protein cofactors, and Fe-S cluster-based chemistry has shaped the cellular metabolism of all living organisms. Over the last 30 years, thanks to molecular biology and genetic approaches, numerous actors for Fe-S cluster assembly and delivery to apotargets have been uncovered. In prokaryotes, *Escherichia coli* is the best-studied for its convenience of growth and its genetic amenability. During evolution, redundant ways to secure the supply of Fe-S clusters to the client proteins have emerged in *E. coli*. Disrupting gene expression is essential for gene function exploration, but redundancy can blur the interpretations as it can mask the role of important biogenesis components. This chapter describes molecular biology and genetic strategies that have permitted to reveal the *E. coli* Fe-S cluster conveying component network, composition, organization, and plasticity. In this chapter, we will describe the following genetic methods to investigate the importance of *E. coli* Fe-S cluster carriers: one-step inactivation of chromosomal genes in *E. coli* using polymerase chain reaction (PCR) products, P1 transduction, arabinose-inducible expression system, mevalonate (MVA) genetic by-pass, sensitivity tests to oxidative stress and iron starvation, β-galactosidase assay, gentamicin survival test, and Hot Fusion cloning method.

Key words Iron-sulfur clusters, A-type carrier, ISC, SUF, Fe-S cluster biogenesis, One-step inactivation of chromosomal genes in *E. coli* using PCR products, Electrotransformation, P1 transduction, Arabinose-inducible expression system, MVA genetic by-pass, Sensitivity tests to oxidative stress and iron starvation, β-Galactosidase assay, Gentamicin survival test, Hot Fusion cloning method

1 Introduction

Iron-sulfur (Fe-S) clusters are thought to have appeared very early in evolution and are still used by most living organisms. Thanks to their chemical reactivity, proteins that host Fe-S clusters are able to perform a wide range of chemical reactions that are implicated in fundamental processes ranging from energy production, central metabolism to DNA replication [1]. Therefore, the maturation of Fe-S proteins is an essential process for eukaryotic and prokaryotic

Patricia C. Dos Santos (ed.), *Fe-S Proteins: Methods and Protocols*, Methods in Molecular Biology, vol. 2353,
https://doi.org/10.1007/978-1-0716-1605-5_1,

organisms [2–5]. To achieve this task, cells employ multiple Fe-S cluster biogenesis components. Two types of biogenesis machineries, ISC (iron-sulfur cluster) and SUF (sulfur utilization factor), are distinguished and are redundant in prokaryotic organisms that possess both machineries, such as *E. coli* [6–11]. In this organism, ISC is the housekeeping machinery, while the SUF is used when the available iron becomes scarce, and when oxidative stress is encountered [8–10]. Eukaryotic photosynthetic organisms also have ISC and SUF machineries, but in this case, physical compartmentation prevents redundancy [12]. Hence, the ISC machinery resides in mitochondria, while the SUF machinery is located in chloroplasts [12].

The maturation of Fe-S proteins by the ISC and SUF machineries can be viewed as a process involving two steps, Fe-S cluster assembly and Fe-S cluster delivery (Fig. 1). The differences between the ISC and SUF machineries of *E. coli* reside at the

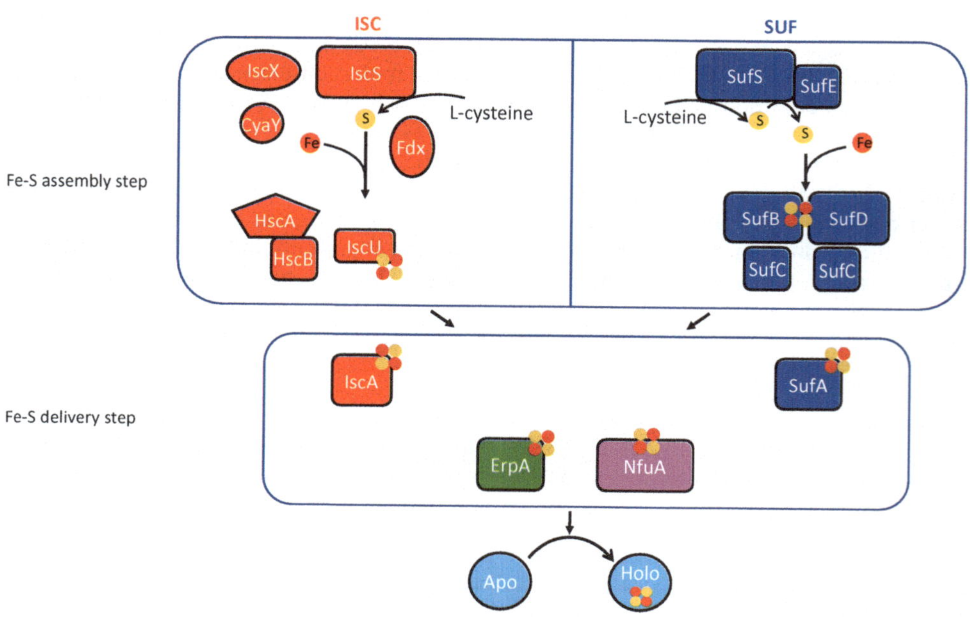

Fig. 1 The *E. coli* ISC and SUF Fe-S cluster biogenesis machineries. Schematic representation of Fe-S cluster biogenesis by the ISC (red components) and the SUF (dark blue components) machineries of *E. coli*. The two main steps, that is, assembly and delivery of the Fe-S clusters, are indicated. The Fe-S clusters are assembled on scaffold components (IscU, $SufBC_2D$ complex) that receive sulfur from cysteine desulfurases (IscS, SufSE) and iron from an as-yet nonidentified source. The iron (red dot) and sulfur (orange dot) elements are represented in free (Fe^{2+}) or in persulfide (S^{2-}) form, and when part of Fe-S clusters. The ferredoxin (Fdx) is proposed to intervene at the sulfur transfer step from IscS to IscU. The CyaY and IscX components, whose role is still debated, intervene at the Fe-S assembly step by interacting with IscS. The preformed Fe-S cluster is transferred to a carrier protein, which delivers it to the final apotarget (turquoise). The ErpA and NfuA carriers, whose encoding genes are not located in the *iscRSUA-hscBA-fdx-iscX* or in the *sufABCDSE* operons, are represented in green and pink, respectively. The HscA/HscB chaperone/cochaperone duo participates in the Fe-S cluster release from IscU

assembly step. In the ISC machinery, the component onto which the iron and sulfur elements are assembled, referred to as a scaffold, is a single protein, IscU, while in the SUF machinery it is a protein complex involving the SufB, SufC, and SufD proteins [13, 14]. In the two systems, sulfur is mobilized from L-cysteine by a cysteine desulfurase, IscS and SufS, and each of these systems requires additional components at this step [15, 16]. In the SUF machinery, an additional sulfur acceptor (SufE) is necessary for sulfur donation to the scaffold, while in the ISC machinery a ferredoxin (Fdx) is required [14]. In *E. coli,* it was shown that the ISC cysteine desulfurase cannot replace the SUF cysteine desulfurase, and vice versa [7, 16]. Thus, despite that the ISC and SUF machineries are redundant, the components of the assembly step are not exchangeable.

In contrast, components of the delivery step have shown some functional redundancy in *E. coli*, especially the A-type carriers (ATCs), IscA, SufA, and ErpA [7, 17]. ATCs are small proteins that bind Fe-S clusters *via* three conserved cysteine residues, and in vitro, the ATC-bound cluster can be transferred directly to a series of 2Fe-2S and 4Fe-4S client proteins [18–22]. Genetic analyses have revealed that, under aerobiosis, *E. coli* IscA and SufA are functionally redundant carriers, as both are potentially able to receive an Fe-S cluster from IscU and $SufBC_2D$ scaffolds and transfer it to ErpA [17, 18]. In contrast, under anaerobiosis, redundancy occurs between ErpA and IscA, which are both potentially able to receive Fe-S clusters from IscU, and transfer them to apotargets [17]. Interestingly, an atypical ATC, NfuA, has been described in *E. coli*, *A. vinelandii,* and *Pseudomonas aeruginosa* [23–25]. NfuA possesses an N-terminal domain with homologies to ATC, but surprisingly without exhibiting the three conserved cysteine residues. The C-terminal domain of NfuA shares identity with the Nfu domain of prokaryotic and eukaryotic proteins that are involved in Fe-S protein maturation [26–30]. The Nfu domain exhibits two conserved cysteine residues coordinating a [4Fe-4S] cluster ligation, very likely at a homodimer interface. In *E. coli*, both genetic and biochemical approaches have revealed that NfuA is not redundant with the ATC, but rather was proposed to act with ErpA to conjointly establish a "stress-resistant" Fe-S cluster delivery route [23, 31].

Our knowledge of the Fe-S cluster biology has considerably improved, thanks to the collaborative multidisciplinary approaches between chemistry, enzymology, structural biology, biophysics, and genetics [32, 33]. In this chapter, *E. coli* is used as an example to illustrate how key genetic approaches have contributed to investigate redundancy at the Fe-S cluster delivery step.

2 Materials

2.1 Common Materials

1. Lysogeny broth (LB): 5 g of yeast extract, 10 g of Bacto Tryptone, 10 g of NaCl, make up to 1 L with deionized water. Autoclave and store at room temperature as much as possible in the dark to avoid the formation of reactive oxygen species [34].
2. Lysogeny broth agar (LB agar): add 15 g of agar to 1 L of LB. Autoclave and store at room temperature.
3. Sterile water: autoclaved deionized water.
4. Phosphate-buffered saline (PBS) buffer: 137 mM NaCl, 2.7 mM KCl, 10 mM Na_2HPO_4, 1.76 mM KH_2PO_4, pH 7.4. Autoclave and store at room temperature.
5. 1 M Tris–HCl buffer, pH 7.5. Autoclave and store at room temperature.
6. Glycerol sterile solution: 80% glycerol solution prepared in deionized water. Autoclave and store at room temperature.
7. Magnesium chloride solution: 1 M magnesium chloride solution in deionized water. Autoclave and store at room temperature.
8. Dithiothreitol (DTT) solution: 1 M dithiothreitol in sterile deionized water. Store as 0.5-mL aliquots at −20 °C.
9. Ampicillin stock: 5 mg/mL of ampicillin in deionized water. Filter-sterilize and store as 15-mL aliquots at −20 °C.
10. Chloramphenicol stock: 2.5 mg/mL of chloramphenicol in 50% ethanol and store as 15-mL aliquots at −20 °C.
11. Kanamycin stock: 3 mg/mL of kanamycin in deionized water. Filter-sterilize and store as 15-mL aliquots at −20 °C.
12. Spectinomycin stock: 2.5 mg/mL of spectinomycin in deionized water. Filter-sterilize and store as 15-mL aliquots at −20 °C.
13. Tetracycline stock: 2.5 mg/mL of tetracycline in 50% ethanol, store as 15-mL aliquots at −20 °C.
14. Selective LB plates with the appropriate antibiotic (depending on the drug cassette and/or the plasmids used) usually at a final concentration of 50 μg/mL of ampicillin, 25 μg/mL of chloramphenicol, 30 μg/mL of kanamycin, 25 μg/mL of spectinomycin, or 25 μg/mL of tetracycline.
15. Sterile 1-L, 250-mL, 100-mL, 50-mL, and 25-mL Erlenmeyer flasks.
16. Sterile 50-mL, 15-mL, 2-mL, and 1.5-mL centrifuge tubes, 0.2-mL PCR tubes.
17. Sterile 96-well flat microplates.

18. Ultraviolet (UV)-visible spectrophotometer and cuvettes.
19. Centrifuge, tubes, and rotors, compatible with spinning volumes of 50 mL, 15 mL, and 2 mL, at approximately 10,000 × *g*.
20. Incubation shaker at an appropriate temperature (usually 37 °C).
21. Incubators at 30 °C, 37 °C, and 43 °C.
22. Thermocycler for PCRs.
23. KOD Hot Start DNA Polymerase and 10× PCR buffer.
24. Deoxynucleotide triphosphate (dNTP) solution: 10 mM deoxynucleotide triphosphate mix solution (deoxyadenosine triphosphate (dATP), deoxythymidine triphosphate (dTTP), deoxycytidine triphosphate (dCTP), deoxyguanosine triphosphate (dGTP), 10 mM each). Store as 100-μL aliquots at −20 °C.
25. Tris-acetate-EDTA (TAE) buffer: 40 mM Tris, 20 mM acetate, 1 mM ethylenediaminetetraacetic acid (EDTA), pH 8.4. Store at room temperature.
26. Agarose gel: 1% agarose in 0.5× TAE solution. Prepare freshly.
27. DNA ladder.
28. DNA loading buffer (×6): 30% glycerol, 0.25% bromophenol blue, and 0.25% xylene cyanol FF. Store as 1-mL aliquots at 4 °C.
29. Agarose gel electrophoresis system.
30. PCR Product Purification Kit and Gel Extraction Kit.
31. Plasmid DNA Miniprep Kit.
32. Sterile glass beads or L-shaped spreader.
33. Restriction enzymes and appropriate 10× buffers.
34. Electronic multichannel pipette (from 5μL to 200μL).
35. Water bath.
36. Heating block at 85 °C.

2.2 Design of In-frame Deletion Mutants

1. Gene knockout DNA oligonucleotide primers (10μM) to PCR-amplify flippase recognition target (FRT)-flanked resistance gene.
2. pKD3: template plasmid pKD3 (Amp^R and Cm^R, R6Kγ origin) in deionized water.
3. pKD4: template plasmid pDK4 (Amp^R and Kan^R, R6Kγ origin) in deionized water.
4. pKD13: template plasmid pKD13 (Amp^R and Cm^R, R6Kγ origin) in deionized water.

5. *Dpn*I restriction enzyme and *Dpn*I 10× buffer.
6. pKD46: λ Red recombinase expression low copy plasmid pKD46 (Amp^R, pSC101 origin, $repA_{101}{}^{ts}$) in deionized water.
7. Materials for electrotransformation (*see* Subheading 2.3).
8. pCP20: recombinase flippase (FLP) expression plasmid pCP20 (Amp^R and Cm^R, pSC101 origin, $repA_{101}{}^{ts}$) in deionized water.

2.3 Electrotransformation

1. DNA solution in deionized water (*see* **Note 1**).
2. Sterile electrotransformation cuvettes (1 mm) and electroporator.

2.4 P1 Transduction

1. 10^9 to 10^{10} plaque-forming unit (pfu)/mL P1 bacteriophage stock solution (*see* Miller JH, 1993 for phage titration [35]).
2. *E. coli* nonlysogenic donor strain containing the genetic marker to be transduced, such as a Keio mutant strain.
3. *E. coli* recipient strain to be transduced, such as the MG1655 strain.
4. Sodium citrate solution: 1 M sodium citrate solution in deionized water. Filter-sterilize and store at room temperature.
5. Calcium chloride solution: 1 M calcium chloride solution in deionized water. Autoclave and store at room temperature.
6. Chloroform.

2.5 Arabinose-Inducible Expression System

1. *E. coli* strain TG1 *aadA7-araC* (*spec*RExBAD).
2. DNA oligonucleotide primers (10μM) to PCR-amplify the selectable repression/expression cassette RExBAD.
3. L-Arabinose solution: 20% L-arabinose solution in deionized water. Filter-sterilize and store at room temperature.
4. D-Glucose solution: 20% D-glucose solution in deionized water. Filter-sterilize and store at room temperature.

2.6 Clustered Regularly Interspaced Short Palindromic RepeatsInterference (CRISPRi)

1. pdCas9: Defective Cas9 expression plasmid (Cm^R, p15A origin) in deionized water.
2. psgRNA: RNA guide expression plasmid (Amp^R, ColE1 origin) in deionized water.
3. DNA oligonucleotide primers (10μM) to PCR-amplify target-specific single guide RNA (sgRNA).
4. aTc solution: 2 mM anhydrotetracycline (aTc) in 50% ethanol. Store as 500-μL aliquots at −20 °C.
5. Quick blunt-end DNA ligation Kit.

2.7 MVA Genetic By-pass

1. L-Arabinose solution: *see* Subheading 2.5.
2. MVA solution: 500 mM mevalonate (MVA) stock solution. MVA is prepared from (±) mevalonolactone by hydrolysis of 1 M mevalonolactone (diluted in deionized water) with 1.05 volume of 1 M KOH (*see* **Note 2**). Store as 400-μL aliquots at −20 °C.

2.8 Sensitivity Tests to Oxidative Stress and Iron Starvation

1. PMS solution: 10 mM phenazine methosulfate (PMS) stock solution in sterile deionized water. Store as 1-mL aliquot at 4 °C (*see* **Note 2**).
2. DIP solution: 100 mM 2,2′-dipyridyl (DIP) stock solution in absolute ethanol. Store as 1-mL aliquot at −20 °C (*see* **Note 2**).
3. PQ solution: 100 mM paraquat (PQ) stock solution in sterile deionized water. Store as 1-mL aliquot at −20 °C (*see* **Note 2**).

2.9 β-Galactosidase Assay

1. Z buffer: 60 mM $Na_2HPO_4 \cdot 7H_2O$, 40 mM $NaH_2PO_4 \cdot H_2O$, 10 mM KCl, 1 mM $MgSO_4$, pH 7. Filter-sterilize and store at 4 °C.
2. SDS solution: 20% sodium dodecyl sulfate (SDS) in deionized water. Store at room temperature.
3. ONPG solution: 4 mg/mL of ortho-nitrophenyl β-D-galactopyranoside (ONPG) in Z buffer (*see* **Note 3**).
4. Sodium carbonate solution: 1 M sodium carbonate in deionized water. Autoclave and store at room temperature.
5. Chloroform.

2.10 Gentamicin Survival Test

Gentamicin solution: 5 mg/mL of gentamicin stock solution in deionized water. Filter-sterilize and store at 4 °C.

2.11 Hot Fusion Cloning Method

1. DNA oligonucleotide primers (10μM) to PCR-amplify the fragment to be cloned.
2. Template vector DNA: 10 ng/μL of plasmid or 50 ng/μL of genomic DNA.
3. PEG-8000: Polyethylene glycol 8000 powder.
4. 2 u/μL KOD Hot Start DNA Polymerase and 10× PCR buffer.
5. 2 u/μL Phusion Hot Start DNA Polymerase: Phusion Hot Start DNA Polymerase.
6. 10 u/μL T5 exonuclease.

3 Methods

3.1 Construction of Single Deletion Mutants

To investigate the in vivo contribution of IscA, SufA, ErpA, and NfuA to the Fe-S cluster biogenesis process, single deletion mutants have been constructed. For this purpose, the method initially developed by Datsenko and Wanner is the one that has been widely used [36]. Briefly, this is a simple and highly efficient method to disrupt chromosomal genes in *E. coli* in which PCR primers provide the homology to the targeted gene. In this procedure, recombination is ensured by the phage λ Red system that includes three genes: *γ*, *β*, and *exo*, whose products are called Gam, Bet, and Exo, respectively. Gam inhibits the *E. coli* RecBCD exonuclease V, allowing Bet and Exo to access the PCR DNA ends to promote recombination [37]. The *γ*, *β*, and *exo* genes are under the control of an inducible promoter on an easily curable, low copy number plasmid (pKD46) [36]. Using this procedure, the Δ*iscA* and the Δ*nfuA* mutants have been constructed [17, 23]. Interestingly, the antibiotic resistance cassette introduced in the target gene locus is flanked by recombinase flippase (FLP) recognition target (FRT) sites and then can be excised from the chromosome (*see* Subheading 3.3). The scar created in the chromosome has stop codons in all six reading frames, an idealized ribosome-binding site, and a start codon for downstream (rightward) gene expression. Thus, this method has been used to create nonpolar gene deletions within operons such as in the *iscRSUA-hscBA-fdx-iscX* and *sufABCDSE* operons. The protocol for inactivation of *E. coli* chromosomal genes described here is the one established by Datsenko and Wanner [36]. Protocols for electrotransformation and P1 transduction are those described in reference books [35, 38]; they have been included in this chapter to give some hints and to provide an A to Z protocol to construct *E. coli* mutants.

3.1.1 One-Step Inactivation of Chromosomal Genes in E. coli Using PCR Products

1. Design the DNA oligonucleotide primer pair that will determine the region to be deleted. For example, the *iscA* gene was deleted from the start codon to the stop codon (Fig. 2). The upstream primer contains at its 5′ extremity the 39-nt-long sequence of the region (H1) immediately upstream the sequence to be deleted and the sequence of the priming site 1 (P1) of the template plasmid (Fig. 2). To design the downstream primer, place the sequence of the priming site 2 (P2) upstream of the 39-nt-long sequence of the region immediately downstream the sequence to be deleted (H2), and use the reverse complementary sequence to obtain the sequence (5′–3′) of the downstream primer to be used in the PCR (Fig. 2).
2. Perform PCR in a 25-μL total volume containing 1 μL of 10 ng/μL template plasmid, 1μL of 10μM upstream primer, 1 μL of 10μM downstream primer, 1μL of 10 mM dNTPs, 0.5 μL of 2 u/μL high-fidelity DNA Polymerase, 2.5 μL of 10× PCR buffer, and sterile water to 25 μL. Template plasmids pKD3 and pKD4 will allow the introduction of

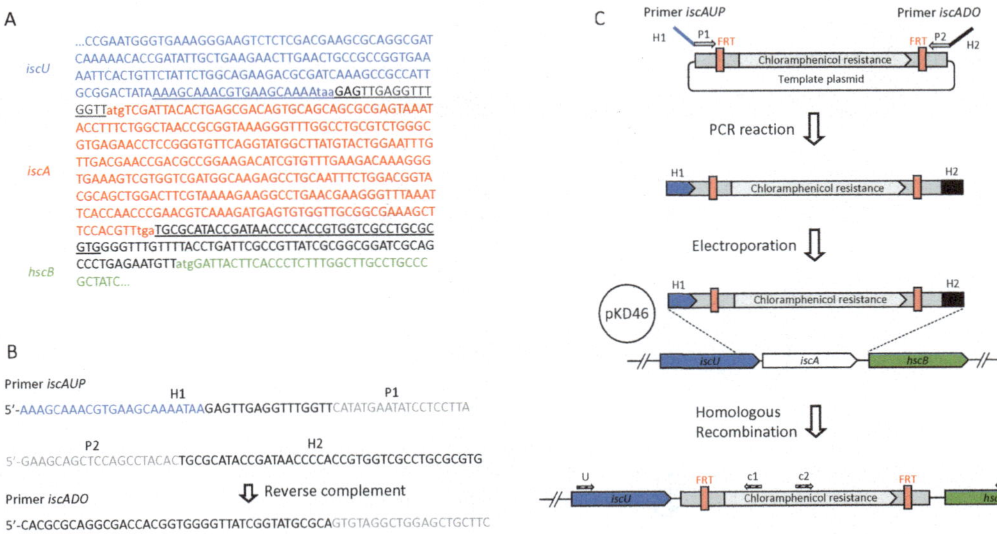

Fig. 2 One-step inactivation of chromosomal genes in *E. coli* using PCR products. (**a**, **b**) Design of the oligonucleotides to delete *iscA* (red). Gene knockout primers (*iscAUP* and *iscADO*) have 19-/20-nt 3′ ends (gray) for priming upstream (P1) and downstream (P2) of the FRT sites flanking the antibiotic resistance gene in the template plasmids (pKD13, pKD4, and pKD3) and 39-nt 5′ ends homologous to upstream (H1) and downstream (H2) chromosomal sequences for targeting the gene deletion. To construct the Δ*iscA* mutant, the pKD3 was used as a template plasmid. The primer *iscAUP* contains the sequence of (1) the region immediately upstream *iscA* corresponding to *iscU* sequence (blue), (2) the *iscU-iscA* intergenic region (black), and (3) the 19-nt 3′ ends for priming upstream (gray) of the FRT sites flanking the kanamycin resistance gene in pKD3. The primer *iscADO* contains the sequence of (1) the *iscA-hscB* intergenic region (black) and (2) the 20-nt 3′ ends for priming downstream (gray) of the FRT sites flanking the kanamycin resistance gene in pKD3. (**c**) PCR is performed using pKD3 plasmid as a template and the *iscAUP-iscADO* primer pair. The strain expressing the λ Red system, from the pKD46 plasmid, is electrotransformed with the PCR-amplified FRT-flanked resistance gene product. Novel junctions created after homologous recombination between the resistance cassette and neighboring upstream (*iscU*) and downstream (*hscB*) genes are verified by PCR with chloramphenicol (c1 and c2) and locus-specific (U and D) oligonucleotide primers. The antibiotic resistance gene can be eliminated using an FLP expression plasmid (pCP20) (*see* Fig. 3)

chloramphenicol and kanamycin resistance cassettes, respectively. Run the following PCR program on the thermocycler: first step: 94 °C for 3 min; second step: 94 °C for 20 s; third step: 15 s at a temperature depending on the melting temperature (T_m) of the oligonucleotides; fourth step: 72 °C during a time that depends on the PCR product length (1 min/kb); fifth step: go back to step 2 for 24 times; sixth step: 72 °C for 5 min and seventh step: 4 °C on hold (*see* **Note 4**).

3. Load an aliquot of the PCR product on an agarose gel (0.8–1.2%) to check that the PCR product has the expected size and that no other bands are present. Purify the PCR product using a commercial PCR Product Purification Kit. Elute DNA with 25μL of sterile deionized water.

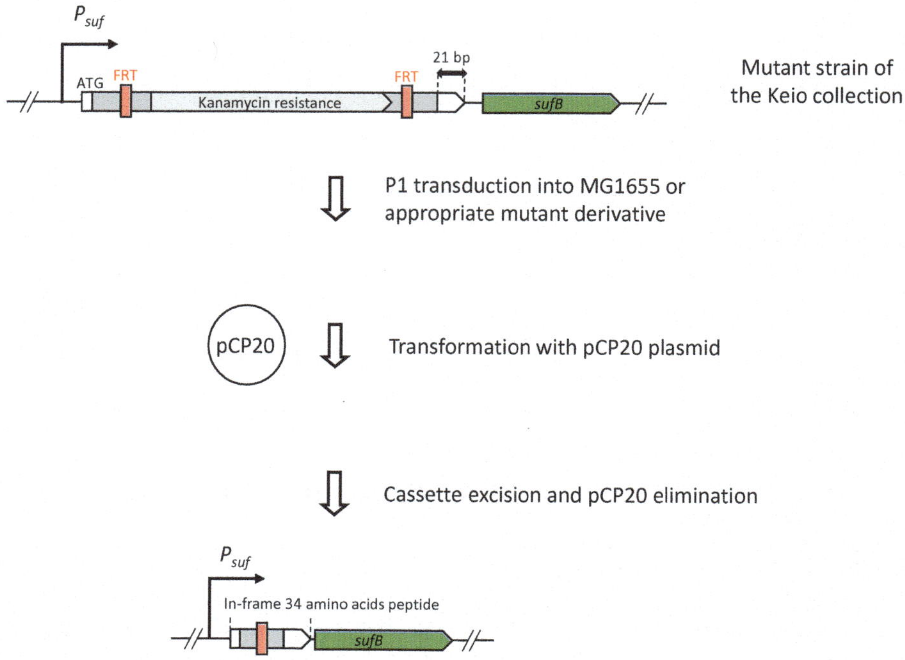

Fig. 3 Elimination of the antibiotic resistance gene of mutants from the Keio collection. The Keio collection is a collection of mutants where open-reading frame coding regions have been replaced with a kanamycin cassette flanked by flippase recognition target (FRT) sites (red boxes) by using a one-step method for inactivation of chromosomal genes (*see* Subheading 3.1). Here, the example of *sufA* is given. To reduce polar effects on downstream gene expression, oligonucleotide primers were designed, so that excision of the resistance cassette with the recombinase flippase would create an in-frame deletion of the second codon through the seventh codon from the C-terminus, leaving the ORF start codon and translational signal for a downstream gene intact. Translation from the initiation codon of the target gene is expected to produce a 34-amino acid scar peptide. Sequences created after excision of the resistance gene are verified by PCR with neighboring gene-specific oligonucleotide primers and by direct DNA sequencing of the region encompassing the scar region

4. To eliminate the template plasmid, the purified PCR product is incubated with *Dpn*I restriction enzyme that will digest the methylated template DNA but not the PCR product. To the 25μL of eluted purified PCR product, add 1μL of 10 u/μL *Dpn*I, 3μL of 10× *Dpn*I buffer, and sterile deionized water to a final volume of 30μL. Incubate for 1 h at 37 °C. The *Dpn*I digestion step is unnecessary when using template plasmids, such as pKD3 and pKD4, because they are conditional (oriRγ) replicons (*see* **Note 5**).
5. Load an aliquot of the *Dpn*I-digested PCR product on an agarose gel (0.8–1.2%) to control the quality and quantity of the recovered DNA. Then, purify the PCR product using a commercial PCR Product Purification Kit. Elute DNA with 25μL of deionized water and measure DNA concentration

using a UV-visible spectrophotometer. Store the purified PCR product at −20 °C or proceed to the next step.

6. Transform the *E. coli* recipient strain (MG1655 or appropriate strains) by chemical methods or by electrotransformation (*see* Subheading 3.1.2) with the λ Red system expression plasmid (pKD46). At **step 13** of Subheading 3.1.2, incubate the culture for 1.5 h at 30 °C because replication of the λ Red helper plasmid pKD46 is temperature sensitive.
7. Select the ampicillin-resistant transformants at 30 °C after incubating the selective plate overnight.
8. Pick single colonies, streak to isolate single colonies onto ampicillin LB agar plates, and incubate overnight at 30 °C.
9. Grow transformants carrying the λ Red helper plasmid in LB medium cultures with ampicillin and L-arabinose at 30 °C to an OD_{600} of 0.6, prior to making them electrocompetent (*see* Subheading 3.1.2).
10. Follow the electrotransformation protocol (Subheading 3.1.2) from **steps 1–9**. At **step 10**, add 50–200 ng of PCR-amplified FRT-flanked resistance gene product (obtained from above **steps 1–5**) to freshly made electrocompetent cells in the cuvette and mix gently. Complete the electrotransformation protocol from **steps 11** to **13**. Centrifuge the culture and resuspend the cells in 100μL of sterile LB in order to spread all the cell suspension onto the appropriate selective LB agar. Incubate overnight at 37 °C.
11. To test for the loss of the pKD46 helper plasmid, pick single colonies and streak them onto nonselective LB agar plates and incubate overnight at 37 °C. Pick single colonies and streak them both on nonselective LB agar and ampicillin LB plate. Clones that have lost the pKD46 helper plasmid will be ampicillin sensitive (*see* **Note 6**).
12. Verify the mutants by PCR to test the presence of new locus- and junction-specific fragments of the predicted sizes. The sequence of the junction is also verified (Fig. 2). Once verified, the mutation is generally moved by P1 transduction into the wt *E. coli* MG1655 strain or appropriate mutant strains to construct single or multiple mutants.

3.1.2 Electrotransformation

1. Using sterile technique, inoculate 5 mL of LB in a 50-mL Erlenmeyer flask with the *E. coli* strain and incubate overnight at the appropriate temperature with shaking (*see* **Note 7**).
2. The following day, dilute the overnight culture 1:100 in a 1-L Erlenmeyer flask containing 100 mL of LB (*see* **Note 8**).

3. Incubate at the appropriate temperature with shaking to reach an OD_{600} of 0.5–0.6. Use a UV-visible spectrophotometer to measure the OD_{600} of the culture.
4. Cool the cell culture on ice for 15 min to stop cell division. Cool the centrifuge rotor at 4 °C.
5. Pellet the cells by centrifugation (4 °C, 5000 × *g*, 10 min). Pour off the supernatant and suspend cells in 1 volume of cold, sterile deionized water. Pellet the cells by centrifugation (4 °C, 5000 × *g*, 10 min).
6. Repeat **step 5** twice and resuspend the cell pellet in cold sterile deionized water to reach a final volume corresponding to 1:100 of the initial culture volume (*see* **Note 9**).
7. Turn on the electroporator and set it to 2.5 kv, 200 Ω, and 50μF.
8. Warm LB at an appropriate temperature.
9. Add 100μL of electrocompetent cells into the chilled electro-transformation cuvette.
10. Add DNA solution (in deionized water) to the cells in the cuvette and gently shake the cuvette to mix without introducing bubbles and make sure that the cells deposit across the bottom of the cuvette (*see* **Note 10**).
11. Wipe water from the exterior of the cuvette, place in the electrotransformation module, and deliver the high-voltage electrical pulse (*see* **Note 11**). The typical time constant is ~2.6 ms.
12. Immediately thereafter, add 900μL of prewarmed LB, mix by pipetting up and down, and transfer to a 2-mL microcentrifuge tube.
13. Incubate for 1 h at the appropriate temperature (*see* **Note 7**) with shaking.
14. Using sterile glass beads or an L-shaped spreader, spread adequate volume of the culture onto the appropriate selective LB agar plate (*see* **Note 12**).
15. Incubate overnight at the appropriate temperature.
16. Pick single colonies and streak to isolate single colonies onto the appropriate selective LB agar plate.

3.1.3 Transduction of Mutations from Genetic Resources

E. coli K-12 is one of the best-characterized organisms in molecular biology and offers key resources for functional genomics, such as a collection of 3985 single-gene knockout mutants in the BW25113 strain background, the Keio collection [39, 40]. In this collection of mutants, open-reading frame (ORF) coding regions have been replaced with a kanamycin cassette flanked by FLP recognition target (FRT) sites by using the one-step method for inactivation of chromosomal genes (*see* Subheading 3.1.1) (Fig. 3). To reduce polar effects on downstream gene expression, DNA oligonucleotide primers were designed, so that excision of the resistance cassette with the FLP recombinase takes place. This would create an in-frame deletion from the 2nd codon from the N-terminus to the seventh codon from the C-terminus, leaving the ORF start codon and translational signal for a downstream gene intact (Fig. 3) (*see* **Note 13**). Further information on the Keio collection is available at http://sal.cs.purdue.edu:8097/GB7/index.jsp or http://ecoli.naist.jp/. The distribution is now being handled by the National Institute of Genetics of Japan (http://www.shigen.nig.ac.jp/ecoli/pec/index.jsp). Strains of this collection can be used in subsequent P1 transduction experiments as described below.

1. To prepare the P1 liquid lysate, first inoculate 5 mL of LB in a 50-mL Erlenmeyer flask with the *E. coli* donor strain (strain from which the P1 lysate is to be made, here a Keio mutant) and incubate overnight at the appropriate temperature (usually 37 °C) with shaking (*see* **Notes 7** and **14**).
2. Dilute the overnight culture of the donor strain 1:100 in a 50-mL Erlenmeyer flask containing 5 mL of LB supplemented with 100μL of calcium chloride solution (final concentration 20 mM) and 100μL of a 10^9 to 10^{10} pfu/mL of P1 bacteriophage stock solution (*see* **Note 15**).
3. Incubate at 37 °C with shaking until the culture lyses (~3.5 h, *see* **Note 16**).
4. Transfer the supernatant to a 15-mL centrifuge tube, add 100μL of chloroform, and vortex thoroughly. After 10 min, vortex again.
5. Centrifuge for 5 min at ~10,000 × *g* or let the tube rest for 20 min.
6. Transfer the supernatant—this is the P1 lysate—in a 15-mL screw-cap tube and add two drops of chloroform. Store at 4 °C (*see* **Note 17**) or proceed to transduction starting at **step 7**.
7. Inoculate 5 mL of LB in a 50-mL Erlenmeyer flask with the *E. coli* recipient strain (strain into which the mutation is to be introduced) and incubate overnight at the appropriate temperature (usually 37 °C, *see* **Note 7**) with shaking.
8. The following day, dilute the overnight culture of the recipient strain 1:100 in a 50-mL Erlenmeyer flask containing 5 mL of LB (*see* **Note 8**).
9. Incubate for 3.5 h at 37 °C with shaking.
10. Add 100μL of calcium chloride solution (final concentration 20 mM) and continue shaking at 37 °C for 30 min.

11. In a 15-mL centrifuge tube, mix 4 mL of this recipient strain culture and 1 mL of the P1 lysate (from the donor strain) (*see* **Note 18**).
12. Incubate for 30 min at 37 °C without shaking.
13. Centrifuge for 3–4 min at 5000 × *g* to pellet the cells.
14. Discard the medium and resuspend the cell pellet in 1 mL of LB supplemented with 5 mM sodium citrate (*see* **Note 19**). Transfer the cell suspension into a 2-mL centrifuge tube.
15. Centrifuge for 1 min at 6000 × *g* to pellet the cells and resuspend the cells in 1 mL of LB supplemented with 5 mM sodium citrate. Repeat this step twice.
16. Incubate the cells in 1 mL of LB supplemented with 5 mM sodium citrate for 1.5 h at 37 °C with shaking.
17. Centrifuge the culture at 5000 × *g* for 2 min, at room temperature, to pellet the cells.
18. Resuspend the cells in 200μL of LB and, using sterile glass beads or an L-shaped spreader, spread the entire suspension onto the appropriate selective LB agar plate supplemented with 2 mM sodium citrate to avoid any remaining P1 (*see* **Note 20**).
19. Incubate the plates overnight at 37 °C.
20. The next day, score the plates. Pick single colonies, streak to isolate single colonies onto the appropriate selective LB agar plates supplemented with 2 mM sodium citrate, and incubate overnight at 37 °C.
21. The next day, pick individual colonies and streak for single colonies on selective LB agar plates.
22. Confirm by PCR that the desired mutation has been introduced at the correct locus and sequence of the appropriate region.

3.2 Overcoming Essentiality of Fe-S Carriers

Deletion of essential genes, such as *erpA*, is, by definition, impossible. Attempts to delete *erpA* in *E. coli* failed, unless the cells had spontaneously duplicated a large chromosomal region (>13 kb in size). In this case, genetically modified strains carried both the inactivated and the wild-type copies of the *erpA* gene [18]. Then, to study the function of *erpA* by genetic approaches, one strategy was to use promoter shut-off experiments by replacing the promoter of *erpA* by a regulable promoter, such as the arabinose-inducible expression system (*see* Subheading 3.2.1). Alternatively, the shut-off of gene expression can also be achieved through the use of a deactivated CRISPR-dCas9 system that halts the transcription of target genes (*see* Subheading 3.2.2). The other strategy was to use genetic by-pass by engineering a strain that provides an Fe-S cluster-independent alternative pathway for the essential pathway relying on Fe-S proteins, as described for the MVA by-pass pathway (*see* Subheading 3.2.3). The use of promoter shut-off experiments and the by-pass strategy has constituted a powerful approach for the functional analysis of *erpA* because it unambiguously showed that *erpA* is an essential gene and they provided tools to study strains lacking ErpA [18]. These approaches have also set up a framework to study all nonviable combinations of Fe-S cluster biogenesis components and have provided the basis for further genetic approaches.

3.2.1 The Arabinose-Inducible Expression System

To allow the functional analysis of any chromosomal genes, including essential genes, such as *erpA*, a selectable repression/expression cassette (RExBAD) has been designed to place a functional P*araBAD* promoter (pBAD) upstream of the target gene at its native loci (Fig. 4) [41]. The introduction of the RExBAD cassette is done by λ Red-mediated homologous recombination.

1. Design the DNA oligonucleotide primer pair that will determine where the P*araBAD* promoter will be introduced.
2. Perform PCR as described in **step 2** of Subheading 3.1.1, except that a colony or purified genomic DNA of the strain TG1 *aadA7-araC*, carrying both the spectinomycin selection marker and the *araC*-P*araBAD* promoter region, was used as a template (Fig. 4) (*see* **Note 21**).
3. Follow the protocol of Subheading 3.1.1, from **steps 3** to **10**.
4. Follow the electrotransformation protocol, from **steps 1** to **9**. At **step 10**, add 50–200 ng of the PCR product, containing a DNA region encoding the functional pBAD promoter (Fig. 4) (obtained from above steps **1** to **2**) to the cells in the cuvette and mix gently. Complete the electrotransformation protocol (Subheading 3.1.2) with **step 11**.
5. Immediately after the electrical pulse, add 900μL of prewarmed LB supplemented with 0.2% arabinose to ensure expression of the target gene, mix by pipetting up and down, and transfer to a 2-mL microcentrifuge tube.
6. Incubate for 1 h at the appropriate temperature (*see* **Note 7**) with shaking.
7. Centrifuge the cell culture and resuspend the cell pellet in 100μL of sterile LB in order to spread all the cell suspension onto spectinomycin LB agar supplemented with 0.2% arabinose.
8. Incubate overnight at the appropriate temperature.
9. Pick single colonies and streak to isolate single colonies onto spectinomycin LB agar supplemented with 0.2% arabinose.

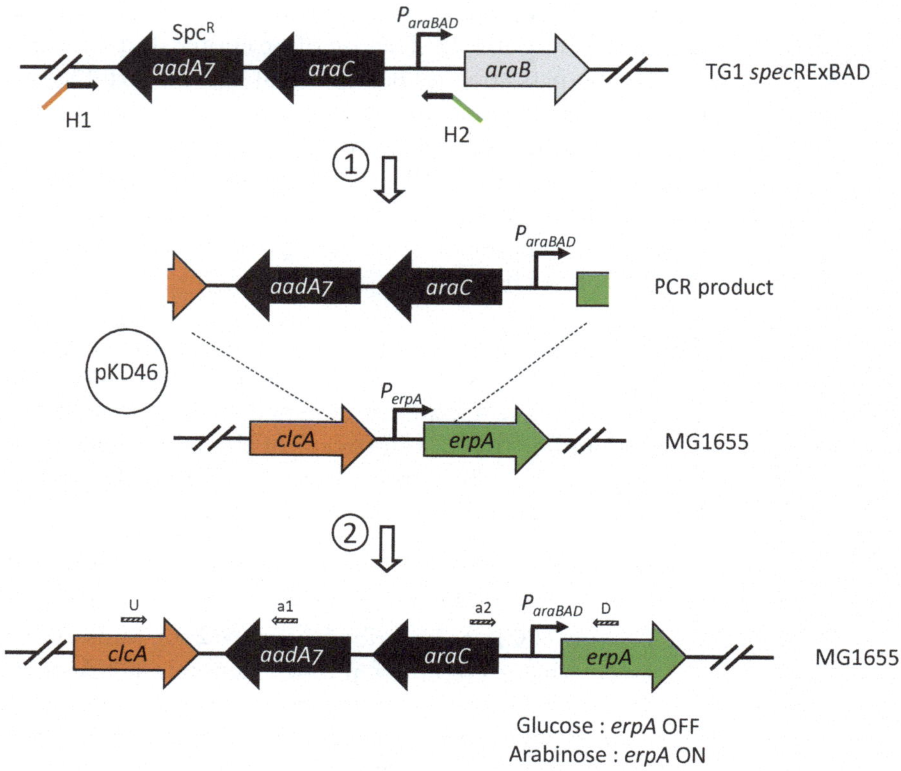

Fig. 4 The RexBAD system to control *erpA* gene expression. The strain TG1 *spec*RExBAD was used as a template to amplify the *aadA7*RExBAD cassette carrying both the spectinomycin selection marker and the *araC*-P*araBAD* promoter region (step 1). The oligonucleotides were designed to introduce a short homologous region to *erpA* (H2) (green line) and to *clcA* (H1) (orange line). Once introduced by λ Red-mediated homologous recombination (dashed lines), the RExBAD cassette permits to place *erpA,* at its native loci, under the control of the repressible/inducible P*araBAD* promoter directly on the chromosome (step 2). Novel junctions created after homologous recombination between the *aadA7*RExBAD cassette and neighboring upstream (*clcA*) and downstream (*erpA*) genes are verified by PCR with cassette-specific primers (a1 and a2) and locus-specific (U and D) oligonucleotide primers

10. To test for the loss of the pKD46 helper plasmid, pick single spectinomycin-resistant colonies, streak them onto nonselective arabinose-containing LB agar plates, and incubate overnight at the appropriate temperature. Pick single colonies and streak them both on nonselective arabinose-containing LB agar and ampicillin arabinose-containing LB plates. Clones that have lost the pKD46 helper plasmid will be ampicillin sensitive (*see* **Note 6**).
11. Verify the mutants by PCR to test the presence of new locus- and junction-specific fragments of the predicted sizes. Check the sequence of the junction (Fig. 4). Once verified, the mutation is generally moved by P1 transduction into the wt *E. coli* MG1655 strain or appropriate mutant strains to construct single or multiple mutant strains, respectively.

12. Using sterile technique, inoculate 5 mL of LB supplemented with 0.2% arabinose, in a 50-mL Erlenmeyer flask with the *E. coli* strain carrying the D-glucose-repressible and L-arabinose-inducible version of the target gene and incubate overnight at the appropriate temperature with shaking (*see* **Note 7**).
13. The following day, dilute the overnight culture 1:100 in a 1-L Erlenmeyer flask containing 100 mL of LB. To modulate the expression of the target gene, the LB medium is supplemented with either 0, 0.4, and 1% of glucose or concentrations of arabinose ranging from 0, 004 to 1%. Incubate at the appropriate temperature with shaking and monitor growth. Whether antibodies are available, immunodetection can be performed to measure the quantity of the target protein before analyzing further the mutant strain.

3.2.2 CRISPR Interference

To repress *erpA* expression, the CRISPR interference (CRISPRi) method has been used [42]. The CRISPRi method used is a repurposing of the *Streptococcus pyogenes* type II CRISPR system that silences foreign DNA elements in an RNA-guided manner [43]. The bacterial CRISPRi system requires only two components, a catalytically inactive Cas9 (dCas9) protein and a customizable single guide RNA (sgRNA) [43]. The dCas9 protein and the sgRNA are expressed from separate plasmids containing an anhydrotetracycline (aTc)-inducible promoter (p_L*tetO*-1) (Fig. 5). The dCas9 protein, when coexpressed with a guide RNA, generates a complex that binds to specific targets by Watson-Crick base pairing between the sgRNA and the DNA target (Fig. 5). Because the dCas9 protein is catalytically dead, the complex does not cleave the DNA, but instead remains DNA-bound. Then, depending on the target genomic locus, the DNA-bound dCas9/sgRNA complex interferes with transcriptional elongation, RNA polymerase binding, or transcription factor binding. The targeting specificity is determined both by base pairing between the sgRNA chimera and the target DNA, and by the binding between the dCas9 protein and a short DNA motif found on the target DNA, called the protospacer adjacent motif (PAM) (Fig. 5). The PAM sequence for the type II CRISPR system of *Streptococcus pyogenes* is 5′-NGG-3′.

1. Select the DNA elements to be targeted, such as the nontemplate DNA strand of the 5′ portion of the target gene, the nontemplate DNA strand of the 5′ untranslated (UTR) region, the −35 box or −10 box of the promoter (either the template or the nontemplate strand), and the binding site of regulatory proteins (either the template or the nontemplate strand). The DNA element of the *erpA* locus to be targeted was the nontemplate DNA strand of the 5′ UTR next to a PAM (Fig. 5).

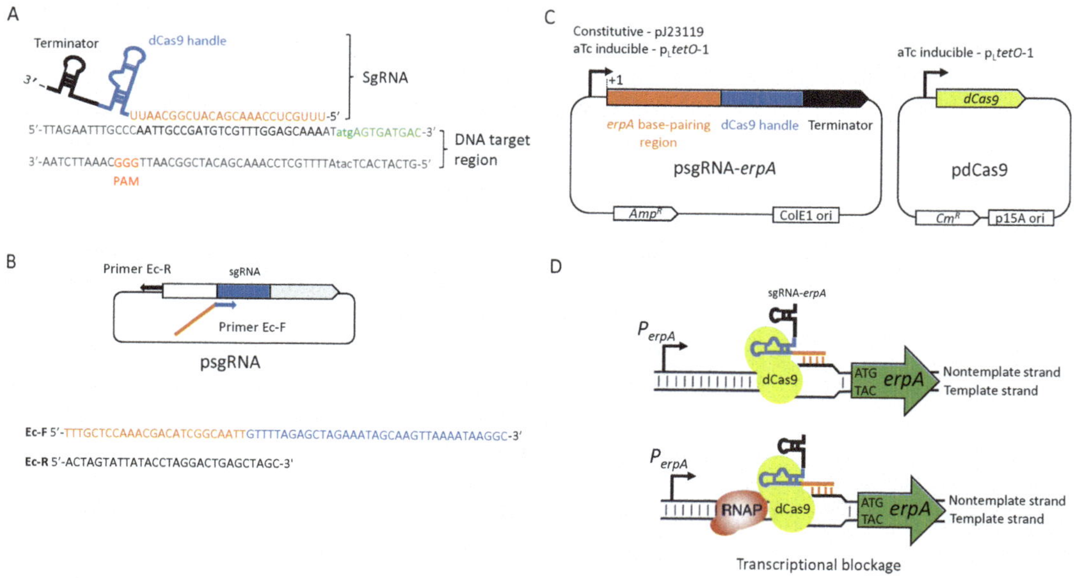

Fig. 5 The CRISPRi system to control *erpA* gene expression. (**a**) Sequence of designed sgRNA specific for *erpA* (sgRNA-*erpA*) and the DNA target region. The sgRNA-*erpA* targets the nontemplate strand at the 5′ untranslated region (UTR) of *erpA*. The *erpA* coding sequence on the nontemplate strand is indicated in green. Base pairing nucleotides in sgRNA-*erpA* are shown in orange. The 42-nt hairpin for dCas9 binding (dCas9 handle) and the 40-nt transcription terminator derived from *S. pyogenes* are indicated in blue and black, respectively. The PAM sequence is shown in red. (**b**) Oligonucleotide primers Ec-R and Ec-F were used to construct, by inverse PCR, the plasmid expressing the sgRNA-*erpA*. (**c**) The sgRNA expression plasmid (psgRNA-*erpA*) contains a promoter (constitutive-pJ23119 or inducible-p_L*tetO*-1) with an annotated transcription site (+1), an ampicillin-selectable marker (Amp^R), and a ColE1 replication origin. The dCas9 expression plasmid (pdCas9) contains an aTc-inducible promoter, a strong ribosome-binding site for dCas9 translation, a chloramphenicol-resistance marker (Cm^R), and a p15A replication origin. (**d**) Binding of the dCas9 (indicated in yellow)/sgRNA-*erpA* complex to the 5′ UTR region of *erpA* can silence gene expression by blocking the elongating RNA polymerases (RNAPs) (indicated in orange)

2. Check the specificity of sgRNA binding in the *E. coli* genome by Basic Local Alignment Search Tool (BLAST) and use only sgRNAs without predicted off-targets.
3. Predict the secondary structure folding of the sgRNA (UNAFold or RNAfold packages), especially for the 42-nt corresponding to the dCas9 handle hairpin that needs to be properly folded.
4. Design an appropriate DNA oligonucleotide primer pair to construct the plasmid expressing the target-specific sgRNA. The *erpA*-specific sgRNA primers were (Ec-F) and (Ec-R). Primers are used to PCR-amplify the whole plasmid (psgRNA) (Fig. 5). Purify the PCR product and digest using *Dpn*I to remove the template psgRNA plasmid.

5. Without further purification or heat inactivation, self-ligate the *Dpn*I digestion reactions using a Quick blunt-end DNA ligation kit.
6. Transform *E. coli* cells by chemical methods or by electrotransformation with the ligation reaction.
7. Select the ampicillin-resistant transformants at 37 °C after incubating the selective plate overnight.
8. Pick single colonies, streak to isolate single colonies onto ampicillin LB agar plates, and incubate overnight at 37 °C.
9. Perform colony PCR on the colonies selected in **step** 7 to check the construction by sequencing. The target-specific sgRNA guide encoded from the plasmid is a chimera and contains three regions: (1) a 20-nt-long base-pairing region for target-specific DNA binding, (2) a 42-nt-long dCas9 handle hairpin for Cas9 protein binding, and (3) a 40-nt-long transcription terminator hairpin.
10. Electrotransform the *E. coli* strain of interest with the pdCas9 and the target-specific psgRNA plasmids simultaneously (*see* Subheading 3.1.2).
11. Select transformants on an LB agar plate containing both ampicillin and kanamycin, after overnight incubation at 37 °C.
12. Pick single colonies, streak to isolate single colonies onto ampicillin and kanamycin LB agar plate, and incubate overnight at 37 °C.
13. Using sterile technique, inoculate 5 mL of LB supplemented with ampicillin and kanamycin, in a 50-mL Erlenmeyer flask with the *E. coli* strain carrying both the pdCas9 and the target-specific psgRNA plasmids. Incubate overnight at the appropriate temperature with shaking (*see* **Note** 7).
14. The following day, dilute the overnight culture 1:100 in a 1-L Erlenmeyer flask containing 100 mL of LB supplemented with ampicillin and kanamycin. To induce expression of the dCas9 protein and the sgRNA, 2μM aTc is used.
15. Incubate at 37 °C with shaking and monitor growth. Whether antibodies are available, immunodetection can be performed to measure the quantity of the target protein before analyzing further the mutant strain.

3.2.3 Replacing the Essential Fe-S Cluster-Dependent Pathway: The MVA Genetic By-Pass

In *E. coli*, the biosynthesis pathway of isopentenyl pyrophosphate (IPP), from 2C-methyl-D-erythritol 4-phosphate (MEP), is an essential process that involves two Fe-S cluster proteins IspG and IspH (Fig. 6). Interestingly, eukaryotes synthesize IPP from mevalonate (MVA) by an Fe-S cluster-independent pathway. Expression in *E. coli* of a chimeric operon encoding the yeast MVA kinase (yMVK), the human phosphomevalonate kinase (hPMK), and the

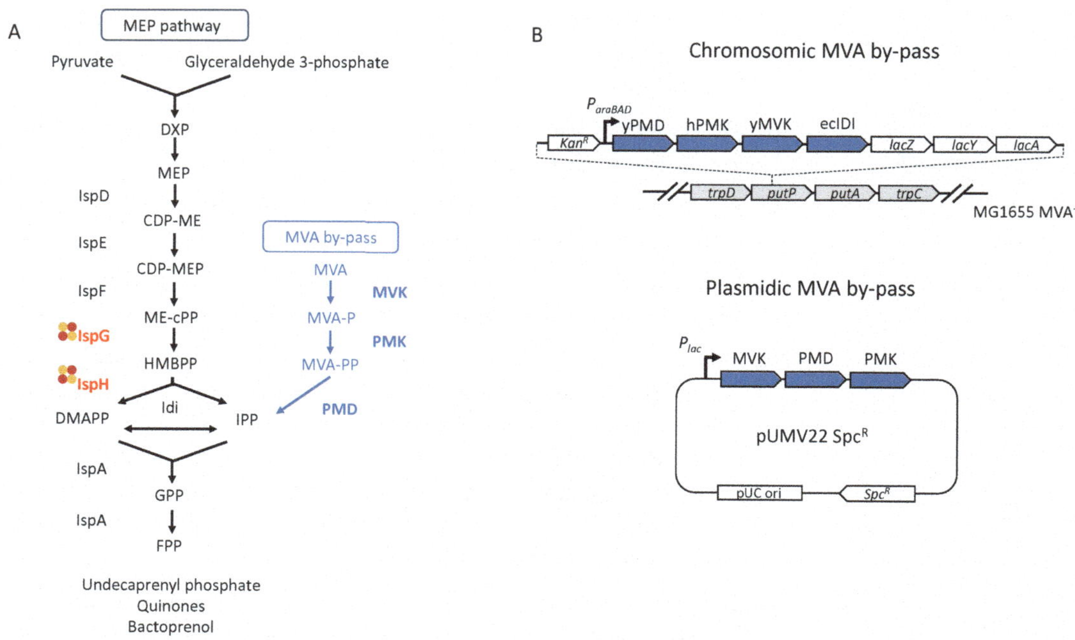

Fig. 6 The methylerythritol phosphate (MEP) pathway and mevalonate phosphate (MVA) by-pass. (**a**) Enzymes intervening at each step of the biosynthetic MEP pathway (black arrows) are indicated. The Fe-S cluster (orange and red dots)-containing proteins, IspH and IspG, are in red. Intermediary products of the MEP pathway are indicated: *DXP* 1-deoxy-D-xylulose 5-phosphate, *MEP* 2C-methyl-D-erythritol 4-phosphate, *CDP-ME* 4-diphosphocytidyl-2C-methyl-D-erythritol, *CDP-MEP* CDP-ME-2-phosphate, *ME-cPP* 2C-methyl-D-erythritol 2,4-cyclodiphosphate, *HMBPP* 1-hydroxyl-2-methyl-2-(*E*)-butenyl 4-diphosphate, *IPP* isopentenyl diphosphate, *DMAPP* dimethylallyl diphosphate, *GPP* geranyl diphosphate, *FPP* farnesyl diphosphate. The enzymes of the MVA by-pass are indicated in blue—mevalonate kinase (MVK), phosphomevalonate kinase (PMK), and mevalonate pyrophosphate decarboxylase (PMD). The substrate—mevalonate (MVA)—and intermediary products—mevalonate 5-phosphate (MVA-P) and mevalonate pyrophosphate (MVA-PP)—of the MVA by-pass are indicated in blue. (**b**) The chromosomic MVA by-pass is ensured by the yeast mevalonate kinase (yMVK), human phosphomevalonate kinase (hPMK), and yeast mevalonate pyrophosphate decarboxylase (yPMD). The resulting *E. coli* MVA$^+$ strains, such as the EcAB1-5 and the MG1655 MVA$^+$ strains, have two copies of genes encoding the isopentenyl-diphosphate Δ-isomerase (eclDI from the synthetic MVA operon and the *idi* gene at its native loci). The synthetic operon is under the control of the p*araBAD* promoter and is located between the transcription terminator of a chimeric kanamycin resistance gene (non-FLP excisable) (Kan^R) and a promoterless *lac* operon (*lacZYA*). The construct has been inserted in the tryptophan operon of the *E. coli* chromosome between *trpC* and *trpD*. The genes *putA* and *putP* are derived from *Salmonella typhimurium put* operon. The plasmidic MVA-bypass is encoded by the pUMV22 Spc^R plasmid, a pUC118 derivative carrying genes for MVA kinase (MVK), MVA-P kinase (PMK), and MVA-PP decarboxylase (PMD) from *Streptomyces* sp. CL190, and a spectinomycin (Spc) resistance marker from the pAPNC213 plasmid

yeast diphosphomevalonate decarboxylase (yPMD), under the control of the pBAD promoter, compensates for inactivation of the MEP pathway (Fig. 6) (*see* **Note 22**) [44]. Thanks to this MVA genetic by-pass, it is possible to construct an *E. coli* strain lacking ErpA, which is able to grow under aerobic conditions [18].

1. Follow the P1 transduction protocol (Subheading 3.1.3), using as a donor the EcAB1-5 strain that carries the chimeric operon (yPMD-hPMD-yMVK) linked to a kanamycin-resistant cassette. Select kanamycin-resistant transductants [18].
2. Confirm by PCR that the chimeric MVA pathway has been introduced at the correct locus.
3. Mutation abolishing Fe-S cluster biogenesis can be introduced in the MVA^+ strain by following either the one-step inactivation protocol (Subheading 3.1.1) or by P1 transduction (Subheading 3.1.3). In both cases, mutants should be selected on a plate containing both 1 mM MVA and 0.2% arabinose, the substrate and the inducer of the MVA-dependent pathway [18].

3.3 Construction of Multigene Deletion Mutants

The MVA genetic by-pass can be subsequently exploited to demonstrate the synthetic lethality of multiple genes, like *iscA* and *sufA*, or offers the possibility to construct the double mutants, like in the case of Δ*erpA* Δ*iscA*, Δ*erpA* Δ*sufA*, and the triple mutantΔ*iscA* Δ*sufA* Δ*erpA* [17].

Combining mutations has a limitation due to the limited number of distinct antibiotic resistance cassettes available. Therefore, the method developed by Datsenko and Wanner (*see* Subheading 3.1.1) to create *E. coli* mutants provides a way to eliminate the antibiotic cassette using the FLP recombinase [36]. Elimination of the resistance genes leaves behind an 82–85-nt scar in place of the disrupted gene (Fig. 3).

1. Transform mutants containing an FRT-flanked resistance cassette (usually Kan^R or Cm^R) with the pCP20 plasmid and select ampicillin-resistant transformants at 30 °C. The pCP20 plasmid shows temperature-sensitive replication and thermal induction of FLP synthesis.
2. Colony-purify the Amp^R transformants onto nonselective LB agar plates at 43 °C.
3. Test isolated colony for the loss of all antibiotic resistances (*see* **Notes 23** and **24**).
4. Confirm by PCR the elimination of the cassette and sequence the scar introduced.

3.4 Varying the Burden Imposed on Fe-S Cluster Biogenesis

Fe-S cluster biogenesis can be impeded in iron scarce conditions and under oxidative stress. In contrast, anaerobiosis is a favorable condition for Fe-S cluster biogenesis. Then, by varying environmental conditions and thus the burden imposed on Fe-S cluster biogenesis, it has been possible to reveal novel Fe-S cluster delivery routes [17, 23]. The nonviable mutants under aerobic conditions, Δ*erpA* and Δ*iscA* Δ*sufA,* were viable under anaerobiosis, while the Δ*erpA* Δ*iscA* double mutants remained nonviable under anaerobiosis. This meant that under anaerobiosis, a condition favorable for Fe-S cluster stability, only one ATC is required, IscA or ErpA, while in aerobiosis, a less favorable condition, two ATCs are necessary, ErpA and either IscA or SufA. For these experiments, anaerobic conditions can be created by using very simple and small devices that are sufficient to incubate Petri dishes, such as the Mitsubishi™ AnaeroPack system.

Conversely to anaerobiosis, it is also possible to impose a burden on Fe-S cluster biogenesis, such as by generating oxidative stress or by creating iron limitation during laboratory growth conditions. This has permitted to reveal the physiological importance of NfuA in Fe-S cluster biogenesis in *E. coli* [23, 31]. To create oxidative stress, redox cyclic compounds are convenient molecules (paraquat, phenazine methosulfate). Hydrogen peroxide (H_2O_2) to endanger Fe-S clusters can also be used. However, because bacteria rapidly consume H_2O_2 that is added to test cultures, specific methods have been described to sustain low-micromolar steady-state concentrations of H_2O_2 which are sufficient to impair cell growth [34]. Alternatively, the *E. coli* Hpx$^-$ strain, devoid of both its catalases and its NADH peroxidase (*katE katG ahpCF*), which accumulates up to 1μM endogenous H_2O_2 when transferred from anaerobic to aerobic medium, can be used [45]. Then, it was shown that the Hpx$^-$ mutant exhibits growth defects if it additionally lacks the *sufBCD* genes, but the Hpx$^-$Δ*sufA* mutant is healthy, suggesting that other carriers can replace SufA in H_2O_2-stressed cells [46].

A classic way to induce iron limitation is to add iron chelators, such as 2,2′-dipyridyl (DIP), in the growth medium. Recently, an "iron-minus" strain has been constructed, which contains a limited known iron level. This strain lacks the Feo-, ZupT-, ferric-citrate-, and siderophore-dependent iron-import systems (Δ*feoABC* Δ*zupT* Δ*tonB*) [47].

Sensitivity tests to paraquat (PQ), phenazine methosulfate (PMS), and 2,2′-dipyridyl (DIP) can be conducted by culturing selected strains in liquid culture or solid medium.

3.4.1 Sensitivity Tests in Liquid Cultures

1. Inoculate 5 mL of LB in a 50-mL Erlenmeyer flask with the *E. coli* strain of interest and incubate overnight at the appropriate temperature (usually 37 °C) with shaking (*see* **Note 7**).
2. The following day, dilute the overnight culture 1:100 in a 250-mL Erlenmeyer flask containing 25 mL of LB (*see* **Note 8**).
3. Incubate at 37 °C with shaking to reach an OD_{600} of 0.2–0.5. Add PQ, PMS, or DIP, usually 250μM, 30μM, and 250μM, respectively. Incubate further at the appropriate temperature with shaking.

4. Follow cell growth at OD_{600} (*see* **Note 25**). It is also possible to determine the colony-forming units per milliliter (cfu/mL) at different incubation times by taking a 100-μL aliquot of the culture (*see* Subheading 3.5.2 from **steps 6** to **10**).

3.4.2 Sensitivity Tests in Solid Medium

1. Inoculate 5 mL of LB in a 50-mL Erlenmeyer flask with the *E. coli* strain of interest and incubate overnight at the appropriate temperature (usually 37 °C) with shaking (*see* **Notes 7** and **26**).
2. Serial dilute overnight cultures in sterile PBS (from 10^{-1} to 10^{-6}).
3. Spot 5μL of each dilution on plates containing the appropriate concentrations tested of PQ, PMS, or DIP (usually 250μM, 30μM, and 250μM, respectively) and incubate overnight at 37 °C.

3.5 Varying the Client Proteins

Phenotypic analysis of Fe-S cluster biogenesis mutants in rich LB medium informs us on the level of IPP and then on the maturation status of the IspG and IspH proteins [18]. Interestingly, the maturation of other client proteins, such as Fe-S transcriptional regulators, can be studied by analyzing the expression of reporter genes using *lacZ* gene fusion. To this purpose, the protocol of β-galactosidase assay from liquid culture is given below. Investigation on how the transcriptional regulators IscR and SoxR acquire their Fe-S clusters has revealed that differences arise in the Fe-S delivery routes [25, 42, 48]. In the case of IscR, several routes occur, including one without the assistance of any ATC, and for SoxR, NfuA and ErpA were dispensable in oxidative stress conditions [25, 42, 48]. These two situations differ from the maturation of IspG/H. Interestingly, IspG and IspH are 4Fe-4S cluster proteins, while IscR and SoxR are 2Fe-2S cluster-containing proteins, suggesting that specialization of carriers depending on the nature of the cluster might exist. Such a hypothesis was also proposed from studies on eukaryotic models [4].

Maturation of the eukaryotic respiratory complex I that is rich in Fe-S clusters has been shown to involve dedicated factors [4]. Recently, in prokaryotes, the same picture seems to emerge with Mrp, a ParA-like Fe-S protein, that is required for the maturation of *E. coli* complex I but not for succinate dehydrogenase [49]. Thanks to the importance of respiratory complexes for aminoglycoside uptake in *E. coli*, a gentamicin sensitivity test, described below, was shown to be a very reliable proxy for the maturation of respiratory complexes [50, 51]. To assess the maturation of respiratory complexes, the gentamycin sensitivity test offers the possibility to run a large number of tests in parallel, then permitting to test numerous mutants, various conditions, and to follow maturation kinetics. With such a test, screening and selection procedure can also be envisioned to search for respiratory complex maturation factors.

3.5.1 β-Galactosidase Assay

1. Inoculate 5 mL of LB in a 50-mL Erlenmeyer flask with the *E. coli* strain containing a *lacZ* gene fusion, such as the P*iscR*-*lacZ* gene fusion, and incubate overnight at the appropriate temperature (usually 37 °C) with shaking (*see* **Note 7**).
2. The following day, dilute the overnight culture 1:100 in a 50-mL Erlenmeyer flask containing 5 mL of LB (*see* **Note 8**).
3. Incubate at 37 °C with shaking to reach a density of 2–5 × 10^8 cells/mL (an OD_{600} of 0.3–0.7). Use a UV-visible spectrophotometer to measure the OD_{600} of the cultures.
4. To prevent further growth, cool for 20 min the cultures in an ice bucket containing a mixture of ice and water.
5. In a 2-mL centrifuge tube, add an aliquot of the cultures (0.1–0.5 mL, referred to as v) and Z buffer to reach 1 mL. To open the cells, add two drops of chloroform and one drop of 0.1% SDS using a Pasteur pipette (*see* **Note 27**). Prepare a blank tube with LB medium, instead of the culture.
6. Vortex the tubes for 10 s and place the tubes in a 28 °C water bath for 5 min.
7. Start the reaction by adding 0.2 mL of ONPG (4 mg/mL) to each tube, start the stopwatch (t_0), and shake the tubes for a few seconds.
8. After sufficient yellow color has developed, stop the reaction by adding 0.5 mL of a 1 M sodium carbonate solution. Record the time of the reaction (t) with the stopwatch.
9. For each tube, record the optical density at 420 nm and at 550 nm (*see* **Note 28**).
10. The β-galactosidase activity expressed in Miller units is calculated as follows:

$$\text{Units of } \beta - \text{galactosidase} = 1000 \times [(OD_{420} - 1.75 \times OD_{550})/(v \times t \times OD_{6}$$

The volume of the culture (v) used in the assay is in mL, and the reaction time (t) is in minutes.

3.5.2 Gentamicin Sensitivity Test

1. Inoculate 5 mL of LB in a 50-mL Erlenmeyer flask with the *E. coli* strain of interest and incubate overnight at the appropriate temperature (usually 37 °C) with shaking (*see* **Note 7**).
2. The following day, dilute the overnight culture 1:100 in a 250-mL Erlenmeyer flask containing 25 mL of LB (*see* **Note 8**).

3. Incubate the culture at 37 °C with shaking to reach an OD_{600} of 0.3.
4. Add gentamicin (usually 5μg/mL) and incubate at 37 °C with shaking. Time 0 is the time at which gentamicin is added.
5. At different time points (usually 0, 1.5, 3, and 4.5 h), take a 100-μL sample of the culture to determine the colony-forming units per milliliter (cfu/mL).
6. To determine the colony-forming units per milliliter, fill sterile 96-well flat microplates with 180μL of PBS.
7. Add 20μL of the culture sample to the first well (10^{-1} dilution) and perform a ten-fold serial dilution of the culture to reach the 10^{-6} dilution using the electronic multichannel pipette (*see* **Note 29**).
8. Spot 5μL of each dilution on LB agar plates and incubate overnight at 37 °C.
9. Count the colonies and determine the cfu/mL as follows: cfu/mL = number of colonies × 200 × dilution factor.
10. Survival curves can be drawn by representing the percentage of survival as a function of time. The cfu/mL at the time point when the drug is added (t_0) can be used as the 100%.

3.6 Multicopy Suppression

Increasing gene dosage by a multicopy suppressor approach has been exploited to investigate the potential redundancies between Fe-S carriers [17, 31, 52–54]. Expression of a carrier at a higher-than-normal level permits to phenotypically complement the defects caused by the lack of another carrier, then revealing potential pathways followed by Fe-S clusters in vivo. Despite that, one can legitimately question whether such a high level can be attainable under natural conditions, and the added value of such an approach is that it reveals potential capacities of some components while pointing the incapacity of others. Thanks to this method, it was shown that the plasmid-driven expression of *sufA*, but not that of *iscA*, suppresses Fe-S cluster defects associated with a mutation in *erpA*, indicating that beyond their overall functional redundancy, some specificities differentiate the two ATCs, IscA and SufA [17]. With the development of convenient and efficient cloning procedures, this approach can be included in a routine strategy to study redundancy [55, 56]. One cloning procedure that does not require ligase is the Hot Fusion protocol for single fragment cloning and detailed below (Fig. 7) [55] (*see* **Note 30**).

1. To obtain the fragment to be cloned, designing a DNA oligonucleotide primer pair with gene-specific sequences along with a portion of the vector sequences, each with a 17–30-bp overlap (Fig. 7).

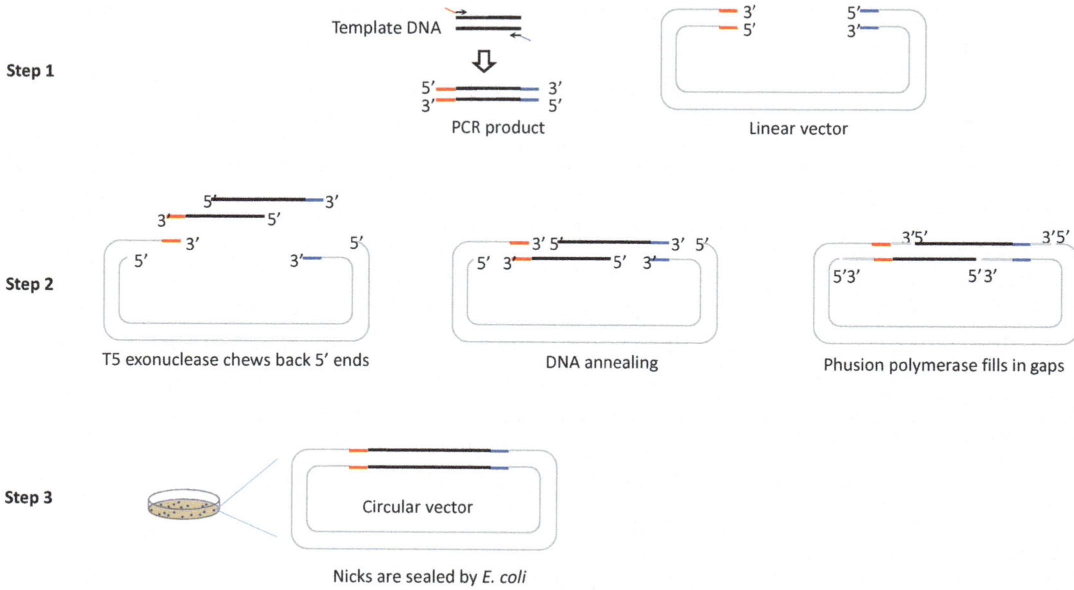

Fig. 7 A diagram of the Hot Fusion process for single fragment cloning. Step 1 is the preparation of a linearized vector and PCR product. Red and blue lines on the vector (gray lines) and PCR product (black lines) indicate the overlapping sequences (17–30 bp). At step 2, the linearized vector and PCR product are added to the Hot Fusion master mix. T5 exonuclease removes nucleotides from the 5′ to 3′ end of double-strand DNA molecules. Phusion DNA polymerase fills in gaps that are overgenerated by T5 exonuclease. At step 3, annealed fragments are transformed into *E. coli*, and nick sites in the nucleotide chain are repaired in *E. coli* cells

2. Perform PCR in a 25-μL total volume reaction mixture containing 1 μL of dNTPs (10 mM), 1μL of 5′ primer (10μM), 1μL of 3′ primer (10μM), 0.5μL of KOD Hot Start DNA Polymerase (2 u/μL), 2.5μL of 10 × PCR buffer, 1μL of DNA template (10 ng/μL for plasmid DNA, 50 ng/μL for genomic DNA), and sterile deionized water to 25μL.
3. Run the following PCR program on the thermocycler: first step: 94 °C for 3 min; second step: 94 °C for 20 s; third step: dedicated temperature depending on the oligonucleotides T_m for 15 s; fourth step: 72 °C for a dedicated time (1 min per Kilobase (kb) to be amplified); fifth step: go back to **step 2** for 24 times; sixth step: 72 °C for 5 min and seventh step: 10 °C on hold (*see* **Note 4**).
4. Run the PCR product on an agarose gel (0.8–1.2%) and purify using PCR Product Purification Kit or Gel Extraction Kit when necessary.
5. Elute DNA with sterile deionized water, measure the DNA concentration using a UV-visible spectrophotometer, and store at −20 °C.

6. Prepare the linearized vector by digesting the DNA vector with appropriate restriction enzymes and dedicated 10× buffers.
7. Depending on the restriction enzyme, heat inactivation is performed at 65 °C or 80 °C for 20 min. When the enzyme cannot be heat-inactivated, the enzyme-digested vector is run on 1% agarose gel and purified using the Gel Extraction Kit. In this latter case, elute DNA with sterile deionized water, measure DNA concentration using a UV-visible spectrophotometer, and store at −20 °C.
8. Prepare 1 mL of 5× preassembly buffer (0.5 M Tris, pH 7.5, 50 mM $MgCl_2$, 1 mM each dNTP, 50 mM DTT, and 25% PEG-8000) by mixing 0.5 mL of 1 M Tris, pH 7.5, 50μL of 1 M $MgCl_2$, 100μL of 10 mM dNTP mix, 50μL of 1 M DTT, 250 mg of PEG-8000, and 150μL of sterile deionized water. Store the buffer at −20 °C.
9. Prepare 400μL of 2× Hot Fusion buffer by mixing 160μL of 5× preassembly buffer, 0.3μL of 10 u/μL T5 exonuclease, 10μL of 2 u/μL Phusion Polymerase, and 230μL of sterile deionized water.
10. Transfer the 2× Hot Fusion buffer as 10-μL aliquots in 0.2-mL PCR tubes. Store at −20 °C.
11. Set up Hot Fusion reaction by adding to the PCR microtubes containing 10μL of 2× Hot Fusion buffer (from **step 10**), 1μL of 30–50 ng/μL of linearized vector (from **step 7**), and 2–3μL of 30–100 ng/μL PCR insert (from **step 5**). Add distilled water to the tube for a final volume of 20μL (*see* **Note 31**).
12. Incubate the tubes in a thermocycler for 1 h at 50 °C.
13. Slowly ramp the temperature down to 20 °C in 5 min (0.1 °C per second) and hold at 10 °C.
14. Use the Hot Fusion reaction for electrotransformation (*see* Subheading 3.1.2) or store at −20 °C, if not used immediately.

3.7 Search for Suppressors

A search of suppressor mutations that revert the mutated phenotype of a carrier mutant has helped in identifying replacement Fe-S cluster carriers. An illustration is provided by the search for suppressors of the Δ*erpA* Δ*iscA* mutant that is rendered viable by the presence of the MVA-dependent pathway (*see* Subheading 3.2.3). Remarkably, all the suppressor mutations exerted their effect via an increased expression of the *sufA* gene [17]. Thus, the suppressor approach illustrates that, in vivo, one barrier to full functional redundancy lies in the control of gene expression. In the case of Fe-S cluster carrier mutants, such as the Δ*erpA* Δ*iscA* mutant, suppressors were selected on plates lacking both arabinose and MVA [17]. To identify the suppressor mutation in the *E. coli* chromosome, whole-genome sequencing is an affordable and convenient method.

3.8 Carriers' Interaction Network

To understand how carriers are organized along the Fe-S clusters' delivery path, as well as to unveil target specificity, mapping protein-protein interactions is an approach that has been used in mammalian, plant, and bacterial model organisms [26, 28, 29, 57]. For this purpose, in *E. coli*, the bacterial two-hybrid (BACTH, for "Bacterial Adenylate Cyclase-based Two-Hybrid") system is a simple, fast, and convenient molecular biology method. This system is based on the interaction-mediated reconstitution of a cyclic adenosine monophosphate (cAMP) signaling cascade [58]. As BACTH uses the production of a stable messenger molecule as a proxy for the protein-protein interaction, a transient physical association can be detected, which is most likely the type of interaction involving Fe-S cluster carriers. Using this method, the atypical A-type domain of NfuA has been shown to be involved in the interaction with both ATC and client proteins [26]. Interestingly, screening and selection procedures can be designed with the BACTH system. For the BACTH protocol, the reader can refer to a recent contribution from the designer of the technique in Methods in Molecular Biology [59].

4 Notes

1. Avoid any salts in the DNA solution, especially for the Subheading 3.1.1 protocol that requires to use a large amount of DNA. In this case, it is best to dialyze DNA by drop dialysis against deionized water, using 0.22-μM Millipore filter, for 1–4 h.
2. The stock solution of MVA is stably frozen for several months. The stock solution of PMS in sterile deionized water is stably frozen and protected from light for several days. The DIP and PQ stock solutions are stable at −20 °C for several weeks.
3. ONPG is prepared at a concentration of 4 mg/mL, preferably in Z buffer. It is best to prepare this solution extemporaneously, although it can be wrapped in aluminum foil and stored at 4 °C for few days (discard if yellow color appears during storage).
4. To obtain a highly concentrated solution of the PCR product, several PCR-independent reactions (2–4 tubes) can be done and pooled before loading the purification column (stay within the DNA-binding capacity of the purification column).
5. The template plasmids are conditional (oriRγ) replicons that require the trans-acting P protein (the *pir* gene product) for

replication. Therefore, these plasmids need to be maintained in an *E. coli* strain carrying the *pir* gene, such as BW25141 [36].

6. If pKD46 is not lost, then colony-purify at 43 °C and test again. It is also recommended to check that the pKD46 plasmid used is really temperature sensitive before the transformation.
7. The optimal *E. coli* growth temperature is 37 °C, unless a temperature-sensitive strain or a strain carrying a temperature-sensitive replication plasmid is used.
8. Different growth culture volumes can be used. However, always respect a ten-fold volume difference between the culture and the Erlenmeyer to ensure good aeration.
9. The final cell concentration is about 2–4 × 10^{10} cells/mL. It is possible to store electrocompetent cells at −80 °C; in this case, use deionized water supplemented with 10% glycerol at the last wash.
10. Dialysis of the PCR product onto a nitrocellulose filter permits to use a large volume of the DNA solution with a low risk of an electric arc. The quantity of DNA added to the cells varies as a function of the type of DNA used (usually 10–50 ng of plasmid or 100–500 ng of PCR product).
11. If an electric arc could be produced, 100μL of sterile cold water can be added to the cell suspension and apply again to the electrical pulse.
12. For transformation by plasmids, spread 100μL of the cell culture onto the appropriate selective LB agar plate. Keep the remaining cell culture at 4 °C (~900μL). If no colonies appear after the overnight incubation, use the remaining cell culture kept at 4 °C to plate again. For transformation by PCR products, centrifuge the cell culture 1 min at 5000 × *g*, resuspend the cell pellet in 100μL of LB, and spread onto the appropriate selective LB agar. Incubate overnight at the appropriate temperature.
13. Be aware that because the *sufC* and *sufD* genes are overlapping by 26-nt, (1) the Keio Δ*sufC* mutant also lacks the start codon of *sufD* and (2) in the Keio Δ*sufD* mutant, the *sufC* gene contains a deletion of the seventh last codons and the stop codon.
14. To perform transduction the same day, start growing the recipient strain (*see* Subheading 3.1.3, **step** 7) at the same time as the donor strain.
15. Make sure the P1 stock titer is at least 10^9 to 10^{10} pfu/mL. It is usually the titer you get when following the Subheading 3.1.3 protocol to prepare the P1 lysate.

16. If the strain is temperature sensitive, use a longer time and a lower temperature. It is important to check that when preparing the P1 lysate, the donor strain exhibited growth before to lyse suddenly.
17. The lysate can be stored at 4 °C for several years.
18. The volume of cell culture can be scaled down to 1 and 0.1 mL, then 250 and 25μL of P1 lysate is used, respectively.
19. Calcium is essential for phage adsorption to the bacteria. Sodium citrate chelates the calcium, thus minimizing secondary infection by residual phage.
20. A control containing only the P1 stock should be done. If the control gives colonies, filter the P1 lysate through a sterile 0.45-μM filter.
21. A *cat*RExBAD cassette, carrying the chloramphenicol selection marker, acetyl transferase (*cat*) gene, instead of the spectinomycin selection marker, is also available [41].
22. A plasmidic version of the MVA genetic by-pass is available using the pUMV22 Spc^R plasmid, which carries the genes for MVA kinase, phosphomevalonate (PMVA) kinase, and diphosphomevalonate (DPMVA) decarboxylase from *Streptomyces* sp. CL190, and the spectinomycin (Spc)-resistant marker gene (Fig. 6) [60, 61].
23. The majority of cells lost the FRT-flanked resistance gene and the FLP helper plasmid simultaneously.
24. FRT-flanked resistance cassettes obtained using pKD3 and pKD4 create an identical scar. The pKD13 scar has no translation signals. Scars could be problematic under certain conditions. Because of the limited homology that is required for gene disruption when using the λ Red system, a new PCR fragment can recombine at the new targeted gene or at the scar of an earlier gene disruption. To avoid such occurrences, single-gene disruptions are made in wild-type hosts and multiple mutants are constructed by standard P1 transduction. However, other chromosomal rearrangements might result from FLP-promoted recombination events between FRT sites at different loci. Although such events are rare, it is best to recheck all gene disruption sites by PCR routinely.
25. If no effect is observed, the treated culture can be used to inoculate fresh LB containing PQ, PMS, or DIP. Alternatively, PQ, PMS, or DIP can be added at the beginning of the culture growth, when the overnight culture is diluted in fresh LB. Depending on the strain sensitivity to PQ, PMS, or DIP, a range of different concentrations can also be tested.
26. Alternatively, the test can be performed using exponentially grown cells.

27. The final volume should always be 1 mL. If high levels of β-galactosidase are being assayed, add 0.1 mL of culture to 0.9 mL of Z buffer; if low levels are being assayed, add 0.5 mL of culture to 0.5 mL of Z buffer.
28. Ideally, the OD_{420} reading (yellow color) should be 0.6–0.9. Centrifugation of the reaction tube in a microfuge for few minutes is possible to remove cell debris and renders OD_{550} reading optional. Ensure that during the transfer of the reaction to spectrophotometer cuvette, no chloroform is taken since it may damage plastic and render the reading false.
29. Depending on the strain sensitivity to gentamicin, no colonies might be observed in the spot corresponding to the 10^{-1} dilution. One solution is to spot 5μL of the nondiluted sample of the cell culture and/or to spread 100μL of the nondiluted cell culture directly on an LB agar plate using sterile glass beads or an L-shaped spreader.
30. The Hot Fusion can be used for multi-DNA fragment cloning by adapting **steps 1** and **2**.
31. Depending on the size of the PCR insert and linearized vector, adjust their quantity to reach a 1:1 molecular ratio.

Acknowledgments

Thanks are due to all members of the Py group and the Barras group (past in Marseille, and present in Paris) who contributed to studies on Fe-S cluster biogenesis. We thank the Ezraty group (Marseille) for sharing material resources and fruitful discussions. Funding sources were from the Center National de la Recherche Scientifique, Aix-Marseille Université, Agence Nationale de la Recherche (JPAIMR «Combinatorial», Blanc «FRACOL»), ERA-CoBioTech « Iron Plug'n Play », and Association Française de l'Ataxie de Friedreich.

References

1. Beinert H (2000) Iron-sulfur proteins: ancient structures, still full of surprises. J Biol Inorg Chem 5:2–15
2. Roche B, Aussel L, Ezraty B et al (2013) Iron/sulfur proteins biogenesis in prokaryotes: formation, regulation and diversity. Biochim Biophys Acta 1827:455–469
3. Lill R, Dutkiewicz R, Freibert SA et al (2015) The role of mitochondria and the CIA machinery in the maturation of cytosolic and nuclear iron-sulfur proteins. Eur J Cell Biol 94:280–291
4. Lill R, Freibert S-A (2020) Mechanisms of mitochondrial iron-sulfur protein biogenesis. Annu Rev Biochem 89:471–499
5. Maio N, Rouault TA (2020) Outlining the complex pathway of mammalian Fe-S cluster biogenesis. Trends Biochem Sci 45:411–426
6. Takahashi Y, Tokumoto U (2002) A third bacterial system for the assembly of iron-sulfur clusters with homologs in archaea and plastids. J Biol Chem 277:28380–28383
7. Tokumoto U, Kitamura S, Fukuyama K et al (2004) Interchangeability and distinct

properties of bacterial Fe-S cluster assembly systems: functional replacement of the isc and suf operons in Escherichia coli with the nifSU-like operon from helicobacter pylori. J Biochem 136:199–209

8. Patzer SI, Hantke K (1999) SufS is a NifS-like protein, and SufD is necessary for stability of the [2Fe-2S] FhuF protein in Escherichia coli. J Bacteriol 181:3307–3309
9. Outten FW, Djaman O, Storz G (2004) A suf operon requirement for Fe-S cluster assembly during iron starvation in Escherichia coli. Mol Microbiol 52:861–872
10. Nachin L, Loiseau L, Expert D et al (2003) SufC: an unorthodox cytoplasmic ABC/ATPase required for [Fe-S] biogenesis under oxidative stress. EMBO J 22:427–437
11. Rincon-Enriquez G, Crété P, Barras F et al (2008) Biogenesis of Fe/S proteins and pathogenicity: IscR plays a key role in allowing Erwinia chrysanthemi to adapt to hostile conditions. Mol Microbiol 67:1257–1273
12. Przybyla-Toscano J, Roland M, Gaymard F et al (2018) Roles and maturation of iron-sulfur proteins in plastids. J Biol Inorg Chem 23:545–566
13. Garcia PS, Gribaldo S, Py B et al (2019) The SUF system: an ABC ATPase-dependent protein complex with a role in Fe-S cluster biogenesis. Res Microbiol 170(8):426–434
14. Baussier C, Fakroun S, Aubert C et al (2020) Making iron-sulfur cluster: structure, regulation and evolution of the bacterial ISC system. Adv Microb Physiol 76:1–39
15. Pérard J, Ollagnier de Choudens S (2018) Iron-sulfur clusters biogenesis by the SUF machinery: close to the molecular mechanism understanding. J Biol Inorg Chem 23:581–596
16. Trotter V, Vinella D, Loiseau L et al (2009) The CsdA cysteine desulphurase promotes Fe/S biogenesis by recruiting Suf components and participates to a new Sulphur transfer pathway by recruiting CsdL (ex-YgdL), a ubiquitin-modifying-like protein. Mol Microbiol 74:1527–1542
17. Vinella D, Brochier-Armanet C, Loiseau L et al (2009) Iron-sulfur (Fe/S) protein biogenesis: phylogenomic and genetic studies of A-type carriers. PLoS Genet 5:e1000497
18. Loiseau L, Gerez C, Bekker M et al (2007) ErpA, an iron sulfur (Fe S) protein of the A-type essential for respiratory metabolism in Escherichia coli. Proc Natl Acad Sci U S A 104:13626–13631
19. Ollagnier-de-Choudens S, Sanakis Y, Fontecave M (2004) SufA/IscA: reactivity studies of a class of scaffold proteins involved in [Fe-S] cluster assembly. J Biol Inorg Chem 9:828–838
20. Ollagnier-de-Choudens S, Mattioli T, Takahashi Y et al (2001) Iron-sulfur cluster assembly: characterization of IscA and evidence for a specific and functional complex with ferredoxin. J Biol Chem 276:22604–22607
21. Ollagnier-de Choudens S, Nachin L, Sanakis Y et al (2003) SufA from Erwinia chrysanthemi. Characterization of a scaffold protein required for iron-sulfur cluster assembly. J Biol Chem 278:17993–18001
22. Gupta V, Sendra M, Naik SG et al (2009) Native Escherichia coli SufA, coexpressed with SufBCDSE, purifies as a [2Fe-2S] protein and acts as an Fe-S transporter to Fe-S target enzymes. J Am Chem Soc 131:6149–6153
23. Angelini S, Gerez C, Ollagnier-de Choudens S et al (2008) NfuA, a new factor required for maturing Fe/S proteins in Escherichia coli under oxidative stress and iron starvation conditions. J Biol Chem 283:14084–14091
24. Bandyopadhyay S, Naik SG, O'Carroll IP et al (2008) A proposed role for the Azotobacter vinelandii NfuA protein as an intermediate iron-sulfur cluster carrier. J Biol Chem 283:14092–14099
25. Romsang A, Duang-Nkern J, Saninjuk K et al (2018) Pseudomonas aeruginosa nfuA: gene regulation and its physiological roles in sustaining growth under stress and anaerobic conditions and maintaining bacterial virulence. PLoS One 13:e0202151
26. Py B, Gerez C, Angelini S et al (2012) Molecular organization, biochemical function, cellular role and evolution of NfuA, an atypical Fe-S carrier. Mol Microbiol 86:155–171
27. Tong W-H, Jameson GNL, Huynh BH et al (2003) Subcellular compartmentalization of human Nfu, an iron-sulfur cluster scaffold protein, and its ability to assemble a [4Fe-4S] cluster. Proc Natl Acad Sci U S A 100:9762–9767
28. Benoit SL, Holland AA, Johnson MK et al (2018) Iron-sulfur protein maturation in helicobacter pylori: identifying a Nfu-type cluster carrier protein and its iron-sulfur protein targets. Mol Microbiol 108:379–396
29. Berger N, Vignols F, Przybyla-Toscano J et al (2020) Identification of client iron-sulfur proteins of the chloroplastic NFU2 transfer protein in Arabidopsis thaliana. J Exp Bot 71:4171–4187
30. Mashruwala AA, Pang YY, Rosario-Cruz Z et al (2015) Nfu facilitates the maturation of iron-sulfur proteins and participates in virulence in

Staphylococcus aureus. Mol Microbiol 95:383–409

31. Py B, Gerez C, Huguenot A et al (2018) The ErpA/NfuA complex builds an oxidation-resistant Fe-S cluster delivery pathway. J Biol Chem 293:7689–7702
32. Py B, Barras F (2015) Genetic approaches of the Fe-S cluster biogenesis process in bacteria: historical account, methodological aspects and future challenges. Biochim Biophys Acta 1853:1429–1435
33. Ollagnier de Choudens S, Barras F (2017) Genetic, biochemical, and biophysical methods for studying FeS proteins and their assembly. Meth Enzymol 595:1–32
34. Li X, Imlay JA (2018) Improved measurements of scant hydrogen peroxide enable experiments that define its threshold of toxicity for Escherichia coli. Free Radic Biol Med 120:217–227
35. Miller JH (1993) A short course in bacterial genetics – a laboratory manual and handbook for Escherichia coli and related bacteria. Cold Spring Harbor 1992. Cold Spring Harbor Laboratory Press, Cold Spring Harbor, New York
36. Datsenko KA, Wanner BL (2000) One-step inactivation of chromosomal genes in Escherichia coli K-12 using PCR products. Proc Natl Acad Sci U S A 97:6640–6645
37. Murphy KC (1998) Use of bacteriophage lambda recombination functions to promote gene replacement in Escherichia coli. J Bacteriol 180:2063–2071
38. Ausubel I, Frederick M (1987) Molecular biology technique. In: Current protocols in molecular biology. Wiley & Sons Inc, New York, New York
39. Baba T, Ara T, Hasegawa M et al (2006) Construction of Escherichia coli K-12 in-frame, single-gene knockout mutants: the Keio collection. Mol Syst Biol 2:2006.0008
40. Yamamoto N, Nakahigashi K, Nakamichi T et al (2009) Update on the Keio collection of Escherichia coli single-gene deletion mutants. Mol Syst Biol 5:335
41. Roux A, Beloin C, Ghigo J-M (2005) Combined inactivation and expression strategy to study gene function under physiological conditions: application to identification of new Escherichia coli adhesins. J Bacteriol 187:1001–1013
42. Gerstel A, Zamarreño Beas J, Duverger Y, Bouveret E, Barras F, Py B (2020) Oxidative stress antagonizes fluoroquinolone drug sensitivity via the SoxR-SUF Fe-S cluster homeostatic axis. PLoS Genet 16(11):e1009198
43. Larson MH, Gilbert LA, Wang X et al (2013) CRISPR interference (CRISPRi) for sequence-specific control of gene expression. Nat Protoc 8:2180–2196
44. Campos N, Rodríguez-Concepción M, Sauret-Güeto S et al (2001) Escherichia coli engineered to synthesize isopentenyl diphosphate and dimethylallyl diphosphate from mevalonate: a novel system for the genetic analysis of the 2-C-methyl-d-erythritol 4-phosphate pathway for isoprenoid biosynthesis. Biochem J 353:59–67
45. Seaver LC, Imlay JA (2001) Alkyl hydroperoxide reductase is the primary scavenger of endogenous hydrogen peroxide in Escherichia coli. J Bacteriol 183:7173–7181
46. Jang S, Imlay JA (2010) Hydrogen peroxide inactivates the Escherichia coli Isc iron-Sulphur assembly system, and OxyR induces the Suf system to compensate. Mol Microbiol 78:1448–1467
47. Martin JE, Waters LS, Storz G et al (2015) The Escherichia coli small protein MntS and exporter MntP optimize the intracellular concentration of manganese. PLoS Genet 11: e1004977
48. Vinella D, Loiseau L, Ollagnier de Choudens S et al (2013) In vivo [Fe-S] cluster acquisition by IscR and NsrR, two stress regulators in Escherichia coli. Mol Microbiol 87:493–508
49. Burschel S, Kreuzer Decovic D, Nuber F et al (2019) Iron-sulfur cluster carrier proteins involved in the assembly of Escherichia coli NADH:ubiquinone oxidoreductase (complex I). Mol Microbiol 111:31–45
50. Roche B, Agrebi R, Huguenot A et al (2015) Turning Escherichia coli into a Frataxin-dependent organism. PLoS Genet 11: e1005134
51. Ezraty B, Vergnes A, Banzhaf M et al (2013) Fe-S cluster biosynthesis controls uptake of aminoglycosides in a ROS-less death pathway. Science 340:1583–1587
52. Pinske C, Sawers RG (2012) Delivery of iron-sulfur clusters to the hydrogen-oxidizing [NiFe]-hydrogenases in Escherichia coli requires the A-type carrier proteins ErpA and IscA. PLoS One 7:e31755
53. Pinske C, Jaroschinsky M, Sawers RG (2013) Levels of control exerted by the Isc iron-sulfur cluster system on biosynthesis of the formate hydrogenlyase complex. Microbiology 159:1179–1189
54. Pinske C, Sawers RG (2012) A-type carrier protein ErpA is essential for formation of an active formate-nitrate respiratory pathway in Escherichia coli K-12. J Bacteriol 194:346–353
55. Fu C, Donovan WP, Shikapwashya-Hasser O et al (2014) Hot fusion: an efficient method to

clone multiple DNA fragments as well as inverted repeats without ligase. PLoS One 9: e115318
56. Hartley JL, Temple GF, Brasch MA (2000) DNA cloning using in vitro site-specific recombination. Genome Res 10:1788–1795
57. Beilschmidt LK, Ollagnier de Choudens S, Fournier M et al (2017) ISCA1 is essential for mitochondrial Fe4S4 biogenesis in vivo. Nat Commun 8:15124
58. Karimova G, Pidoux J, Ullmann A et al (1998) A bacterial two-hybrid system based on a reconstituted signal transduction pathway. Proc Natl Acad Sci U S A 95:5752–5756
59. Karimova G, Gauliard E, Davi M et al (2017) Protein-protein interaction: bacterial two-hybrid. Methods Mol Biol 1615:159–176
60. Takagi M, Kuzuyama T, Takahashi S et al (2000) A gene cluster for the mevalonate pathway from Streptomyces sp. strain CL190. J Bacteriol 182:4153–4157
61. Tanaka N, Kanazawa M, Tonosaki K et al (2016) Novel features of the ISC machinery revealed by characterization of Escherichia coli mutants that survive without iron-sulfur clusters. Mol Microbiol 99:835–848

Chapter 2

Study of Fe-S Cluster Proteins in *Methanococcus maripaludis*, a Model Archaeal Organism

Cuiping Zhao, Christina A. Roberts, Ian J. Drake, and Yuchen Liu

Abstract

Iron-sulfur (Fe-S) clusters are among the oldest and most versatile cofactors present in all domains of life. Many bacterial and eukaryotic Fe-S proteins have been well-characterized, whereas the archaeal ones are less studied. Fe-S proteins are particularly abundant and play essential roles in methanogenic archaea. *Methanococcus maripaludis* is a model methanogen with available genetic tools. Here, we describe the techniques for anaerobic cultivation of *M. maripaludis* with formate, liposome-mediated transformation, expression and anoxic affinity purification of Fe-S proteins, Fe-S cluster reconstitution, and analysis of Fe-S proteins by UV-visible absorption spectroscopy.

Key words Iron-sulfur clusters, Methanogen, *Methanococcus*, Archaea, Anaerobic cultivation, Liposome-mediated transformation, Anoxic protein purification, Fe-S cluster reconstitution, UV-visible absorption spectroscopy

1 Introduction

Fe-S clusters are one of the most ancient and versatile cofactors present in all three domains of life [1, 2]. The proteins containing Fe-S clusters (Fe-S proteins) are essential in various fundamental cellular processes, including central metabolism, respiration, photosynthesis, DNA repair and replication, protein synthesis, and gene regulation [3, 4]. Fe-S clusters are often vulnerable to oxidants, including O_2 [5], and thus, it is important to maintain strict anoxic conditions during the entire process of Fe-S protein purification and analysis. Although many Fe-S proteins have been characterized in bacteria and eukaryotes, the studies of archaeal proteins are often hindered by difficulties in their cultivations and limited genetic tools.

Methanogens, or methanogenic archaea, are obligate anaerobes that obtain most energy from methane production [6–9]. Methanogens possess a large number of Fe-S proteins [10], many of which are directly involved in methanogenesis. For example, the

Patricia C. Dos Santos (ed.), *Fe-S Proteins: Methods and Protocols*, Methods in Molecular Biology, vol. 2353, https://doi.org/10.1007/978-1-0716-1605-5_2, © Springer Science+Business Media, LLC, part of Springer Nature 2021

methanogenesis pathway enzyme formyl-methanofuran dehydrogenase (Fmd) complex contains 46 [4Fe-4S] clusters [11], and the heterodisulfide reductase-[NiFe] hydrogenase (Hdr-Mvh) complex possesses 28 [4Fe-4S] clusters [12]. Therefore, studying Fe-S proteins in methanogens is critical for elucidating their metabolism, which contributes to our understanding of archaeal diversity and evolution.

Methanococcus maripaludis is a valuable model organism for archaea studies. It has rapid (doubling time ~ 2 h) and reliable growth on either H_2 or formate as the energy source [13]. Many genetic tools have been developed in this organism, including DNA delivery, gene deletion, and recombinant protein expression from shuttle vectors [14–16]. High coverage transcriptomic and proteomic databases are also available [17–21]. Moreover, the essentiality of each gene was analyzed by a whole-genome transposon mutagenesis study [22]. This chapter introduces specific techniques for the cultivation and genetic manipulation of this archaeon for the expression of Fe-S proteins. Furthermore, methods for affinity purification, Fe-S cluster reconstitution, and characterization of O_2-sensitive Fe-S cluster proteins by UV-visible absorption spectroscopy are also described.

2 Materials

2.1 Strain and Growth Media

1. *M. maripaludis* strain S0001 (available upon request) [23].
2. MCYF liquid medium: Mix medium components (Table 1) in a 500-mL glass bottle (*see* **Notes 1–3**). Cap with a rubber stopper and sparge with a stream of O_2-free N_2 for 1 h to remove dissolved O_2 (Fig. 1a). Tightly screw an open-top plastic cap and transfer the bottle into the anaerobic chamber with an atmosphere of 95% (v/v) N_2 and 5% (v/v) H_2. Add 0.1 g of L-cysteine hydrochloride and 0.1 g sodium 2-mercaptoethanosulfonate per 100-mL medium as reductants. Wait until the medium becomes colorless. Inside the chamber, aliquot 5-mL medium into 18 × 150 mm anaerobic culture tubes or 200-mL medium into 1-L culture bottles (*see* **Notes 4** and **5**). Seal tubes or bottles with 20-mm butyl rubber stoppers and crimp with aluminum seals (Fig. 1b). Outside the chamber, flush the medium tubes with N_2/CO_2 (4:1, v/v) using a gassing station and then pressurize to 5 psi. Sterilize by autoclave at 121 °C for 20 min (*see* **Note 6**).
3. Sodium sulfide stock solution: Dissolve 2.5 g $Na_2S{\cdot}9H_2O$ in 100-mL anoxic distilled H_2O containing one pellet of NaOH inside the anaerobic chamber (*see* **Note 7**). Aliquot 5-mL solution into anaerobic culture tubes (*see* **Note 4**). Seal tubes with 20-mm butyl rubber stoppers and crimp with aluminum

Table 1
Composition of the MCYF medium

Component	Amount per 100 mL
General salts solution: 0.67 g/L KCl, 5.5 g/L $MgCl_2 \cdot 6H_2O$, 6.9 g/L $MgSO_4 \cdot 7H_2O$, 1 g/L NH_4Cl, 0.28 g/L $CaCl_2 \cdot 2H_2O$	50 mL
Glycyl-glycine solution (1 M, pH 8.0)	20 mL
K_2HPO_4 solution (14 g/L)	1 mL
Trace mineral solution: 1.5 g/L nitrilotriacetic acid, 0.1 g/L $MnCl_4 \cdot 4H_2O$, 0.2 g/L $Fe(NH_4)_2(SO_4)_2 \cdot 6H_2O$, 0.1 g/L $CoCl_2 \cdot 6H_2O$, 0.1 g/L $ZnSO_4 \cdot 7H_2O$, 0.01 g/L $CuSO_4 \cdot 5H_2O$, 0.025 g/L $NiCl_2 \cdot 6H_2O$, 0.2 g/L Na_2SeO_3, 0.1 g/L $Na_2MoO_4 \cdot 2H_2O$, 0.1 g/L $Na_2WO_4 \cdot 2H_2O$	1 mL
Trace vitamin solution: 2 mg/L biotin, 2 mg/L folic acid, 10 mg/L pyridoxine hydrochloride, 5 mg/L thiamine hydrochloride, 5 mg/L riboflavin, 5 mg/L nicotinic acid, 5 mg/L DL-calcium pantothenate, 0.1 mg/L vitamin B_{12}, 5 mg/L 4-aminobenzoic acid, 5 mg/L lipoic acid	1 mL
Iron stock solution: 2 g/L $Fe(NH4)_2(SO_4)_2 \cdot 6H_2O$ dissolved in distilled water containing 500 μL/L (v/v) concentrated HCl	0.5 mL
Resazurin (1 g/L)	0.1 mL
Sodium acetate solution: 136 g/L $CH_3COONa \cdot 3H_2O$	1 mL
L-Alanine solution (100 mM)	1 mL
$NaHCO_3$	0.5 g
Sodium formate	2.7 g
Yeast extract	0.2 g
Casamino acid	0.2 g
Distilled H_2O	Bring total volume to 100 mL

seals. Sterilize by autoclave at 121 °C for 20 min. These tubes can be stored inside the anaerobic chamber for about 2 months. Discard tubes with precipitates.

4. Puromycin stock solutions: Dissolve 10 mg of puromycin dihydrochloride in 10-mL anoxic distilled water to make a 1 mg/mL (400×) stock solution. Sterilize by filtration. Dilute eightfold by adding 710 μL 400 × stock solution to 5-mL sterilized MCYF medium in a sealed anaerobic culture tube to make a 125 μg/mL (50×) stock solution. Store stock solutions at −20 °C.

5. MCYF solid medium: Mix MCYF medium components (Table 1) in a 500-mL glass bottle and sparge with N_2 for 1 h. Transfer the medium bottle into the anaerobic chamber and add 0.1 g of L-cysteine hydrochloride, 0.1 g of sodium

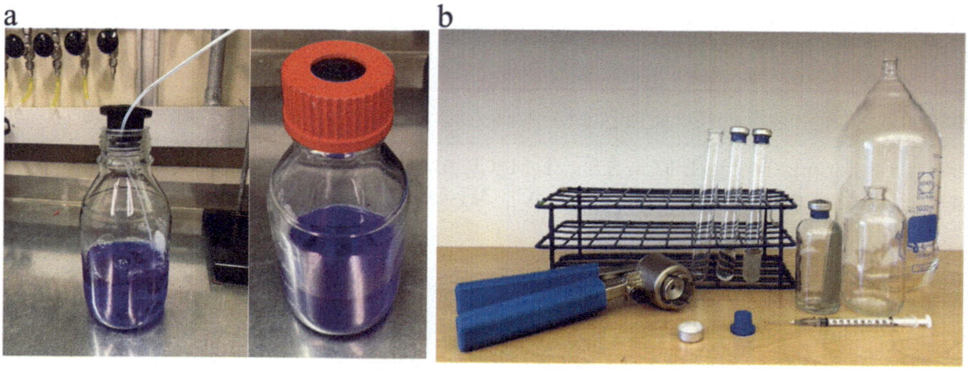

Fig. 1 Preparation of MCYF media. (**a**) A 500-mL glass bottle containing MCYF medium is capped with a rubber stopper and inserted with long plastic tubing for sparging with N_2. After sparging, the bottle is closed with an open-top screw cap before transfer into the anaerobic chamber. (**b**) The 18 × 150-mm anaerobic culture tubes and 1-L bottles are used for 5-mL and 200-mL liquid cultures, respectively. The 60-mL serum bottle with 15-mL solid agar medium is used for plating. The culture tubes or bottles are sealed with the blue butyl stoppers and aluminum seals using a 20-mm crimper

2-mercaptoethanosulfonate, and 1 g of agar per 100-mL medium. Mix thoroughly and aliquot 15-mL suspension into 60-mL serum bottles (Fig. 1b). Seal bottles with 20-mm butyl rubber stoppers and crimp with aluminum seals. Outside the chamber, flush bottles with N_2/CO_2 (4:1, v/v) and then pressurized to 5 psi. Sterilize by autoclave at 121 °C for 20 min. After bottles cool down to 48–55 °C, add 0.3 mL of sodium sulfide stock solution and 0.3 mL of 125 μg/mL puromycin stock solution. Mix immediately by inversion and then place bottles horizontally to solidify the agar medium on one side of the bottle (Fig. 1b).

6. Glycerol storage solution: Prepare 20 mL of MCYF medium by mixing components listed in Table 1. Supplement with 30-mL glycerol. Sparge the solution with N_2 for 1 h and then transfer it into the anaerobic chamber. Add 0.05 g of L-cysteine hydrochloride and 0.05 g sodium 2-mercaptoethanosulfonate per 50-mL MCYF-glycerol mixture. Wait until the mixture becomes colorless. Aliquot 0.8-mL mixture into 2-mL serum bottles. Seal bottles with 13-mm rubber stoppers and aluminum seals. Autoclave at 121 °C for 20 min. Store inside the anaerobic chamber at room temperature until use.

2.2 Liposome-Mediated Transformation of *M. maripaludis*

1. Anaerobic chamber with an atmosphere of 95% (v/v) N_2 and 5% (v/v) H_2.
2. Shaking incubator.
3. Spectrophotometer with vial/test tube holder.
4. Microcentrifuge in the anaerobic chamber.
5. Disposable syringes (1 mL and 5 mL).

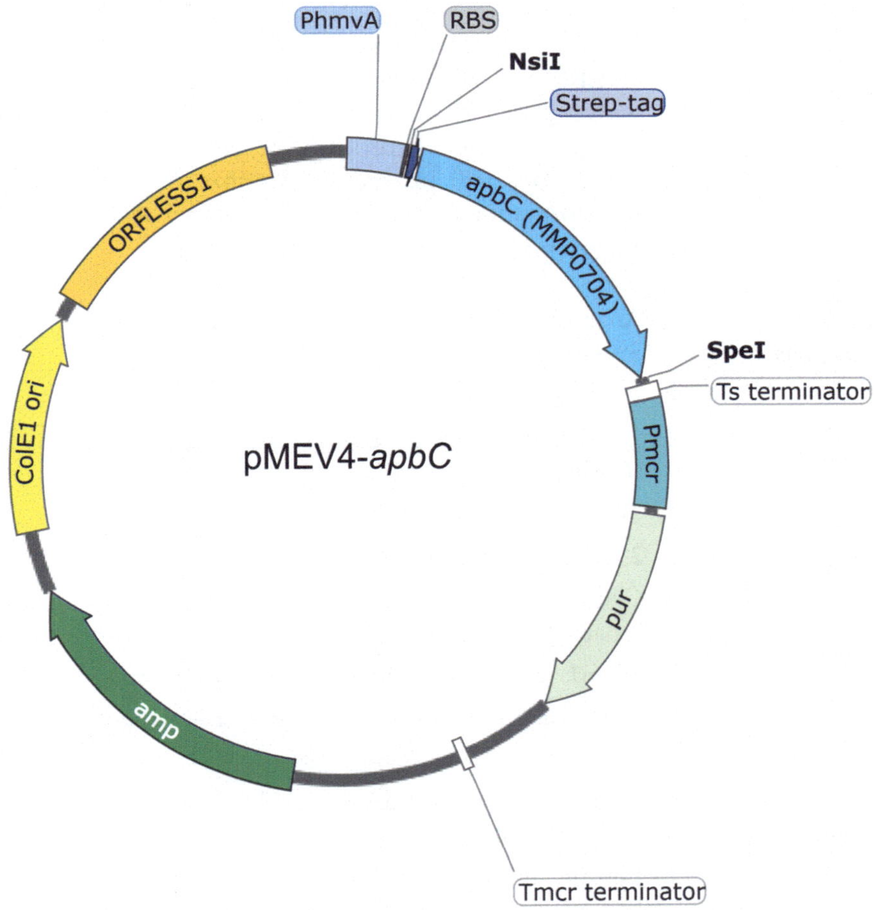

Fig. 2 Map of the pMEV4 shuttle vector with the Strep-tagged *apbC* (locus tag: MMP0704) gene. The Strep-tagged *apbC* gene was cloned in-frame into the NsiI and SpeI sites for expression under the control of the P_{hmvA} promoter. The plasmid (5338 bp) contains ColE1 ori, an origin of replication in *E. coli*; ORFLESS1, a possible origin of replication in *M. maripaludis*; P_{hmvA}, a histone gene promoter from *Methanococcus voltae*; RBS, ribosome-binding site (aggtga); Ts terminator, a synthetic terminator (gatatagcaaaagtgggacttaagttcc-cacttttgctatt); P_{mcr}, the methyl coenzyme M reductase gene promoter from *M. voltae*; pur, a codon-optimized puromycin transacetylase gene from *Streptomyces alboniger* for positive selection in *M. maripaludis*; T_{mcr} terminator, the methyl coenzyme M reductase gene terminator from *M. voltae*; amp, the ampicillin resistance cassette for positive selection in *E. coli*

6. Syringe needles (22-gauge, L 1 in.).
7. pMEV4 shuttle vector [24].
8. pMEV4-*apbC* plasmid (Fig. 2) [25].

9. Transformation buffer: Prepare a solution containing 50 mM sodium 4-(2-Hydroxyethyl)piperazine-1-ethanesulfonic acid (HEPES), 0.85 M sucrose, 1 mg/L resazurin, pH 7.5. Sparge with N_2 for 1 h and transfer into the anaerobic chamber. Add 0.1 g L-cysteine hydrochloride and 0.015 g dithiothreitol (DTT) per 100 mL buffer. Sterilize by filtration. Store the transformation buffer inside the chamber at room temperature.
10. DOTAP liposomal transfection reagent.

2.3 Anoxic Fe-S Protein Purification

1. Anaerobic chamber with an atmosphere of 95% (v/v) N_2 and 5% (v/v) H_2.
2. High-speed centrifuge.
3. 1-L centrifugation bottles with gastight cap assembly (O-rings).
4. 50-mL centrifuge tubes.
5. 2-mL serum bottles.
6. Centrifugal filter units for concentrating proteins (loading capacity = 15 mL).
7. 20-mL gravity-flow chromatography columns.
8. Strep-Tac tin Super flow protein purification resin.
9. DNase I.
10. Ethylenediamine tetra acetic acid (EDTA)-free Protease Inhibitor Cocktail.
11. Binding buffer: 50 mM sodium HEPES (pH 7.5), 300 mM NaCl, 5 mM $MgCl_2$. Degas by sparging with N_2. Transfer into the anaerobic chamber and keep stirring the buffer overnight to equilibrate with the anoxic atmosphere.
12. Elution buffer: 50 mM sodium HEPES (pH 7.5), 300 mM NaCl, 5 mM $MgCl_2$, 2.5 mM desthiobiotin. Degas by sparging with N_2. Transfer into the anaerobic chamber and keep stirring the buffer overnight to equilibrate with the anoxic atmosphere.
13. Glycerol.

2.4 UV-Visible Absorption Spectroscopy

1. UV-visible spectrophotometer in the anaerobic chamber.
2. Ultra-micro UV cuvettes (path length = 1 cm).
3. 100 mM DTH: 100 mM sodium dithionite (DTH) dissolved in the binding buffer (*see* **Note 8**).

2.5 In Vitro Reconstitution of Fe-S Clusters

1. Microcentrifuge in the anaerobic chamber.
2. PD MiniTrap G-25 desalting column.
3. Centrifugal filter units for concentrating proteins (loading capacity = 0.5 mL).

4. Buffer A: 50 mM sodium HEPES (pH 7.5), 300 mM NaCl, 25% (v/v) glycerol. Degas by sparging with N_2. Transfer into the anaerobic chamber and keep stirring the buffer overnight to equilibrate with the anoxic atmosphere.
5. Buffer B: 50 mM sodium HEPES (pH 7.5), 300 mM NaCl. Degas by sparging with N_2. Transfer into the anaerobic chamber and keep stirring the buffer overnight to equilibrate with the anoxic atmosphere.
6. 1 M DTH: 1 M sodium dithionite in buffer A (*see* **Note 8**).
7. EDTA: 0.5 M disodium EDTA (pH 8.0). Degas by sparging with N_2.
8. Ferrous ammonium sulfate solution: 100 mM Fe $(NH_4)_4(SO_4)_2{\cdot}6H_2O$ in anoxic water.
9. Sodium sulfide solution: 100 mM $Na_2S{\cdot}9H_2O$ in anoxic water (*see* **Note 7**).
10. DTT solution: 1 M dithiothreitol in anoxic water.

3 Methods

3.1 Genetic Engineering of M. maripaludis for Expression of Strep-Tagged Fe-S Proteins

To engineer *M. maripaludis* for expression of a recombinant archaeal Fe-S protein, the gene encoding the target protein is cloned into the pMEV4 shuttle vector (*see* **Note 9**) with an N-terminal Strep-tag. This chapter uses the Fe-S proteins ApbC (locus tag: MMP0704) [25] and ThiI (locus tag: MMP1354) [26] as examples to illustrate the techniques. The final plasmid construct, for example, pMEV4-*apbC* (Fig. 2), is introduced into the *M. maripaludis* strain S0001 (*see* **Note 10**) via a liposome-mediated transformation method.

1. Before inoculation of the *M. maripaludis* culture, add anoxically 0.1 mL of sodium sulfide stock solution per 5 mL of MCYF medium using syringes as the sulfur source for growth (*see* **Note 11**). The final sodium sulfide concentration is 2 mM.
2. To inoculate *M. maripaludis* strain S0001 from a frozen glycerol culture stock, transfer the stock to a −20 °C cooler. Scratch the surface of the frozen sample with a syringe needle and then insert the needle directly through the butyl stopper into a 5 mL of MCYF medium tube (from **step 1**). Rinse the syringe several times with the medium to suspend the cells. Incubate at 37 °C with 100 rpm orbital shaking until OD_{600nm} reaches ~0.7–1.0 (usually 2–3 days). This culture can be stored on the bench at room temperature for up to 4 weeks.
3. To prepare fresh cultures for transformation, inoculate 0.1 mL of the *M. maripaludis* S0001 culture (from **step 2**) into 5 mL of MCYF medium containing 2 mM sodium sulfide (*see* **Note**

11) anoxically with syringes. Incubate at 37 °C with 100 rpm orbital shaking until OD_{600nm} reaches ~0.7–1.0 (usually overnight).

4. Transfer culture tubes into the anaerobic chamber. Collect cells by centrifugation at 5000 × *g* for 5 min at room temperature.
5. Resuspend cells in 0.5 mL of transformation buffer. Spin down cells by centrifugation at 5000 × *g* for 2 min at room temperature. Discard the supernatant and resuspend cells in 0.5 mL of transformation buffer.
6. For each transformation reaction, mix 1 μg of plasmid DNA in 10 μL of H_2O with 15 μL of DOTAP liposomal transfection reagent and 55 μL transformation buffer. Incubate the mixture at room temperature inside the anaerobic chamber for 30 min.
7. Add the DNA-liposome mixture to the cell suspension (from **step 5**). Mix gently and incubate for 2 h at room temperature inside the chamber.
8. Transfer the entire transformation mixture to 5 mL of MCYF medium with 2 mM sodium sulfide (*see* **Note 11**) in an anaerobic culture tube. Move cultures outside the anaerobic chamber and incubate at 37 °C overnight with 100 rpm orbital shaking.
9. Make serial dilutions (10×, 100×, and 1000×) of the transformation cultures with the MCYF medium.
10. Transfer 1 mL diluted culture to a 60-mL serum bottle containing 15 mL of MCYF solid agar medium with 2 mM sodium sulfide and 2.5 μg/mL puromycin anoxically using syringes. Swirl the bottle to spread the liquid evenly on the agar surface. Place the bottle horizontally with the agar surface facing up for 1 h.
11. Incubate the serum bottles at 37 °C. Keep them upright and avoid shaking during the incubation. It usually takes 5–7 days for colonies to appear.
12. Transfer the serum bottles into the anaerobic chamber. Remove stoppers and pick up single colonies using syringes with needles. Inoculate each single colony into 5 mL of MCYF medium containing 2 mM sodium sulfide and 2.5 μg/mL puromycin (*see* **Note 12**).
13. Grow cultures at 37 °C with 100 rpm orbital shaking till OD_{600nm} reaches 0.6–1.0 (usually 2–3 days). Screen cultures for desired Fe-S protein expressions by Western blot with a Strep-tag antibody.
14. To prepare glycerol stock cultures, inside the anaerobic chamber, add 0.8 mL of cell culture to 0.8 mL of glycerol storage solution in 2-mL serum bottles using syringes. Mix

thoroughly. The final concentration of glycerol is ~30% (v/v). Freeze and store glycerol stock cultures in a −80 °C freezer.

3.2 Cultivation Conditions for Fe-S Protein Expression

1. Prepare fresh starter cultures in 5 mL of MCYF medium supplemented with 2 mM sodium sulfide and 2.5 μg/mL puromycin as described in Subheading 3.1 (**steps 1–3**).
2. Transfer 5-mL starter culture (OD_{600nm} ~ 0.7–1.0) into each 200-mL MCYF medium supplemented with 2 mM sodium sulfide and 2.5 μg/mL puromycin in 1-L culture bottles (Fig. 1b) using syringes.
3. Incubate at 37 °C with 100 rpm orbital shaking. It usually takes 1–2 days for OD_{600nm} to reach ~1.0.

3.3 Anoxic Purification of Strep-Tagged Fe-S Protein

In order to prevent cluster degradation by O_2, many Fe-S cluster proteins should be purified and analyzed under strictly anoxic conditions.

1. Harvest 1 L cell culture from Subheading 3.2 under anoxic conditions. Inside the anaerobic chamber, transfer liquid cultures to centrifuge bottles. Close tightly with O-ring screw caps and wrap caps with a parafilm to minimize O_2 contamination. Transfer bottles outside the chamber and collect cells by centrifugation at 4000 × *g* for 10 min at 4 °C.
2. Transfer centrifuge tubes back to the anaerobic chamber and discard the supernatant. Resuspend the cell pellet with residual medium (~1 mL) and transfer the suspension to a 50-mL centrifuge tube. The cell suspension can be stored in a −80 °C freezer until use. When transferring centrifuge tubes outside the chamber, wrap caps with a parafilm to minimize O_2 contamination.
3. Inside the anaerobic chamber, add 20 mL of binding buffer to the cell suspension and mix by inverting the tube.
4. Lyse cells by two freeze/thaw cycles (*see* **Note 13**). In each cycle, freeze the cell suspension in a −80 °C freezer for 15 min and then thaw at room temperature.
5. Inside the anaerobic chamber, add one pellet of EDTA-free Protease Inhibitor Cocktail and 10 U DNase I to ~20 mL of cell suspension. Incubate for 30 min at room temperature.
6. Centrifuge the cell lysate at 23,000 × *g* for 20 min at 4 °C.
7. Transfer centrifuge tubes back to the anaerobic chamber. Load the supernatant onto a 20-mL gravity-flow chromatography column containing 1-mL Strep-Tactin Superflow protein purification resin equilibrated with the binding buffer.
8. Wash the column with 20 mL of binding buffer.
9. Elute the Strep-tagged protein with 10 mL of elution buffer.

10. Concentrate the entire 10 mL of eluate using a centrifugal filter unit to a final volume of ~0.2–0.3 mL. Concentrated Fe-S proteins are usually brownish in color (*see* **Note 14**). Analyze the concentrated protein by SDS-polyacrylamide gel electrophoresis (PAGE) [26].
11. Supplement the purified protein with anoxic glycerol to a final glycerol concentration of 20–40% (v/v). Store protein samples in 2-mL serum bottles sealed with rubber stoppers and aluminum seals at −80 °C until use.

3.4 UV-Visible Absorption Spectroscopy

Fe-S clusters display broad, characteristic UV-visible absorption spectra that may provide a hint of the cluster type. The spectra of [2Fe-2S] proteins often exhibit several bands at around 320, 410, 450–460 nm and a shoulder around 550 nm [27]. Spectra of [4Fe-4S] and [3Fe-4S] proteins display a broad band at around 420 nm (Fig. 3). The Fe-S proteins reduced with DTH show decreased UV-visible absorptions.

1. Transfer ≥70 μL of as-purified protein (~40–200 μM) into an ultra-micro UV cuvette inside the anaerobic chamber (*see* **Note 4**).
2. Use a spectrophotometer inside the anaerobic chamber to determine the UV-visible absorption spectrum of the as-purified protein.

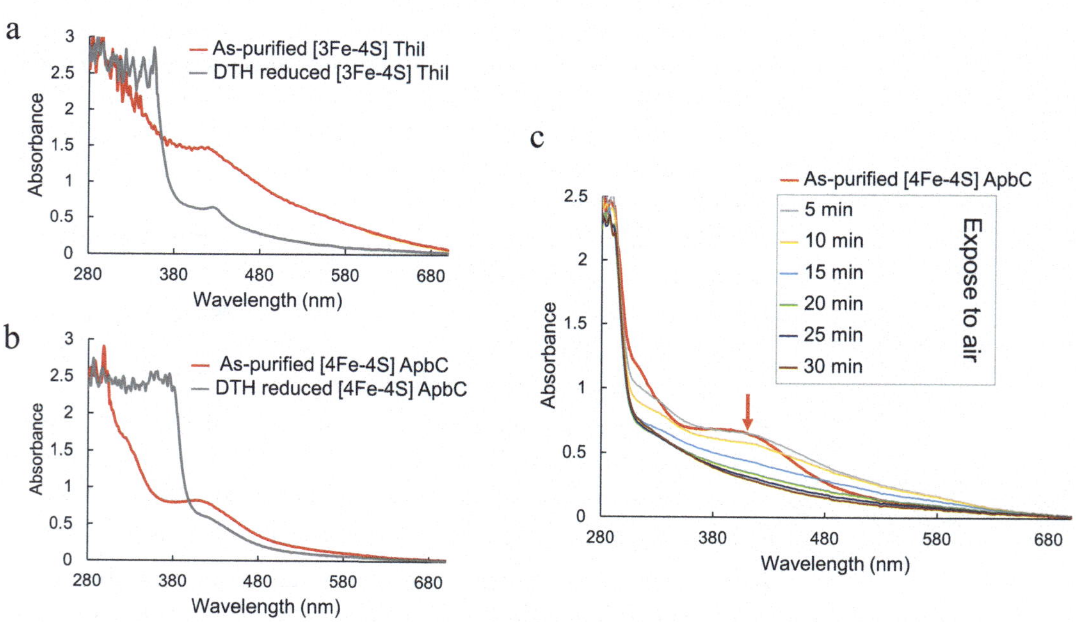

Fig. 3 UV-visible spectra of representative [3Fe-4S] and [4Fe-4S] proteins anoxically purified from *M. maripaludis*. (**a**) Spectra of the [3Fe-4S] protein ThiI (locus tag: MMP1354) [26]. (**b**) Spectra of the [4Fe-4S] protein ApbC (locus tag: MMP0704) [25]. (**c**) Spectra of the [4Fe-4S] protein ApbC (locus tag: MMP0704) upon air exposure [25]. The red arrow indicates the decrease of the absorbance at ~420 nm after the protein was exposed to air

3. Add freshly prepared 100 mM DTH stock solution (*see* **Note 8**) to give a final concentration of 5 mM. Mix well and incubate for 5 min at room temperature to reduce the protein inside the cuvette.
4. Repeat **step 2** to determine the UV-visible absorption spectrum of the reduced protein.
5. To determine the O_2-sensitivity of an Fe-S protein, seal the cuvette with a rubber stopper and transfer it outside of the anaerobic chamber. Record its initial UV-visible absorption spectrum using a spectrophotometer. Remove the rubber stopper and pipette up and down to mix the protein with air. Measure UV-visible absorption spectra every 5 min (Fig. 3c).

3.5 *In Vitro* Reconstitution of Fe-S Clusters

When the archaeal Fe-S cluster protein is isolated mainly in its apo-form (e.g., purified under oxic conditions or recombinantly expressed and purified from *E. coli*), the clusters can often be reconstructed chemically by incubation with ferrous ammonium sulfate as the Fe source and sodium sulfide as the sulfur source under anoxic conditions. The following experiments are performed anoxically in the anaerobic chamber.

1. To prepare the apo-protein, incubate the as-purified protein (~ 400 μM in 400 μL buffer A) with 10 mM DTH and 10 mM EDTA at 4 °C for 4 h.
2. Pass the protein sample through a PD Mini Trap G-25 desalting column pre-equilibrated with buffer A to remove chelated Fe and free sulfide species.
3. Incubate the apo-protein with 5 mM DTT at room temperature for 1 h.
4. Add four- to eight-fold molar excess of Fe from the 100 mM Fe $(NH_4)_4(SO_4)_2$ solution drop by drop with gentle stirring. Then, add sodium sulfide at an equal molar concentration as Fe from the 100 mM solution drop by drop with gentle stirring. Incubate the mixture at room temperature for 2 h.
5. Remove precipitates by centrifugation at 14,000 × *g* for 2 min.
6. Remove unbound Fe and sulfide bypassing the supernatant through a PD Mini Trap G-25 desalting column pre-equilibrated with buffer B (*see* **Note 15**).
7. Concentrate the reconstituted protein using a centrifugal filter unit to a final volume of ~0.2 mL. Supplement the reconstituted protein with anoxic glycerol to a final glycerol concentration of 30% (v/v).
8. Confirm the reconstituted Fe-S cluster by determination of the Fe content and UV-visible absorption spectroscopy.

4 Notes

1. To prepare the trace mineral solution, dissolve nitrilotriacetic acid and neutralize to pH 6.5 with KOH. Then, add minerals in order, allowing each chemical to dissolve before adding the next one. Adjust the final pH to 7.0.
2. Store the trace vitamin solution in the dark at 4 °C.
3. Adjusting the pH of the 1 M glycyl-glycine solution with NaOH can facilitate dissolving.
4. For anoxic work, all plastic and glass materials (tubes, bottles, pipettes, syringes, rubber stoppers, etc.) should stay inside the anaerobic chamber for at least 24 h to allow complete deaeration.
5. The MCYF medium contains 0.4 M sodium formate. During growth, methanogens convert formate to CO_2 and CH_4 gases that cause an increase in pressure. For each milliliter of the medium, the culture can produce up to 0.4 mmol of gas. Depending on the pH, some of the CO_2 is dissolved in the medium. Nevertheless, the glassware used must have sufficient headspace and be able to withstand this increase in pressure.
6. After autoclave, the MCYF medium may form white precipitates, which will be re-dissolved after cooling down to room temperature.
7. Weigh out slightly more than 2.5 g $Na_2S{\cdot}9H_2O$ in the fume hood. The $Na_2S{\cdot}9H_2O$ crystals may have yellowish color on the surface. Rinse the crystals briefly with distilled H_2O and then quickly dry the crystals with a paper towel. Re-weigh the crystals and adjust the volume of the distilled water accordingly to give a final concentration of 2.5% (w/v) $Na_2S{\cdot}9H_2O$.
8. The DTH stock solution must be freshly prepared just before use. Store solid DTH away from the light inside a jar with a desiccant at room temperature. Weigh out DTH and transfer into a serum bottle subsequently closed with a rubber stopper and aluminum seal. Degas through the airlock with a needle puncturing through the stopper and transfer into the anaerobic chamber. Dissolve in anoxic buffers to make stock solutions.
9. The pMEV4 shuttle vector is derived from the pAW42 vector by adding a synthetic terminator (Ts) to the end of the inserted gene [24].
10. The *M. maripaludis* strain S0001 is derived from integrating ORF1 into the *M. maripaludis* strain Mm900 genome. ORF1 supports the replication of the pMEV4 shuttle vector [23].

11. For better growth, add sodium sulfide 1 day before culture inoculation. Incubate the medium with sulfide at 37 °C overnight with orbital shaking.
12. Release gas inside serum bottles with syringe needles before transferring them into the chamber. Keep bottles upright. Gently remove butyl stoppers and avoid contamination. Discard liquid inside serum bottles without contacting the agar surface. Use a 1-mL syringe with a needle to pick up colonies. It is easier to reach some colonies with bent needles. After picking up a colony, insert the needle directly through the butyl stopper into a 5-mL MCYF medium tube and rinse the syringe several times with the medium to suspend the cells.
13. Alternatively, cells can be lysed by mixing with 2-mL BugBuster® 10X Protein Extraction Reagent.
14. The Fe content can be determined using the QuantiChrom Iron Assay Kit (BioAssay Systems).
15. Sometimes Fe^{3+}-containing black precipitates are formed during the reconstitution after adding sulfide, usually due to residual oxygen. The precipitates are often not fully removed by the G25 desalting column. Re-purifying the protein with the Strep-Tactin resin can help remove the black precipitates.

References

1. Eck RV, Dayhoff MO (1966) Evolution of the structure of ferredoxin based on living relics of primitive amino acid sequences. Science 152 (3720):363–366. https://doi.org/10.1126/science.152.3720.363
2. Meyer J (2008) Iron-sulfur protein folds, iron-sulfur chemistry, and evolution. J Biol Inorg Chem 13(2):157–170. https://doi.org/10.1007/s00775-007-0318-7
3. Johnson DC, Dean DR, Smith AD, Johnson MK (2005) Structure, function, and formation of biological iron-sulfur clusters. Annu Rev Biochem 74:247–281. https://doi.org/10.1146/annurev.biochem.74.082803.133518
4. Pain D, Dancis A (2016) Roles of Fe-S proteins: from cofactor synthesis to iron homeostasis to protein synthesis. Curr Opin Genet Dev 38:45–51. https://doi.org/10.1016/j.gde.2016.03.006
5. Imlay JA (2006) Iron-Sulphur clusters and the problem with oxygen. Mol Microbiol 59 (4):1073–1082. https://doi.org/10.1111/j.1365-2958.2006.05028.x
6. Liu Y, Whitman WB (2008) Metabolic, phylogenetic, and ecological diversity of the methanogenic archaea. Ann N Y Acad Sci 1125:171–189. https://doi.org/10.1196/annals.1419.019
7. Buan NR (2018) Methanogens: pushing the boundaries of biology. Emerg Top Life Sci 2 (4):629–646. https://doi.org/10.1042/ETLS20180031
8. Lyu Z, Shao N, Akinyemi T, Whitman WB (2018) Methanogenesis. Curr Biol 28(13): R727–R732. https://doi.org/10.1016/j.cub.2018.05.021
9. Lyu Z, Liu Y (2019) Diversity and taxonomy of methanogens. In: Stams AJM, Sousa D (eds) Biogenesis of hydrocarbons. Springer International Publishing, Cham, pp 19–77. https://doi.org/10.1007/978-3-319-53114-4_5-1
10. Liu Y, Sieprawska-Lupa M, Whitman WB, White RH (2010) Cysteine is not the sulfur source for iron-sulfur cluster and methionine biosynthesis in the methanogenic archaeon *Methanococcus maripaludis*. J Biol Chem 285 (42):31923–31929. https://doi.org/10.1074/jbc.M110.152447
11. Wagner T, Ermler U, Shima S (2016) The methanogenic CO_2 reducing-and-fixing enzyme is bifunctional and contains 46 [4Fe-4S] clusters. Science 354

(6308):114–117. https://doi.org/10.1126/science.aaf9284
12. Wagner T, Koch J, Ermler U, Shima S (2017) Methanogenic heterodisulfide reductase (HdrABC-MvhAGD) uses two noncubane [4Fe-4S] clusters for reduction. Science 357 (6352):699–703. https://doi.org/10.1126/science.aan0425
13. Long F, Wang L, Lupa B, Whitman WB (2017) A flexible system for cultivation of *Methanococcus* and other formate-utilizing methanogens. Archaea 2017:7046026. https://doi.org/10.1155/2017/7046026
14. Leigh JA, Albers SV, Atomi H, Allers T (2011) Model organisms for genetics in the domain archaea: methanogens, halophiles, *Thermococcales* and *Sulfolobales*. FEMS Microbiol Rev 35 (4):577–608. https://doi.org/10.1111/j.1574-6976.2011.00265.x
15. Sarmiento F, Leigh JA, Whitman WB (2011) Genetic systems for hydrogenotrophic methanogens. Methods Enzymol 494:43–73. https://doi.org/10.1016/B978-0-12-385112-3.00003-2
16. Lyu Z, Whitman WB (2019) Transplanting the pathway engineering toolbox to methanogens. Curr Opin Biotechnol 59:46–54. https://doi.org/10.1016/j.copbio.2019.02.009
17. Xia Q, Hendrickson EL, Zhang Y, Wang T, Taub F, Moore BC, Porat I, Whitman WB, Hackett M, Leigh JA (2006) Quantitative proteomics of the archaeon *Methanococcus maripaludis* validated by microarray analysis and real time PCR. Mol Cell Proteomics 5 (5):868–881. https://doi.org/10.1074/mcp.M500369-MCP200
18. Hendrickson EL, Haydock AK, Moore BC, Whitman WB, Leigh JA (2007) Functionally distinct genes regulated by hydrogen limitation and growth rate in methanogenic archaea. Proc Natl Acad Sci U S A 104(21):8930–8934. https://doi.org/10.1073/pnas.0701157104
19. Hendrickson EL, Liu Y, Rosas-Sandoval G, Porat I, Söll D, Whitman WB, Leigh JA (2008) Global responses of *Methanococcus maripaludis* to specific nutrient limitations and growth rate. J Bacteriol 190 (6):2198–2205. https://doi.org/10.1128/JB.01805-07
20. Xia Q, Wang T, Hendrickson EL, Lie TJ, Hackett M, Leigh JA (2009) Quantitative proteomics of nutrient limitation in the hydrogenotrophic methanogen *Methanococcus maripaludis*. BMC Microbiol 9:149. https://doi.org/10.1186/1471-2180-9-149
21. Yoon SH, Reiss DJ, Bare JC, Tenenbaum D, Pan M, Slagel J, Moritz RL, Lim S, Hackett M, Menon AL, Adams MWW, Barnebey A, Yannone SM, Leigh JA, Baliga NS (2011) Parallel evolution of transcriptome architecture during genome reorganization. Genome Res 21 (11):1892–1904. https://doi.org/10.1101/gr.122218.111
22. Sarmiento F, Mrázek J, Whitman WB (2013) Genome-scale analysis of gene function in the hydrogenotrophic methanogenic archaeon *Methanococcus maripaludis*. Proc Natl Acad Sci U S A 110(12):4726–4731. https://doi.org/10.1073/pnas.1220225110
23. Walters AD, Smith SE, Chong JP (2011) Shuttle vector system for *Methanococcus maripaludis* with improved transformation efficiency. Appl Environ Microbiol 77(7):2549–2551. https://doi.org/10.1128/AEM.02919-10
24. Lyu Z, Chou C-W, Shi H, Wang L, Ghebreab R, Phillips D, Yan Y, Duin EC, Whitman WB (2018) Assembly of methyl coenzyme M reductase in the methanogenic archaeon *Methanococcus maripaludis*. J Bacteriol 200 (7):e00746–e00717. https://doi.org/10.1128/JB.00746-17
25. Zhao C, Lyu Z, Long F, Akinyemi T, Manakongtreecheep K, Söll D, Whitman WB, Vinyard DJ, Liu Y (2020) The Nbp35/ApbC homolog acts as a nonessential [4Fe-4S] transfer protein in methanogenic archaea. FEBS Lett 594(5):924–932. https://doi.org/10.1002/1873-3468.13673
26. Liu Y, Vinyard DJ, Reesbeck ME, Suzuki T, Manakongtreecheep K, Holland PL, Brudvig GW, Söll D (2016) A [3Fe-4S] cluster is required for tRNA thiolation in archaea and eukaryotes. Proc Natl Acad Sci U S A 113 (45):12703–12708. https://doi.org/10.1073/pnas.1615732113
27. Ollagnier de Choudens S, Barras F (2017) Genetic, biochemical, and biophysical methods for studying FeS proteins and their assembly. Methods Enzymol 595:1–32. https://doi.org/10.1016/bs.mie.2017.07.015

Chapter 3

Genetic Approaches to Uncover Gene Products Involved in Iron-Sulfur Protein Maturation: High-Throughput Genomic Screening Using Transposon Sequencing

Valerie J. Carabetta, Karla Esquilin-Lebron, Ehud Zelzion, and Jeffrey M. Boyd

Abstract

Iron-sulfur (Fe-S) clusters are one of the most ubiquitous and versatile prosthetic groups exploited by nature. Fe-S clusters aid in conducting redox reactions, carbon activation, and environmental sensing. This chapter presents an overview of the genetic approaches that have been useful for identifying and characterizing bacterial factors involved in Fe-S protein assembly. Traditional genetic screens that assess viability or conditional auxotrophies and bioinformatic approaches have identified the majority of the described genes utilized for Fe-S protein assembly. Herein, we expand upon this list of genetic methods by detailing the use of transposon sequencing (TnSeq) to identify gene products that are necessary for the proper function of metabolic pathways that require Fe-S enzymes. TnSeq utilizes the power of genomics and massively parallel DNA sequencing to allow researchers to quantify the necessity of individual gene products for a specific growth condition. This allows for the identification of gene products or gene networks that have a role in a given metabolic process but are not essential for the process. An advantage of this approach is that it allows researchers to identify mutants that have partial phenotypes that are often missed using traditional plate-based selections. Applying TnSeq to address questions of Fe-S protein maturation will result in a more comprehensive understanding of genetic interactions and factors utilized in Fe-S biogenesis and Fe-S protein assembly.

Key words Bacteria, Iron-sulfur cluster, Iron, Tn-seq, Transposon sequencing

1 Introduction

Fe-S clusters are cofactors found in almost all living organisms. The synthesis of Fe-S clusters and the maturation of Fe-S proteins are essential processes [1, 2]. Fe-S proteins are utilized for a wide range of physiological processes. In bacteria, there are three primary systems that have evolved to build Fe-S clusters. The iron-sulfur cluster (ISC), sulfur mobilization (SUF), and nitrogen fixation (NIF) biosynthetic systems are phylogenetically and biochemically distinct, but they all share a similar strategy to build the Fe-S

Patricia C. Dos Santos (ed.), *Fe-S Proteins: Methods and Protocols*, Methods in Molecular Biology, vol. 2353,
https://doi.org/10.1007/978-1-0716-1605-5_3,

clusters (reviewed in [3]). In general, Fe-S cluster biogenesis occurs in three steps. First, sulfur (S^0) is acquired from L-cysteine by a cysteine desulfurase and combined with iron (Fe^{2+}) and electrons (e^-) on a proteinaceous scaffold [4]. Once assembled, the Fe-S cluster can be transferred directly to an apo-protein or via an intermediate, Fe-S cluster carrier protein. This results in the activation of the target protein [5, 6].

Despite the progress that has been made in understanding Fe-S cluster assembly and maturation, a number of important questions remain. For instance, in most species, the Fe or electron donors are yet to be identified, and we still do not have a complete picture of how Fe-S cluster synthesis is regulated. It is known that many organisms utilize multiple Fe-S cluster carrier proteins. It has not yet been established if there is underlying specificity of carriers for target proteins or if cells prioritize the maturation of certain Fe-S proteins over others. Moreover, a number of factors have been identified that are utilized in Fe-S protein maturation, but their biochemical functions are unknown [7, 8].

Microorganisms have been exploited to characterize gene products involved in maturating Fe-S proteins because of their genetic tractability, high growth rate, and metabolic diversity. The first genes utilized for Fe-S protein maturation were discovered by scoring mutant strains of the diazotroph *Azotobacter vinelandii* for growth in the presence and absence of a reduced nitrogen source. *A. vinelandii* strains lacking the *nifSU* genes, which were later shown to encode a cysteine desulfurase and a Fe-S cluster synthesis scaffold, respectively, were not able to grow in the absence of a reduced nitrogen source [9–11]. The requirement for Fe-S clusters for nitrogenase function, and the sensitivity of the cofactors to oxygen, places a high demand on the Fe-S cluster assembly machinery when reduced nitrogen is not provided in the medium. The process of nitrogen fixation requires a dedicated maturation system (Nif) for the synthesis of simple and complex Fe-S clusters [12]. Scoring for diazotrophic growth has also been successfully applied to study the redundancy between maturation factors [13].

Escherichia coli can synthesize nicotinic acid, which is required for survival. NadA is utilized in nicotinic acid synthesis, and an *E. coli* mutant strain lacking the *isc* operon is auxotrophic for nicotinic acid [14]. Using a Δ*isc* mutant, researchers isolated strains that contained a second-site suppressor mutation providing nicotinic acid prototrophy. The suppressor strains also showed increased growth compared to the parent when alternate vitamins or amino acids that require an Fe-S enzyme for synthesis were absent from the media. The mutations mapped to the promoter region of the *suf* operon and ultimately resulted in the discovery of the third system for Fe-S cluster synthesis.

Salmonella enterica synthesizes thiamine using the SAM-radical enzymes ThiH and ThiC, which have solvent exposed and oxygen labile Fe-S clusters [15, 16]. Mutations that decrease carbon flux through the purine synthesis pathway, which provides precursors for thiamine synthesis, have allowed researchers to identify a number of genes that participate in thiamine synthesis. Many of these mutations are in genes that encode proteins that are not dedicated members of the core thiamine biosynthetic machinery and have provided hints as to how metabolic pathways are tethered together (reviewed in [17]). Some of thiamine auxotrophic mutants had a general defect in maturating Fe-S proteins [17], and at least, one was biochemically demonstrated to be a Fe-S cluster carrier [18]. Thiamine-dependent growth phenotypes have been used to analyze heterologous expression of homologues [19] and site-directed variants [20] of Fe-S cluster assembly factors.

S. enterica also has the ability to grow using tricarballylate as a carbon and energy source. Tricarballylate is transported into the cells where it is reduced to *cis*-aconitate by the flavoprotein TcuA [21]. The Fe-S protein TcuB is thought to provide electrons for the reduction of TcuA [22]. Strains lacking specific Fe-S maturation factors did not grow on medium containing tricarballylate as a sole carbon source but were proficient for growth on medium containing glucose. It has been hypothesized that the high demand for Fe-S assembly machinery is the result of poor TcuB maturation, resulting in sustained tricarballylate import and decreased reduction. This results in cytosolic accumulation, which inhibits isocitrate dehydrogenase [23]. Tricarballylate growth has been used to conduct suppressor analysis and study functional redundancy between Fe-S cluster assembly systems.

Other phenotypes utilized to identify and characterize Fe-S cluster biosynthesis systems include growth in the presence of elevated oxygen [24] or Fe limitation [25]. It was recently demonstrated [26] that Fe-S cluster synthesis is essential because it is required to maturate the Fe-S enzymes IspG and IspH, which are utilized isoprenoid synthesis in *E. coli* and *Bacillus subtilis* [27, 28]. The need for IspH and IspG can be bypassed by the heterologous expression of a mevalonate-dependent pathway for isoprenoid synthesis, which contains no Fe-S proteins, and supplementation of the growth media with mevalonate. Removing mevalonate from the growth medium provides a strong selective pressure. Researchers have used these genetic systems to isolate strains with second-site mutations that provided information about ISC- and SUF-dependent Fe-S cluster synthesis.

The examples provided above and others discussed in [29] allowed researchers to answer questions by scoring for a conditional auxotrophy or viability. Although powerful, especially when applied to study second-site mutations, they do not allow researchers to identify genes that may contribute to a process but do not have a strong solid media phenotype. Transposon sequencing (Tn-Seq) utilizes the power of genomics and massively parallel DNA sequencing to quantify the necessity of individual gene products for cellular fitness under specific growth conditions [30]. By comparing the fitness needs of loci under two growth conditions, such as with and without thiamine supplementation, we can gain a systems level understanding of the genes that contribute to a given metabolic process, which allows us to build models of gene interaction. There are a couple of potential drawbacks to using Tn-Seq. First, the network analyses do not allow for the incorporation of essential genes. Second, the genes of interest require mutant construction or acquisition for further study.

Herein, we outline the use of Tn-Seq to determine the importance of specific gene products for the fitness of the Gram-positive bacterium *Bacillus subtilis*. Applying Tn-Seq to address questions of Fe-S protein maturation will result in a more comprehensive understanding of genetic interactions and factors utilized in Fe-S cluster biogenesis and Fe-S protein assembly.

2 Materials

2.1 Genomic DNA Preparation

1. DNeasy Blood and Tissue Kit (Qiagen).
2. Enzymatic lysis buffer: 20 mM Tris–HCl, pH 8.0, 2 mM EDTA, 1.2% Triton X-100. Tris–HCl and EDTA stock solutions should be autoclaved prior to addition. A stock solution of Triton-X should be filter sterilized.
3. 100 mg/mL RNase A.
4. Lysozyme.

2.2 Transposon Library Generation

1. Plasmid pCJ41 (*see* **Note 1**).
2. MarC9 transposase (*see* **Note 2**).
3. 3 M sodium acetate, pH 5.2.
4. Ethanol (100% and 70%).
5. 2× Buffer A: 41 mM HEPES, pH 7.9, 20% glycerol, 200 mM NaCl, 20 mM $MgCl_2$, 0.5 mg/mL bovine serum albumin (BSA), 4.0 mM dithiothreitol (DTT). Prepare a small volume from autoclaved or filtered stock solutions fresh before use.
6. 10× Buffer B: 500 mM Tris–HCl, pH 7.8, 100 mM $MgCl_2$, 10 mM DTT.
7. 1 mg/mL bovine serum albumin.
8. dNTP solution mix: 10 mM per nucleotide solution.
9. T4 DNA polymerase: 3000 U/mL T4 DNA polymerase.
10. DNA ligase: 10,000 U/mL *E. coli* DNA ligase.
11. NAD solution: 2.6 mM nicotine adenine dinucleotide (NAD).

2.3 Competent Cells

1. Competence medium: 1X SS base: 6 g/L (44 mM) KH_2PO_4, 14 g/L (80.4 mM) K_2HPO_4, 2 g/L (38 mM) $(NH_4)_2SO_4$, 1 g/L (4 mM) sodium citrate. Autoclave to sterilize. To 500 mL of 1× SS medium, add 5 mL of 10% yeast extract (autoclaved twice prior to addition), 5 mL of 2% casein hydrolysate (autoclaved), 5 mL of 50% glucose (autoclaved), 2.5 mL of 0.5 M $MgCl_2$, and auxotrophic requirements (final concentration 50 μg/mL).
2. 0.5 M $CaCl_2$.
3. Luria-Bertani (LB) broth agar and liquid media.
4. Spectinomycin solution: 100 mg/mL spectinomycin in dH_2O. Solution should be filtered sterilized before use. When added to plates, use at a final concentration of 100 μg/mL.
5. S7 + metals: 5 mM potassium phosphate, pH 7.0, 10 mM $(NH_4)_2SO_4$, 100 mM MOPS (adjusted to pH 7.0 with KOH), 20 mM sodium glutamate, pH 7.0, 2 mM $MgCl_2$, 0.7 mM $CaCl_2$, 50 μM $MnCl_2$, 5 μM $FeCl_2$, 1 μM $ZnCl_2$, 2 μM thiamine, 100 mM glucose + required auxotrophic requirements (50 μg/mL). All salt solutions should be autoclaved. MOPS and thiamine should be filter sterilized. To recover the transposon library, leave out all carbon sources (glucose, glutamate, and amino acids).
6. Glycerol.

2.4 Preparation of DNA for Illumina Sequencing

1. Primers (*see* Table 1).
2. MmeI restriction endonuclease.
3. Phenol:chloroform:isoamyl alcohol (25:24:1).
4. 400,000 U/mL T4 DNA ligase
5. Phusion DNA polymerase.
6. Agarose.
7. Electrophoresis equipment.
8. MinElute Gel Extraction kit.

3 Methods

3.1 Preparation of Genomic DNA

Genomic DNA can be prepared using any standard method or kit that produces a large quantity of a highly pure product. The following protocol contains instructions for use with the DNeasy Blood and Tissue kit (Qiagen), using the Gram-positive bacteria protocol.

Table 1
Primer sequences utilized for Tn-Seq

Name	Sequence (5′ to 3′)	Purification
AP-A	GTTCAGAGTTCTACAGTCCGACGATCACACNN	PAGE
AP-B	5′-Phos/GTGTGATCGTCGGACTGTAGAACTCTG AACCTGTC/3′Phos	PAGE
oCJ22	CAAGCAGAAGACGGCATACGAGATNNNNNNTGT GTGAGACCGGG GACTTATCATCCAACCTGT	PAGE
GEX-PCR	AATGATACGGCGACCACCGAGATCTACACGTTCA GAGTTCTA CAGTCCGA	PAGE
Prim-seq	TACACGTTCAGAGTTCTACAGTCCGACGATCACAC	None
Bar-seq	ACAGGTTGGATGATAAGTCCCCGGTCTCACACA	None

Primer oCJ22 is specific for the inverted repeat sequences of the *magellan6* transposon and attaches a barcode (NNNNNN, where N is any randomly selected nucleotide) and Illumina-specific sequences. The barcode can be any combination of nucleotides and should be 4–6 nt in length. Ordering multiple different barcoded primers allows for multiplexing when performing Illumina sequencing. Primer GEX-PCR is specific for the ligated adaptor and adds Illumina-specific sequences. The primers oCJ22, GEX-PCR, and Prim-seq are optimized for use with the MiSeq instrumentation. It is recommended to speak with an Illumina specialist when using new or different instrumentation to check the compatibility of the sequences (For use with the HiSeq platform, *see* [32])

1. Inoculate 5 mL of liquid LB medium with a single colony of the desired *Bacillus* strain and incubate overnight at 37 °C with aeration (shaking).
2. Centrifuge 1.5 mL of cells in an Eppendorf tube for 5 min at 11,000 × *g* and remove and discard the resulting supernatant. Repeat this step again (same tube) to collect a total of 3 mL of cells.
3. Resuspend the pellet in 180 μL of enzymatic lysis buffer that has 40 mg/mL lysozyme (final concentration) freshly added. Incubate at 37 °C for 30 min.
4. Add 4 μL of RNaseA (stock concentration of 100 mg/mL) and mix well. Incubate for 15 min at room temperature.
5. Add 25 μL of Proteinase K and 200 μL of Buffer AL (provided in the DNeasy kit). Vortex samples to mix. Incubate at 56 °C for 30 min.
6. Add 200 μL of 100% ethanol and vortex to mix.
7. Transfer the sample to a DNeasy spin column. Centrifuge for 1 min at 6000 × *g*. Discard the flow-through, and transfer the column to a new collection tube.
8. Wash the column by adding 500 μL of Buffer AW1 (provided in the DNeasy kit) and centrifuge for 1 min at 6000 × *g*. Discard the flow-through and transfer column to a clean collection tube.

9. Wash column a second time by adding 500 μL of Buffer AW2 (provided in the DNeasy kit) and centrifuge for 3 min at maximum speed. Discard the flow-through and transfer column to a clean Eppendorf tube for sample elution.
10. Add 200 μL Buffer AE (provided in the DNeasy kit) to the spin column. Incubate for 1 min at room temperature. Centrifuge for 1 min at 6000 × *g* to collect the genomic DNA. Samples may be stored at −20 °C for future use.

3.2 Generation of Transposon Libraries In Vitro

More information on transposon library preparation can be found for *Streptococcus pneumoniae* [30, 31], *Bacillus subtilis* [32], and *Staphylococcus aureus* [33, 34]. The *B. subtilis* detailed protocol is presented below. Each reaction can be used for two library transformations. It is recommended to perform at least five separate reactions in total.

1. Mix together in an Eppendorf tube, 1.3 μg of pCJ41 (*magellan6X* transposon donor, *see* **Note 1**), 5 μg of target gDNA, 10 μL of 2× Buffer A, and dH_2O to 18 μL. Finally, add 2 μL of MarC9 transposase (final concentration of 15–20 μg/mL, *see* **Note 2**). Incubate the reaction overnight at 30 °C.
2. The following morning, add 2 μL of 3 M sodium acetate (pH 5.2), mix, and then add 50 μL of ice-cold 100% ethanol. Incubate on ice for at least 20 min, but can go up to an hour.
3. Centrifuge samples at 4 °C, at maximum speed for 30 min.
4. Remove the supernatant (with a pipette) and add 100 μL of ice-cold 70% EtOH. Centrifuge samples at 4 °C, at maximum speed for 15 min. Remove supernatant with a pipette and allow to dry at 37 °C.
5. Resuspend the pellet by adding 2 μL of 10× Buffer B, 2 μL of 1 mg/mL BSA, and 11 μL of H_2O. Incubate at 37 °C for 3–4 h.
6. Add 4 μL of a 2.5 mM dNTP solution and 1 μL of T4 DNA Polymerase (3000 U/mL) and incubate for 20 min at 12 °C (*see* **Note 3**).
7. Following incubation, heat inactivate the polymerase by incubating the reactions at 75 °C for 15 min. Cool the reactions on ice for 3 min.
8. Add 0.2 μL of 2.6 mM NAD and 1 μL of *E. coli* DNA ligase (10,000 U/mL, NEB), and incubate overnight at 16 °C.
9. The transposed DNA, in theory, can be stored at −20 °C for future use; however, the effects of freeze–thaw cycles are unknown. It is recommended to proceed immediately with the transformation procedure described in the following.

3.3 Preparation of the In Vivo Transposon Library

1. Streak out the desired *Bacillus* strain on LB agar plates with appropriate antibiotics (if necessary), and incubate overnight at 30 °C.
2. The following morning, inoculate two flasks with 10 mL of competence medium with enough bacteria for a starting Klett of 15–20 (If using OD_{600}, start at an OD of ~0.05).
3. Incubate cells at 37 °C with aeration for 2.5 h. From this point, take Klett readings every 10 min. When the Klett value changes by 2–5 units every 10 min, your cells are in the exponential phase. Dilute each culture 1:10 into 10 mL of prewarmed competence medium supplemented with 0.5 mM $CaCl_2$ (final concentration).
4. Incubate cells while shaking aerobically for 90 min.
5. Centrifuge cells for 10 min at 4500 × *g* at 4 °C to collect cells. Resuspend each pellet in 1 mL of spent competence medium (supernatant).
6. Divide the competent cells into ten 200 μL of aliquots (in test tubes, or 15 mL conical tubes), and add 10 μL of the in vitro transposed gDNA (2.5 μg each, or ½ of the reaction mixture) to each tube.
7. Incubate the reactions for 90 min at 37 °C with aeration (shaking or rotation).
8. Add 1 mL of prewarmed LB to each tube and incubate for an additional 30 min.
9. Plate cells on LB agar supplemented with 100 μg/mL spectinomycin (250 μL/plate, to select for the transposon). Incubate plates overnight at 37 °C.
10. OPTIONAL: Plate ten-fold serial dilutions to determine the number of transposants in your library. Do not recover these cells as these clones have more space to grow on the plate and may become overrepresented in the library.
11. Recover the library. Add 2 mL of S7 with metals (no carbon source) to each plate, and use a 1 mL pipette to disrupt the colonies and recover the fluid from the plate. Pool all of the recovered cells. This approach yields a library of ~4.0×10^5 independent clones.
12. Mix the library with ¼ of the volume with 50% glycerol, then freeze in 100 μL aliquots for storage at −80 °C (*see* **Note 4**).

3.4 Selection for Mutants Involved in Fe-S Cluster Biosynthesis

There are numerous biosynthetic pathways that encode Fe-S cluster proteins (Table 2). Any individual pathway or multiple pathways, including additional ones not listed, could be chosen for the screen. The base medium for all of the following studies is S7 with metals. Alternate selection pressures, including culturing in Fe deplete and replete conditions or including a small molecule that catalyzes ROS

Table 2
Select genes encoding Fe-S enzymes and the consequences of mutation

Gene	Phenotype	Reference
nadA	Nicotinic acid auxotroph	[23]
leuC	Leucine auxotroph	[7]
ilvD	Isoleucine and valine auxotroph	[5]
lipA	Lipoic acid auxotroph	[46]
bioB	Biotin auxotroph	[47]
thiC	Thiamine auxotroph	[15]
thiH	Thiamine auxotroph	[48]
citB	Glutamate auxotroph	[8]

formation, could also be examined. As in all genetic selections, care should be taken to minimalize variables to prevent pleotropic effects caused by the selective pressure.

Before beginning the selection process, it is recommended to perform simple growth experiments to adjust concentrations of added supplements and to verify that the selection process will be successful. To do this, deletion mutants can be either acquired (i.e., from the Bacillus Genetic Stock Center) or constructed. For example, the *leuC* deletion mutant requires leucine for growth and does not grow in its absence, which was confirmed by growth in a different medium (Fig. 1). When all nutritional requirements are determined, known Fe-S cluster biosynthetic genes are tested in the supplemented S7 and minimal medium (Fig. 2). Deletions of known, nonessential assembly factors, such as *nfu* [5], *sufA* [35], or *sufT* [36], were grown in each medium, and growth defects of each mutant were apparent. Consequently, these mutants should be underrepresented in the population following selection, and thus serve as a positive internal selection control. Once the medium conditions are chosen, the following steps are done to perform the selection. It is recommended to perform this selection at least three independent times for appropriate statistical analyses.

1. Thaw a 100 μL of aliquot of the transposon library. Inoculate 20 mL of LB medium.
2. Incubate at 37 °C with shaking until the culture reaches an OD_{600} of 1.0. It will likely take between 2 and 3.5 h.
3. Collect two 1 mL samples of cells and centrifuge at 11,000 × *g* for 5 min. Wash each pellet three times in either supplemented or minimal S7 medium.
4. Resuspend each pellet in supplemented or minimal medium, and add to 100 mL of the same medium (1:100 dilution). It is

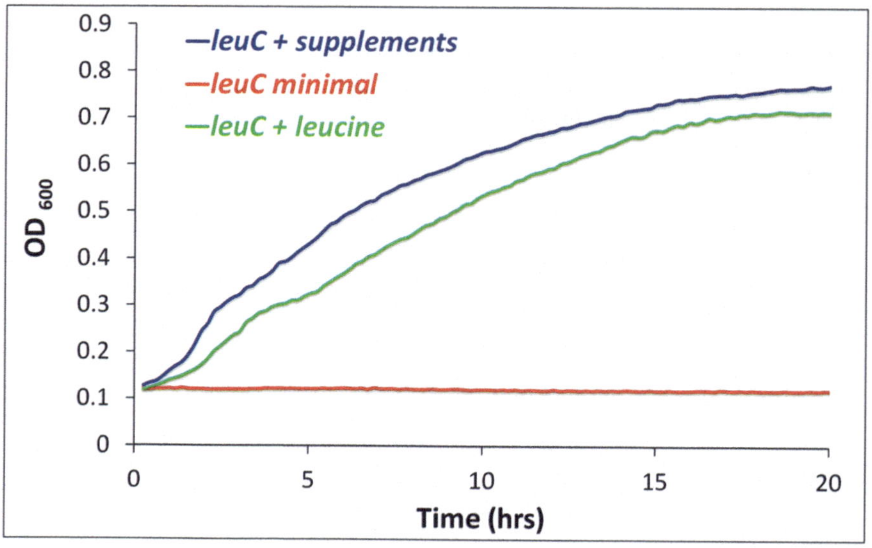

Fig. 1 Sample growth curves with the *leuC* deletion mutant. The *leuC* mutant was cultured in S7 medium with the addition of all selected supplements listed in Table 2 (blue), S7 medium with leucine (green), and minimal S7 medium (red). The *leuC* mutant was acquired from the Bacillus Genetic Stock Center. Growth curves were acquired by measuring OD_{600} in an EnVision multimode 96-well plate reader. Data are presented as the average of two growth experiments

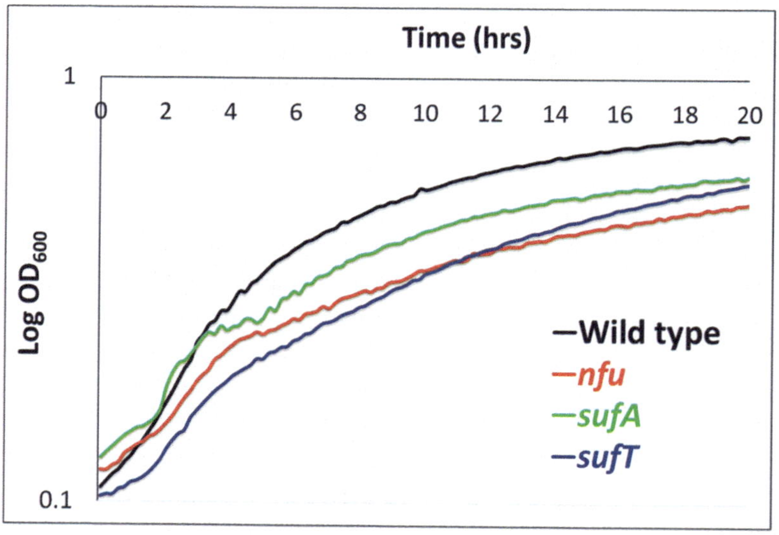

Fig. 2 Sample growth curves of wild type (black), *nfu* (red), *sufA* (green), and *sufT* (blue) *B. subtilis* deletion mutants. Strains were obtained from the Bacillus Genetic Stock Center. The strains were grown, in duplicate, in minimal S7 medium. Growth curves were acquired by measuring OD_{600} in an EnVision multimode 96-well plate reader

recommended to use 1 mL of cells to have a good representation of the mutant clones in the library. Let cells grow overnight (~24 h) at 37 °C with shaking.

5. The next morning, record OD readings. Back dilute the culture 1:100 in the same medium and let grow for another 24 h at 37 °C with shaking.
6. Record OD readings, and collect six 1.5 mL of aliquots of each culture in Eppendorf tubes. Centrifuge at 11,000 × *g* for 10 min. The pellets may be frozen at −20 °C for future processing. Typically, only 2–3 of these tubes will need to be processed to ensure enough gDNA, but collecting extra aliquots of cells is recommended.
7. Isolate gDNA from the pellets as described in Subheading 3.1. A total of 6 μg of gDNA are required to proceed.

3.5 Preparation of DNA for Illumina Sequencing

The following protocol was adapted from Johnson and Grossman [32]. *See* Table 1 for the primer sequences.

3.5.1 Precipitate Genomic DNA

1. Add 20 μL of 3 M sodium acetate pH 5.2 and 500 μL of ice-cold 100% ethanol to 200 μL of gDNA.
2. Incubate at −20 °C (or on ice) for at least 30 min.
3. Centrifuge at 20,000 × *g* at 4 °C for 30 min.
4. Discard the supernatant and wash with 500 μL of ice-cold 70% ethanol. Centrifuge at 20,000 × *g* at 4 °C for 10 min.
5. Discard the supernatant and dry at 37 °C for 10 min.
6. Once dry, resuspend in 60 μL of 10 mM Tris–HCl, pH 8.5.
7. Determine the concentration and purity of the DNA using a Nanodrop or other DNA quantification methods. If you have less than 6 μg of total DNA for one condition, prepare more gDNA from the frozen pellets.

3.5.2 Anneal the Adapter

1. Prepare the oligonucleotides AP-A and AP-B as 200 mM solutions in 1 mM Tris, pH 8.5 (*see* **Note 5**).
2. Mix equal volumes of the oligonucleotides and place in 96 °C heat block for 2 min.
3. Remove the heat block, and let cool to room temperature for 15 min. This will allow the oligonucleotides to anneal.
4. The annealed adapter will be at a 100 μM concentration. The adapter can be stored at −20 °C.

3.5.3 Digest gDNA with MmeI

1. Set up the following reaction: 6 μg gDNA, 3 μL 2000 U/mL MmeI, 0.66 μL SAM (32 mM), 30 μL CutSmart buffer. Adjust volume to 300 μL total volume with dH_2O.

2. Incubate for 2.5 h at 37 °C.
3. Add 0.6 μL of calf intestinal phosphatase (6 units) and incubate for an additional 1 h at 37 °C.
4. Add 300 μL of phenol:chloroform:isoamyl alcohol (25:24:1), vortex to mix, and centrifuge at maximum speed for 10 min. Recover the top aqueous layer into a fresh Eppendorf tube.
5. Ethanol precipitate the DNA as described in Subheading 3.5.1, with one modification. Resuspend in 27.5 μL of 2 mM Tris–HCl, pH 8.5 and incubate for at least 2 h at 37 °C to dissolve the pellet.

3.5.4 Adapter Ligation

1. Prepare the following reaction: 27.5 μL of MmeI-digested gDNA, 3.5 μL of 10X T4 DNA ligase buffer, 2.0 μL of 100 μM annealed adapter, and 1.5 μL of 400,000 U/mL T4 DNA ligase.
2. Incubate overnight at 16 °C.
3. The following morning, purify the reaction products using a PCR purification kit as per manufacturer's instructions. Elute in 50 μL of warm elution buffer (preheated to 37 °C, 10 mM Tris–HCl, pH 8.5).

3.5.5 PCR Amplification of the Fragment of Interest

For PCR, any high-fidelity polymerase can be used. This protocol uses Phusion High Fidelity DNA Polymerase, which we have found works well with AT-rich genomes.

1. Set up the following reaction (per condition): 5 μL of 10X buffer, 2–5 μL of adaptor ligation (can vary template concentration), 0.5 μL of 25 mM dNTP mix, 1.0 μL of 10 μM GEX-PCR primer, 1.0 μL of 10 μM oCJ22-NNNN, 0.5 μL of Phusion, and adjust volume to 50 μL with dH_2O.
2. Set up the PCR conditions as follows. Adjust temperatures and times according to manufacturer's instructions (*see* **Note 6**). After an initial denaturation step of 98 °C for 30 s, the cycle conditions used are 98 °C for 10 s, 58 °C for 25 s, and 72 °C for 10 s for total of 18 cycles. A 10-min final extension at 72 °C is included, followed by a hold step at 4 °C.
3. Load the entire PCR sample (mixed with loading dye) onto a 1.8% agarose gel, with a 100-bp ladder. Excise the appropriate band (size ~130 bp, Fig. 3) from the gel using the MinElute gel extraction kit (Qiagen) following manufacturer's instructions. When following the instructions, perform the optional column wash step with buffer QG (provided with kit) to ensure the removal of residual agarose. Elute with 14 μL of elution buffer (10 mM Tris, pH 8.5).
4. This DNA sample is now ready for Illumina sequencing. Submit the custom sequencing primers Prim-seq (primary

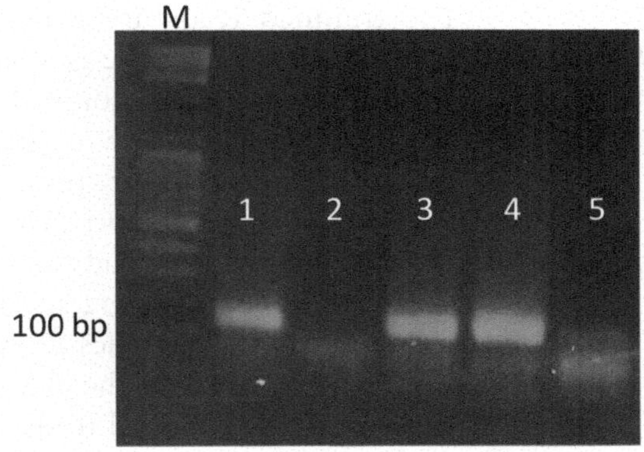

Fig. 3 Example of PCR products separated on a 1.8% agarose gel. Lane M: 100 bp ladder (New England Biolabs). The 100 bp band is marked. Lanes 1, 3–5 are experimental PCR samples. Lanes 1, 3, and 4 contain the desired product, which runs slightly above the 100 bp band. This is the band that should be extracted. Lane 2 is a negative control (adaptor only). This control is important to include to properly localize the correctly sized band for purification. The bottom two bands are artifacts

sequencing primer to determine the genomic locations) and Bar-seq (sequences barcode). Our sequencing facility utilized the NextSeq 500 and MiSeq instruments.

3.6 Bioinformatic Analysis

This protocol is written with the assumption that the genome of the organism of interest has been fully sequenced and annotated with genes, rRNA, tRNA, pseudogenes, etc.

3.6.1 Demultiplexing and Trimming of Raw Reads

Because of the way the Tn-Seq libraries are constructed, the raw reads include barcode sequences, sequencing adaptors and remaining transposon sequences, which need to be removed before any downstream analyses.

1. The first 8 bp of each read form a barcode added before the transposon insertion to enable the pooling of different conditions together. With a simple bash/perl/python script that utilizes the barcode sequence, the reads can be split, so there is one file per each sample. This can also be achieved with the Cutadapt program [37].
2. Remove Illumina sequencing adapters from the 3′-end of the reads. There are many open-source programs that deal with sequencing adapter trimming, such as Cutadapt and Trimmomatic [38].
3. Trim the remaining transposon sequence and filter out any sequence that is longer than 17 bp or shorter than 16 bp. All of the programs mentioned above can be used for this step as well.

3.6.2 Determine the Number of Insertions in Each TA Site

The trimmed reads are genomic sequences found next to the insertion sites (TA). The goal is to align the trimmed reads to the reference genome and count the number of reads aligned to each TA site. There are many programs that can perform this task, but the two most popular programs are bowtie [39] and BWA [40]. The reads are short (16–17 bp), so do not allow any mismatches between the reads and the genome in the alignment process. Each alignment program uses different parameters to control the number of mismatches allowed; consult with the manual of the program in use. From the alignment output, extract the coverage (e.g., number of aligned reads) at each TA site, including the sites with zero read counts and save it as a text file. If your experiment includes more than one condition, such a file should be generated for each condition.

3.6.3 Predicting Putative Essential Genes

The goal here is to identify genes or regions in the genome that are essential or advantageous for growth under specific selection conditions. The basic idea is that mutants with insertions in conditionally essential genes would not be viable or underrepresented in the population.

The different methods used to identify essential genes can be divided into two general categories: (1) annotation dependent: the total count of disrupted sites per annotation feature (i.e., gene) are counted and compared with the rest of the genes to determine which genes are essential and which are not, and (2) annotation independent: the total counts of disrupted sites are counted independently of annotation features. One such method uses a Hidden Markov Model to determine the essentiality of each disrupted site, which are then crossed with the gene annotations to identify which genes are essential. There are several programs that can be used for determining essential genes in Tn-Seq experiments; some options are TRANSIT [41], ESSENTIALS [42], Tn-seq Explorer [43], and TnseqDiff [44]. Some of these programs contain methods for both annotation-dependent and annotation-independent methods such as TRANSIT. The required input files for these programs are usually a gene annotation file and the read alignment file (one for each library). The output, in most cases, is a tab-delimited file that holds all the genes as rows while the columns represent the results (i.e., essential or nonessential). Some programs add probability values to their results and are able to compare the essential genes between two conditions and quantify the statistical significance.

4 Notes

1. The plasmid pCJ41 serves as the source of the *magellan6X* transposon [30, 32]. This plasmid was modified from the original plasmid pR412-*magellan6* [30], to remove the

B. subtilis BsuM restriction site. The inverted repeats on both sides of the transposon contain a MmeI restriction site. This enzyme cleaves the DNA downstream of the restriction site. Following digestion, approximately 16 bp of genomic DNA will be attached to the ends of the transposon, facilitating the mapping of the insertion sites.

2. MarC9 must be purified first, as described [30, 45]. The plasmid containing the MarC9 transposase is available on request [30].
3. The temperature is important for this reaction. The T4 DNA polymerase fills in four nucleotide single-stranded gaps on each end of the transposon insertion. This enzyme also has 3′ to 5′ exonuclease activity that increases at temperatures above 12 °C. If a cooling block is not available, a heat block placed in a cold room can be used to achieve this temperature.
4. To regenerate a previously used library: Take a 100 μL aliquot of the transposon library and inoculate 20 mL of LB medium. Let the cells grow, with shaking, at 37 °C until stationary phase. Measure the OD_{600}. Centrifuge the cells for 10 min at 5000 × *g* at 4 °C, and discard the supernatant. Resuspend the cells in 10 mL of S7 medium (no carbon source) and centrifuge as before. Resuspend the pellet in S7 medium (no carbon source) to achieve an OD_{600} of 10. Add glycerol to a final concentration of 12.5%, and freeze at −80 °C in 100 μL aliquots.
5. AP-A contains a random two base overhang (NN). MmeI digestion leaves a random 3 base overhang, so this allows for ligation to occur. AP-B must be phosphorylated at both ends. The 5′ phosphorylation allows for ligation to the dephosphorylated genomic DNA fragments. The 3′ phosphorylation prevents the formation of self-ligated adapter products.
6. Touchdown PCR can be used in this step. Drop the annealing temperature from 66 °C to 58 °C in successive cycles. Do not change the total number of cycles.

Acknowledgments

This work was funded by a CAREER award (J.M.B.) from the National Science Foundation.

References

1. Roberts CA, Al-Tameemi HM, Mashruwala AA, Rosario-Cruz Z, Chauhan U, Sause WE, Torres VJ, Belden WJ, Boyd JM (2017) The Suf iron-sulfur cluster biosynthetic system is essential in *Staphylococcus aureus*, and decreased Suf function results in global metabolic defects and reduced survival in human

neutrophils. Infect Immun 85(6):e00100-17. https://doi.org/10.1128/IAI.00100-17

2. Boyd ES, Thomas KM, Dai Y, Boyd JM, Outten FW (2014) Interplay between oxygen and Fe-S cluster biogenesis: insights from the Suf pathway. Biochemistry 53(37):5834–5847. https://doi.org/10.1021/bi500488r
3. Chahal HK, Boyd JM, Outten FW (2012) Fe-S cluster biogenesis in archaea and bacteria. In: Culotta V, Scott RA (eds) Metals in Cells. Wiley, West Sussex, UK
4. Py B, Barras F (2010) Building Fe-S proteins: bacterial strategies. Nat Rev Microbiol 8 (6):436–446. https://doi.org/10.1038/nrmicro2356
5. Mashruwala AA, Pang YY, Rosario-Cruz Z, Chahal HK, Benson MA, Mike LA, Skaar EP, Torres VJ, Nauseef WM, Boyd JM (2015) Nfu facilitates the maturation of iron-sulfur proteins and participates in virulence in Staphylococcus aureus. Mol Microbiol 95(3):383–409. https://doi.org/10.1111/mmi.12860
6. Chahal HK, Outten FW (2012) Separate FeS scaffold and carrier functions for SufB(2)C (2) and SufA during in vitro maturation of [2Fe2S] Fdx. J Inorg Biochem 116:126–134. https://doi.org/10.1016/j.jinorgbio.2012.06.008
7. Mashruwala AA, Bhatt S, Poudel S, Boyd ES, Boyd JM (2016) The DUF59 containing protein SufT is involved in the maturation of iron-sulfur (FeS) proteins during conditions of high FeS cofactor demand in Staphylococcus aureus. PLoS Genet 12(8):e1006233. https://doi.org/10.1371/journal.pgen.1006233
8. Rosario-Cruz Z, Chahal HK, Mike LA, Skaar EP, Boyd JM (2015) Bacillithiol has a role in Fe-S cluster biogenesis in Staphylococcus aureus. Mol Microbiol 98(2):218–242. https://doi.org/10.1111/mmi.13115
9. Jacobson MR, Cash VL, Weiss MC, Laird NF, Newton WE, Dean DR (1989) Biochemical and genetic analysis of the *nifUSVWZM* cluster from Azotobacter vinelandii. Mol Gen Genet 219(1–2):49–57
10. Yuvaniyama P, Agar JN, Cash VL, Johnson MK, Dean DR (2000) NifS-directed assembly of a transient [2Fe-2S] cluster within the NifU protein. Proc Natl Acad Sci U S A 97 (2):599–604
11. Zheng L, White RH, Cash VL, Jack RF, Dean DR (1993) Cysteine desulfurase activity indicates a role for NIFS in metallocluster biosynthesis. Proc Natl Acad Sci U S A 90 (7):2754–2758
12. Seefeldt LC, Hoffman BM, Dean DR (2009) Mechanism of Mo-dependent nitrogenase. Annu Rev Biochem 78:701–722. https://doi.org/10.1146/annurev.biochem.78.070907.103812
13. Dos Santos PC, Johnson DC, Ragle BE, Unciuleac MC, Dean DR (2007) Controlled expression of nif and isc iron-sulfur protein maturation components reveals target specificity and limited functional replacement between the two systems. J Bacteriol 189 (7):2854–2862
14. Takahashi Y, Tokumoto U (2002) A third bacterial system for the assembly of iron-sulfur clusters with homologs in archaea and plastids. J Biol Chem 277(32):28380–28393
15. Martinez-Gomez NC, Downs DM (2008) ThiC is an [Fe-S] cluster protein that requires AdoMet to generate the 4-amino-5-hydroxymethyl-2-methylpyrimidine moiety in thiamin synthesis. Biochemistry 47(35):9054–9056
16. Leonardi R, Fairhurst SA, Kriek M, Lowe DJ, Roach PL (2003) Thiamine biosynthesis in Escherichia coli: isolation and initial characterisation of the ThiGH complex. FEBS Lett 539 (1–3):95–99
17. Skovran E, Downs DM (2003) Lack of the ApbC or ApbE protein results in a defect in Fe-S cluster metabolism in Salmonella enterica serovar typhimurium. J Bacteriol 185:98–106
18. Boyd JM, Pierik AJ, Netz DJ, Lill R, Downs DM (2008) Bacterial ApbC can bind and effectively transfer iron-sulfur clusters. Biochemistry 47(31):8195–8202
19. Boyd JM, Drevland RM, Downs DM, Graham DE (2009) Archaeal ApbC/Nbp35 homologs function as iron-sulfur cluster carrier proteins. J Bacteriol 191(5):1490–1497
20. Boyd JM, Sondelski JL, Downs DM (2009) Bacterial ApbC protein has two biochemical activities that are required for *in vivo* function. J Biol Chem 284(1):110–118
21. Lewis JA, Escalante-Semerena JC (2006) The FAD-dependent tricarballylate dehydrogenase (TcuA) enzyme of Salmonella enterica converts tricarballylate into cis-aconitate. J Bacteriol 188(15):5479–5486
22. Lewis JA, Escalante-Semerena JC (2007) Tricarballylate catabolism in Salmonella enterica. The TcuB protein uses 4Fe-4S clusters and heme to transfer electrons from FADH2 in the tricarballylate dehydrogenase (TcuA) enzyme to electron acceptors in the cell membrane. Biochemistry 46(31):9107–9115
23. Boyd JM, Teoh WP, Downs DM (2012) Decreased transport restores growth of a Salmonella enterica *apbC* mutant on tricarballylate. J Bacteriol 194(3):576–583. https://doi.org/10.1128/JB.05988-11

24. Bandyopadhyay S, Naik SG, O'Carroll IP, Huynh BH, Dean DR, Johnson MK, Dos Santos PC (2008) A proposed role for the Azotobacter vinelandii NfuA protein as an intermediate iron-sulfur cluster carrier. J Biol Chem 283(20):14092–14099
25. Angelini S, Gerez C, Ollagnier-de Choudens S, Sanakis Y, Fontecave M, Barras F, Py B (2008) NfuA, a new factor required for maturing Fe/S proteins in Escherichia coli under oxidative stress and iron starvation conditions. J Biol Chem 283(20):14084–14091
26. Loiseau L, Gerez C, Bekker M, Ollagnier-de Choudens S, Py B, Sanakis Y, Teixeira de Mattos J, Fontecave M, Barras F (2007) ErpA, an iron sulfur (Fe S) protein of the A-type essential for respiratory metabolism in Escherichia coli. Proc Natl Acad Sci U S A 104 (34):13626–13631
27. Yokoyama N, Nonaka C, Ohashi Y, Shioda M, Terahata T, Chen W, Sakamoto K, Maruyama C, Saito T, Yuda E, Tanaka N, Fujishiro T, Kuzuyama T, Asai K, Takahashi Y (2018) Distinct roles for U-type proteins in iron-sulfur cluster biosynthesis revealed by genetic analysis of the *Bacillus subtilis sufCDSUB* operon. Mol Microbiol 107(6):688–703. https://doi.org/10.1111/mmi.13907
28. Tanaka N, Kanazawa M, Tonosaki K, Yokoyama N, Kuzuyama T, Takahashi Y (2016) Novel features of the ISC machinery revealed by characterization of Escherichia coli mutants that survive without iron-sulfur clusters. Mol Microbiol 99(5):835–848. https://doi.org/10.1111/mmi.13271
29. Ollagnier de Choudens S, Barras F (2017) Genetic, biochemical, and biophysical methods for studying FeS proteins and their assembly. Methods Enzymol 595:1–32. https://doi.org/10.1016/bs.mie.2017.07.015
30. van Opijnen T, Bodi KL, Camilli A (2009) Tn-seq: high-throughput parallel sequencing for fitness and genetic interaction studies in microorganisms. Nat Methods 6 (10):767–772. https://doi.org/10.1038/nmeth.1377
31. van Opijnen T, Camilli A (2010) Genome-wide fitness and genetic interactions determined by Tn-seq, a high-throughput massively parallel sequencing method for microorganisms. Curr Protoc Microbiol. chapter 1: Unit1E 3. https://doi.org/10.1002/9780471729259.mc01e03s19
32. Johnson CM, Grossman AD (2014) Identification of host genes that affect acquisition of an integrative and conjugative element in *Bacillus subtilis*. Mol Microbiol 93(6):1284–1301. https://doi.org/10.1111/mmi.12736
33. Valentino MD, Foulston L, Sadaka A, Kos VN, Villet RA, Santa Maria J Jr, Lazinski DW, Camilli A, Walker S, Hooper DC, Gilmore MS (2014) Genes contributing to Staphylococcus aureus fitness in abscess- and infection-related ecologies. mBio 5(5):e01729–e01714. https://doi.org/10.1128/mBio.01729-14
34. Grosser MR, Paluscio E, Thurlow LR, Dillon MM, Cooper VS, Kawula TH, Richardson AR (2018) Genetic requirements for Staphylococcus aureus nitric oxide resistance and virulence. PLoS Pathog 14(3):e1006907. https://doi.org/10.1371/journal.ppat.1006907
35. Ollagnier-de Choudens S, Nachin L, Sanakis Y, Loiseau L, Barras F, Fontecave M (2003) SufA from Erwinia chrysanthemi. Characterization of a scaffold protein required for iron-sulfur cluster assembly. J Biol Chem 278 (20):17993–18001. https://doi.org/10.1074/jbc.M300285200
36. Mashruwala AA, Boyd JM (2017) Investigating the role(s) of SufT and the domain of unknown function 59 (DUF59) in the maturation of iron-sulfur proteins. Curr Genet 64 (1):9–16. https://doi.org/10.1007/s00294-017-0716-5
37. Martin M (2011) Cutadapt removes adapter sequences from high-throughput sequencing reads. EMBnet J 17(1):10–12. https://doi.org/10.14806/ej.17.1.200
38. Bolger AM, Lohse M, Usadel B (2014) Trimmomatic: a flexible trimmer for Illumina sequence data. Bioinformatics 30 (15):2114–2120. https://doi.org/10.1093/bioinformatics/btu170
39. Langmead B, Salzberg SL (2012) Fast gapped-read alignment with bowtie 2. Nat Methods 9 (4):357–359. https://doi.org/10.1038/nmeth.1923
40. Li H, Durbin R (2009) Fast and accurate short read alignment with burrows-wheeler transform. Bioinformatics 25(14):1754–1760. https://doi.org/10.1093/bioinformatics/btp324
41. DeJesus MA, Ambadipudi C, Baker R, Sassetti C, Ioerger TR (2015) TRANSIT--A software tool for Himar1 TnSeq analysis. PLoS Comput Biol 11(10):e1004401. https://doi.org/10.1371/journal.pcbi.1004401
42. Zomer A, Burghout P, Bootsma HJ, Hermans PW, van Hijum SA (2012) ESSENTIALS: software for rapid analysis of high throughput transposon insertion sequencing data. PLoS One 7(8):e43012. https://doi.org/10.1371/journal.pone.0043012

43. Solaimanpour S, Sarmiento F, Mrazek J (2015) Tn-seq explorer: a tool for analysis of high-throughput sequencing data of transposon mutant libraries. PLoS One 10(5):e0126070. https://doi.org/10.1371/journal.pone.0126070

44. Zhao L, Anderson MT, Wu W, TM HL, Bachman MA (2017) TnseqDiff: identification of conditionally essential genes in transposon sequencing studies. BMC Bioinformatics 18 (1):326. https://doi.org/10.1186/s12859-017-1745-2

45. Lampe DJ, Churchill ME, Robertson HM (1996) A purified mariner transposase is sufficient to mediate transposition in vitro. EMBO J 15(19):5470–5479

46. Mashruwala AA, Roberts CA, Bhatt S, May KL, Carroll RK, Shaw LN, Boyd JM (2016) Staphylococcus aureus SufT: an essential iron-sulfur cluster assembly factor in cells experiencing a high-demand for lipoic acid. Mol Microbiol 102(6):1099–1119. https://doi.org/10.1111/mmi.13539

47. Bali AP, Lennox-Hvenekilde D, Myling-Petersen N, Buerger J, Salomonsen B, Gronenberg LS, Sommer MOA, Genee HJ (2020) Improved biotin, thiamine, and lipoic acid biosynthesis by engineering the global regulator IscR. Metab Eng 60:97–109. https://doi.org/10.1016/j.ymben.2020.03.005

48. Skovran E, Downs DM (2000) Metabolic defects caused by mutations in the *isc* gene cluster in Salmonella enterica serovar typhimurium: implications for thiamine synthesis. J Bacteriol 182(14):3896–3903

Chapter 4

Methods for Heterologous Overproduction of Fe-S Proteins

Elliot I. Corless and Edwin Antony

Abstract

Proteins carrying iron-sulfur ([Fe-S]) clusters are critical to the basic metabolism of all organisms. Structural and biochemical investigations of many such [Fe-S] cluster proteins depend on recombinant overproduction using heterologous bacterial hosts such as *Escherichia coli*. Here, we describe a detailed procedure for the overproduction and purification of two oxygen-sensitive component proteins of the dark-operative protochlorophyllide oxidoreductase (DPOR) complex. The method relies on an engineered *Escherichia coli* cell line carrying a correction in its genome to restore the loss of a key [Fe-S] cluster biogenesis pathway. The method can also be potentially adapted for the overproduction of other Fe-S proteins.

Key words [Fe-S], [4Fe-4S], Cluster, Anaerobic, DPOR, Electron transfer, Fe-S protein expression

1 Introduction

More than a third of the enzymes in biology are metalloproteins and contain one or more metal clusters required for function. Iron-sulfur ([Fe-S]) clusters are particularly important for catalyzing electron transfer (ET) redox reactions necessary for many basic metabolic processes, including respiration, photosynthesis, DNA replication, DNA repair, and nitrogen fixation [1–3]. In many cases, biophysical and structural studies of the respective catalytic enzymes can be facilitated and/or simplified by recombinant overproduction using heterologous host expression systems such as *E. coli*. Several proteins that contain [Fe-S] clusters also require anaerobic conditions for stability.

The dark-operative protochlorophyllide oxidoreductase (DPOR) enzyme complex is one such oxygen-sensitive metalloprotein. DPOR catalyzes the reduction of protochlorophyllide (Pchlide) to chlorophyllide (Chlide), a key reduction event in the biosynthesis of bacteriochlorophyll [4]. DPOR is made of two-component proteins: BchL and BchNB. BchL is a homodimer carrying a [4Fe-4S] cluster and serves as the electron donor component protein. BchNB is the electron acceptor component and

Patricia C. Dos Santos (ed.), *Fe-S Proteins: Methods and Protocols*, Methods in Molecular Biology, vol. 2353,
https://doi.org/10.1007/978-1-0716-1605-5_4, © Springer Science+Business Media, LLC, part of Springer Nature 2021

functions as a heterotetramer composed of two BchN and BchB subunits, which are arranged as an $\alpha_2\beta_2$ tetramer carrying two [4Fe-4S] clusters and active sites for Pchlide binding [4–7]. In this chapter, we describe two key modifications to the overproduction and purification of these two protein component proteins, which result in enhanced protein yields. We use an engineered "SufFeScient" *E. coli* cell line that carries a constitutively expressed *Suf* [Fe-S] biogenesis pathway [8]. In addition, we describe a recently discovered translational initiation tag that increases protein yields [9]. While DPOR was used as a model system, this method is not limited to Bch proteins. It can be applied to other oxygen-sensitive [Fe-S] metalloproteins of interest.

2 Materials

2.1 Reagents and Solutions

1. Luria Broth (LB): 25 g/L LB powder in deionized (DI) H_2O and autoclaved in an appropriately sized flask(s). Baffled flasks produce better aeration (*see* **Note 1**).
2. Fe(III) citrate solution: 50 mM Fe(III) Citrate solution (12.25 g/L) in DI H_2O. This solution is freshly prepared by heating the mixture using a hot plate while stirring. Alternatively, the mixture can be autoclaved (*see* **Note 2**).
3. L-Cysteine solution: 50 mM L-Cysteine monohydrate (6.06 g/L) in DI H_2O.
4. Ampicillin solution: 100 mg/mL ampicillin solution prepared in DI H_2O, followed by filter sterilization. Prepared as 1000× stock and used as the final (1×) concentration of 100 μg/mL ampicillin.
5. Kanamycin solution: 50 mg/mL kanamycin solution prepared in DI H_2O, followed by filter sterilization. Prepared as 1000× stock and used as the final (1×) concentration of 50 μg/mL kanamycin.
6. IPTG solution: 100 mM (23.83 g/L) Isopropyl-β-D-1-thiogalactopyranoside (IPTG) solution prepared freshly in filter-sterilized DI H_2O before inducing the cultures. Final concentration added into the culture is 0.3 M.
7. Dithionite: 1 M (174 g/L) sodium dithionite (sodium hydrosulfite) prepared freshly using degassed water or STD buffer, or added as a powder to all buffer solutions.
8. Standard "STD" buffer: 100 mM HEPES (pH 7.5), 150 mM NaCl, containing freshly added sodium dithionite (1.7 mM).
9. Q-sepharose "Q equilibration" buffer: 100 mM HEPES (pH 7.5), containing freshly added sodium dithionite (1.7 mM).

10. Q-sepharose "Q elution" buffer: 100 mM HEPES (pH 7.5), 1 M NaCl, containing freshly added sodium dithionite (1.7 mM).
11. "Storage" buffer: 100 mM HEPES (pH 7.5), 150 mM NaCl, 10 mM $MgCl_2$, containing freshly added sodium dithionite (1.7 mM).
12. ATP solution: ~100 mM ATP stock solution prepared in STD buffer in the glovebox; the pH is adjusted to 7.5 and stored at −20 °C. The concentration of ATP is measured spectroscopically using extinction coefficient 15,400 $M^{-1}.cm^{-1}$.
13. Protochlorophyllide: Protochlorophyllide is generated as described [6, 8].

2.2 Equipment

1. 1 L gasket-sealed centrifuge bottle(s).
2. 50 mL gasket-sealed high-speed centrifuge tubes.
3. Airtight septum-sealed bottles.
4. Appropriate septa and aluminum crimp seals.
5. Anaerobic chamber/glove box.
6. Sonifier/sonicator.
7. Nickel-NTA resin.
8. Q-sepharose anion exchange resin.
9. 30-kDa molecular weight cutoff spin concentrator(s)
10. S200 gel filtration column with good resolution of proteins between ~15 kDa and 150 kDa.

3 Methods

3.1 Transformation of BchL and BchNB Plasmids into E. coli BL21(DE3) and SufFeScient Cells

DPOR components are expressed as individual open reading frames that are cloned into separate plasmids. BchL is cloned into an RSF-Duet plasmid and carries an N-terminal poly-His (6×) affinity tag. BchL is purified as an individual protein complex and is a constitutive homodimer. BchN and BchB are cloned into RSF-Duet and pET-Duet plasmids, respectively. BchB carries an N-terminal poly-His (6×) affinity tag. BchN and BchB are co-expressed and purified as a heterotetrameric complex. In this case, appending affinity tags onto the C-terminus of BchB perturbs activity. Similarly, adding an affinity tag to either the N- or C-terminus of BchN interferes with enzymatic activity.

We recommend two options for the choice of BL21(DE3) cells. While DPOR component proteins can be overexpressed using a standard BL21(DE3) cell line [10], we recently developed a modified BL21(DE3) strain with advantages for overproduction of [Fe-S] cluster-containing proteins [8]. This *SufFeScient* BL21

(DE3) cell line carries a restored *Suf* operon, one of the two operons for the production and incorporation of [FeS] clusters in *E. coli*. In addition, the *SufFeScient* strain carries a deletion in the Fur-binding site within the *sufA* promoter region, which constitutively increases *suf* expression by four-fold under aerobic growth conditions [8]. The *SufFeScient* cells produce a remarkable ~300% increase in the yield of BchL and near-maximal incorporation of [4Fe-4S] clusters in the synthesized proteins. However, this strain does not enhance the yields of the BchNB protein complex [8]. This strain is available by request.

1. Transform plasmid DNA using heat-shock method. Mix aliquots of ~50 ng for BchL or ~300 ng of each plasmid for BchN and BchB with 50 μL of chemically competent *E. coli* cells and incubate with plasmid DNA for 30 min on ice.
2. Subject cells to heat-shock at 42 °C for 45 s, followed by incubation on ice for 2 min.
3. Add autoclaved LB medium (250 μL) to the cells and incubate for 60 min at 37 °C with mixing for 60 min.
4. Plate cells on LB-agar plates containing appropriate antibiotic (s) and incubate at 37 °C overnight. Plasmids described in these procedures can be obtained by request.

3.2 "MKIV" Translation-Initiation Tag Enhances Overall Protein Overproduction

A recent study showed that the efficiency of protein translation is dictated by the context of the first four amino acids incorporated by the ribosome at the initiating start codon [9]. One of the optimal sequences for the highest translation efficiency is M-K-I-V.

1. Introduce the MKIV coding sequence ahead of the N-terminal affinity tag. The tag enhances the overproduction of DPOR protein components by around 50% (*see* **Note 3**).
2. Transform the engineered plasmid(s) suitable for expression of the desired protein with the MKIV tag using the method described in Subheading 3.1.

3.3 Overproduction of DPOR Component Proteins

1. Use transformants from the plate (from Subheading 3.1) to generate a 10-mL starter culture with the appropriate antibiotic(s) and grow overnight at 37 °C.
2. Media for the larger growth are prepared using autoclaved LB media supplemented with 1 mM iron(III) citrate and 1 mM L-cysteine. The L-cysteine stock solution is prepared similarly to the iron(III) citrate stock solution but does not require heating to create the stock solution. Add appropriate antibiotics to the media as required.
3. Start the larger cultures by adding 1 mL of the overnight starter to each liter of media. Incubate cultures at 37 °C with shaking at 200 rpm until the OD_{600} reaches 0.6.

4. Induce protein overproduction by adding 300 μM IPTG and incubate cells overnight (~17 h) at 25 °C with shaking at 150 rpm (*see* **Note 4**).
5. Transfer induced cells to 1 L centrifuge bottles which contain a rubber gasket in their cap.
6. After transfer to centrifuge bottle(s), add 1.7 mM (~0.3 g) of sodium dithionite (DT) powder to the cell suspension and immediately cap and turn end over end to dissolve the sodium dithionite.
7. Incubate cells in the centrifuge bottles at 17 °C for an additional 3 h to promote/complete [Fe-S] cluster incorporation (no shaking required).
8. Pellet cells via centrifugation (4392 × *g*, 20 min, 4 °C).
9. Perform all subsequent steps either in airtight septum-sealed serum bottles or in an anaerobic chamber (glove box). All buffers are thoroughly degassed (*see* **Note 5**) and contain fresh reductant (sodium dithionite) to maintain a reducing environment and to scavenge any residual oxygen (*see* **Note 6**).
10. Place the pelleted cells into the glove box. Decant the spent media and discard it. Resuspend the pelleted cells in STD buffer (25 mL/L of growth) and transfer it to an appropriately sized glass serum bottle. The bottle is capped with an airtight septum and crimped using a crimper tool. The cells are then ready for lysis or can be stored at −20 °C.

3.4 Purification of DPOR Component Proteins

3.4.1 Lysis and Affinity Purification

1. Prepare all buffers and thoroughly degas them on the day of lysis and purification.
2. Add lysozyme to the cells (0.5 mg/mL; *see* **Notes 7** and **8**) and stir for 30 min.
3. Sonicate cells inside the glovebox on ice at 50% duty cycles, 3 cycles of 1 min on and 1 min off, respectively.
4. Transfer the cell lysate to centrifuge tubes with rubber gaskets in their cap. Clarify lysates via centrifugation (37,157 × *g*, 60 min). Inside the glovebox, transfer the supernatant to a glass serum bottle and crimp it to seal.
5. Preequilibrate the column with 10 column volumes (CVs) of STD buffer (*see* **Note 9**). Load clarified lysates onto a nickel-nitrilotriacetic acid (Ni^{2+} NTA) column using a peristaltic pump.
6. Wash nonspecifically bound proteins using 10 CVs of "wash" buffer.
7. Elute bound proteins into a sealed serum bottle using 5 CVs of "elution" buffer.

8. Assess the purity of eluted BchL and BchNB component proteins using SDS-PAGE analysis. Following affinity chromatography of BchL and BchNB are purified further using different approaches as described in Subheadings 3.4.2 and 3.4.3, respectively.

3.4.2 Additional Steps for the Purification of BchL Using Ion-Exchange and Size-Exclusion Chromatographies

1. Use a two-step procedure for the equilibration of the Q-sepharose column (*see* **Notes 10** and **11**). The equilibration is initiated with high salt "Q elution" buffer (1 M) and is followed by the low (or zero) salt "Q equilibration" buffer (*see* **Note 12**).
2. After equilibration, load the pooled BchL fractions onto the column using an FPLC (*see* **Note 13**) and wash with 10 CVs of STD buffer.
3. Elute the desired protein with a linear gradient (150 mM to 1 M NaCl), collecting fractions inside the glovebox (*see* **Notes 14** and **15**).
4. Analyze fractions by SDS-PAGE.
5. Pool BchL-containing fractions and concentrate to an appropriate volume for size exclusion chromatography (SEC). The pooled protein is concentrated using a 30-kDa molecular weight cutoff spin concentrator. Transfer the protein to the concentrator in the glove box and cap. Wrap the cap with parafilm before removing it from the glovebox. Perform the concentration by centrifugation at ($4122 \times g$) in a table-top centrifuge (*see* **Notes 16** and **17**).
6. Load the concentrated BchL protein onto the preequilibrated S200 column (*see* **Note 18**).
7. Collect fractions as described earlier.
8. Pool fractions containing BchL and concentrate using a 30-kDa molecular weight cutoff spin concentrator. Buffer exchange the protein solution with the "storage buffer" via dilution and re-concentration (*see* **Note 19**).
9. After concentration, aliquot the purified protein in the glovebox into 1.2 mL cryotubes and seal. Move the tubes out from the glovebox, and flash freeze in liquid nitrogen and store them under liquid nitrogen (*see* **Note 20**).

3.4.3 Additional Steps for Purification of BchNB

The BchNB heterotetramer is a large complex (206 kDa) and thus separates from trace impurities in size exclusion chromatography (SEC).

1. Concentrate the protein fractionated from the affinity chromatography step using a 30-kDa molecular weight cutoff spin concentrator.

2. Purify the BchNB complex using SEC as described earlier for BchL.
3. Pool fractions containing BchNB from SEC. Concentrate, buffer exchange, flash freeze, and store the protein under liquid nitrogen (as described for BchL earlier).

3.5 Determination of Protein Concentration

Reduced sodium dithionite and [Fe-S] clusters exhibit strong absorbance at 280 nm. Since the dithionite signal overlaps with protein absorbance, protein concentration cannot be measured using intrinsic tryptophan-based signals and must be measured using the Bradford method [11].

1. Prepare a standard curve using bovine serum albumin or lysozyme to be used as a reference.
2. Measure the protein concentration using the Bradford method.

3.6 Experiments to Measure DPOR Activity

The activity of DPOR can be assayed by monitoring the progress of the reaction over time.

1. Dilute the individual reaction components BchNB, BchL, ATP, and the substrate protochlorophyllide to 5 μM, 20 μM, 3 mM, and 20 μM, respectively, with freshly prepared STD buffer inside the glovebox.
2. Prepare a negative control reaction, which does not contain ATP.
3. Take 200- μL aliquots from the reaction at various time points. Because DPOR is a slow enzyme, 10, 20, 30, 40, and 60 min are appropriate if the experiment is performed at room temperature.
4. Quench each 200 μL aliquot with the addition of 800 μL of 100% acetone. This causes the protein components to precipitate out of solution and can be separated from the Pchlide and Chlorophyllide (Chlide) components via centrifugation (13,226 × *g*, 4 min).
5. Transfer the 80% acetone mixture to a cuvette and record the absorbance spectra of the extracted pigments. Pchlide has a strong absorbance peak at 626 nm, and the product (Chlide) has a new absorbance peak at 666 nm.
6. Use the absorbance peaks to determine the quantity of Chlide that has been produced using the molar extinction coefficient 30,400 M^{-1} cm^{-1}. Purified DPOR components generate activity ~200 pmol min^{-1} mg^{-1}(total protein).

4 Notes

1. Do not use surfactants during growth as they interfere with the oxidation of [Fe-S] clusters.
2. The iron(III) citrate stock can be autoclaved rather than heating on a heat block. Adding the iron(III) citrate to LB media before autoclaving is not recommended. The L-cysteine stock can be filter sterilized but should not be autoclaved.
3. We note here that this tag will enhance the overall overproduction of the protein, but an increase in yield (soluble protein) depends on the intrinsic solubility. The tag can be utilized as an additional tool to enhance the yield of any protein of interest.
4. The noted IPTG concentration and overnight induction conditions are optimized for DPOR component proteins.
5. The degassing method described here relied on a custom-built Schlenk line. All buffers were subjected to at least three 20-min cycles of vacuum and nitrogen pressure (~7 PSI) while stirring. Buffers can also be degassed by stirring in the glovebox and utilizing the oxygen scavenging catalyst in the glove box. The latter strategy is much slower and relies on a well-maintained catalyst stack and oxygen meter.
6. Determining the actual concentration of reduced sodium dithionite in the solution is difficult as the powder exists as a mixture of reduced and oxidized species. It is therefore recommended to prepare a fresh stock of sodium dithionite every day prior to any experiments or protein preparations.
7. Dry lysozyme powder can be added to the cells in the glove box to maintain anaerobicity, or it can be added outside the glovebox assuming it is added immediately after removing from −20 °C, and the sealed container is degassed on the Schlenk line for at least 3 brief (~3 min) cycles of vacuum and nitrogen. If the latter strategy is used, it is recommended to add additional fresh sodium dithionite to the frozen suspension before thawing.
8. Thawing can also be performed in the glove box instead of under positive nitrogen pressure. Depending on the thermal stability of the protein of interest, it may be ideal for thawing on ice or may be accomplished more quickly by thawing in room temperature water.
9. It is also possible to perform Ni^{2+} NTA purification (or Strep-Tag) via gravity inside the glovebox, but this process tends to be slower. It is also possible (or even preferable) to move the peristaltic pump into the glovebox to maintain anaerobicity.

10. The quantity of resin necessary for this step must be empirically determined.
11. High salt must be used first when equilibrating a Q-sepharose column, as the negatively charged dithionite of the sodium dithionite salt will be bound to the top of the column without the 1.0 M NaCl. If the high salt equilibration step is skipped, most of the column will not be completely reduced.
12. The salt concentration at which a protein of interest will bind to Q-sepharose varies based on the surface charge of the protein. The theoretical PI of the protein may help determine binding strength to an ion-exchange column.
13. Proteins can be loaded onto a column using a variety of methods. It is possible to use a peristaltic pump inside the glove box. Outside the glovebox, positive nitrogen pressure can be used to load the proteins. Alternatively, if an FPLC is used, the protein can be loaded using an injection port.
14. The starting and ending concentrations of NaCl for the gradient shown here for ideal separation must be determined empirically. This particular gradient works well for BchL.
15. Fraction size is determined by the size of the preparation, the volume of resin used, and the slope of the gradient chosen. For the DPOR preparation described here, 2.0 mL fractions were collected.
16. The time and speed at which to centrifuge depend on the overall concentration, the ratio of the cutoff weight of the spin contractor to the size of the protein, and the viscosity of the solution.
17. Large-scale concentrations can be accomplished using a Stirred Cell (Amicon) or similar concentrator, but care must be taken to vent the cell with excess nitrogen and always maintain an outflow. This methodology is more effective when using argon in place of nitrogen, as it is heavier than air and creates a better barrier.
18. Lower than recommended equilibration volumes can safely be used for equilibration if the column has been stored in a reducing buffer, and it is possible to visualize the degree of equilibration by dripping the outflow into a methyl viologen (0.2 g/mL) solution, which will produce purple coloration when exposed to reductant.
19. This step may not be necessary depending on the storage requirements of a given protein.
20. Liquid nitrogen flash freezing can be done outside of the glove box but should be done immediately after removal. It is possible to bring liquid nitrogen into the glovebox depending on the degree of inlet chamber purging. A vacuum cycle that is too

strong will cause the nitrogen to boil and spill out of the vessel and may cause damage to the glove box. It is recommended to slightly deflate the chamber to allow for evaporation if liquid nitrogen is brought into the chamber.

References

1. Fontecave M (2006) Iron-sulfur clusters: ever-expanding roles. Nat Chem Biol 2 (4):171–174. https://doi.org/10.1038/nchembio0406-171
2. Beinert H (2000) Iron-sulfur proteins: ancient structures, still full of surprises. J Biol Inorg Chem 5:2–15
3. Kiley PJ, Beinert H (2003) The role of Fe-S proteins in sensing and regulation in bacteria. Curr Opin Microbiol 6(2):181–185
4. Reinbothe C, El Bakkouri M, Buhr F et al (2010) Chlorophyll biosynthesis: spotlight on protochlorophyllide reduction. Trends Plant Sci 15:614–624. https://doi.org/10.1016/j.tplants.2010.07.002
5. Fujita Y, Takagi H, Hase T (1996) Identification of the chlB gene and the gene product essential for the light-independent chlorophyll biosynthesis in the cyanobacterium Plectonema boryanum. Plant Cell Physiol 37 (3):313–323. https://doi.org/10.1093/oxfordjournals.pcp.a028948
6. Fujita Y, Bauer CE (2000) Reconstitution of light-independent protochlorophyllide reductase from purified Bchl and BchN-BchB subunits: in vitro confirmation of nitrogenase-like features of a bacteriochlorophyll biosynthesis enzyme. J Biol Chem 275:23583–23588. https://doi.org/10.1074/jbc.M002904200
7. Nomata J, Swem LR, Bauer CE et al (2005) Overexpression and characterization of dark-operative protochlorophyllide reductase from Rhodobacter capsulatus. Biochim Biophys Acta Bioenerg 1708:229–237. https://doi.org/10.1016/j.bbabio.2005.02.002
8. Corless EI, Mettert EL, Kiley PJ et al (2020) Elevated expression of a functional suf pathway in *Escherichia coli* BL21(DE3) enhances recombinant production of an iron-sulfur cluster-containing protein. J Bacteriol. https://doi.org/10.1128/JB.00496-19
9. Verma M, Choi J, Cottrell KA et al (2019) A short translational ramp determines the efficiency of protein synthesis. Nat Commun 10:5774. https://doi.org/10.1038/s41467-019-13810-1
10. Moser J, Bröcker MJ (2011) Methods for nitrogenase-like dark operative protochlorophyllide oxidoreductase. Methods Mol Biol 766:129–143. https://doi.org/10.1007/978-1-61779-194-9_9
11. Bradford MM (1976) A rapid and sensitive method for the quantitation of microgram quantities of protein utilizing the principle of protein-dye binding. Anal Biochem 72:248–254. https://doi.org/10.1016/0003-2697(76)90527-3

Chapter 5

Enzymatic and Chemical In Vitro Reconstitution of Iron-Sulfur Cluster Proteins

Mauro Marengo, Rita Puglisi, Simonetta Oliaro-Bosso, Annalisa Pastore, and Salvatore Adinolfi

Abstract

Iron-sulfur (Fe-S) clusters are key cofactors for proteins involved in essential cellular processes such as DNA replication and repair, ribosome biogenesis, tRNA thio-modification, and co-enzyme synthesis. Fe-S clusters can assemble spontaneously from inorganic compounds, but their biogenesis requires dedicated machineries to circumvent the toxic nature of iron and sulfur. To address how these machines work, different laboratories have applied various biochemical and biophysical approaches, both in vivo and in vitro. Fe-S cluster enzymatic and chemical formation in vitro is the most efficient way to follow Fe-S cluster biogenesis in a controlled environment and investigate each component of the machinery at the molecular level. In this review, we detail and discuss an efficient protocol for an in vitro Fe-S cluster enzymatic and chemical formation, which we successfully developed to study Fe-S cluster formation. We underline the applications of this approach to the study of an essential biological system.

Key words Iron-sulfur cluster, Biogenesis, Anaerobicity, Glove box, Fe-S protein, Cluster reconstitution

1 Introduction

Inorganic iron-sulfur (Fe-S) clusters are present in nearly all organisms, spanning all three domains of life: eubacteria, archaebacteria, and eukarya. Fe-S clusters are considered to be among the most ancient catalysts. This concept is supported by the elevated levels of iron and sulfur in the primordial environment and by the unique characteristics of these elements [1, 2]. In the cell, Fe-S clusters are usually transported by proteins that coordinate them through Cys and His residues. Fe-S proteins perform a variety of biological functions, such as enzyme activity, protein regulation, and translation, DNA replication and repair, ribosome biogenesis, tRNA thio-modification, and coenzyme synthesis [3–14].

While many cell types present different strategies to import their cofactors or their precursors, Fe-S clusters biogenesis relies

Patricia C. Dos Santos (ed.), *Fe-S Proteins: Methods and Protocols*, Methods in Molecular Biology, vol. 2353, https://doi.org/10.1007/978-1-0716-1605-5_5,

on the intracellular environment for their iron and sulfur sources. This requirement is dictated by the oxygen sensitivity and the instability of Fe-S clusters in aqueous solution, especially in their free forms [15]. Fe–S clusters were initially believed to assemble spontaneously, as they can be synthesized nonenzymatically in vitro in the presence of iron cations and sulfide and proteins with a proper number of correctly positioned Cys ligands [16]. However, in vivo, both iron and sulfur are toxic elements highly reactive, and their presence requires tight regulation. Therefore, all living organisms utilizing these cofactors must rely on a biosynthesis apparatus to mobilize the iron and sulfur and to promote the assembly and insertion of Fe-S clusters into final acceptor proteins.

It has been well documented that the Fe-S cluster biogenesis requires tightly coordinated biosynthetic pathways. Dean and coworkers showed for the first time the requirement of a specialized biosynthetic pathway dedicated to the assembly of Fe-S clusters in *A. vinelandii* [17]. This pathway involves several proteins associated with a complex machinery devoted not only to the building of the Fe-S clusters, but also to their delivery to the recipient proteins [18]. In the 1990s, exhaustive genetic, biochemical, and cell-biology evidence were gathered to prove that the maturation of Fe-S proteins in living cells is a tightly controlled process that involves not only one but several complex protein machineries. Three different systems have been identified in bacteria: the nitrogen fixation NIF machinery, which is responsible for the specific maturation of nitrogenase in diazotrophic bacteria, the iron-sulfur cluster assembly machinery ISC, and the methanoarchaeal sulfur mobilization SUF machinery, which generates housekeeping Fe-S proteins [19].

Genomic analyses of these machineries revealed a great variability across diverse phylogenetic groups while relying on similar biosynthetic steps. Indeed the number and type of operons coding for these machineries vary from one bacterium to another. Some organisms such as *Erwinia crysanthemi* contain all three systems [20], others such as *Escherichia coli* have only two [21] while *Mycobacterium tuberculosis* has only one [22]. Some other organisms, such as *Helicobacter* pylori, contain only some components of one of these systems [23]. Fe-S cluster assembly has been intensively investigated in the model bacterium *Escherichia coli*, which contains both the ISC and SUF operons operating during normal and stress conditions, respectively [24, 25]. These two operons share common features for the formation of Fe-S clusters that can be divided into three main steps. First, cysteine mobilization is required and carried out by a cysteine desulfurase pyridoxal 5-′-phosphate (PLP)-dependent enzyme (IscS). This enzyme generates an S-precursor in the form of a cysteine-bound persulfide by desulfurization of L-cysteine, thus producing alanine as a by-product. Second, the cluster is assembled on a scaffolding protein (IscU) through coordination mainly with cysteines. These residues are also involved in sulfur transfer by accepting the sulfur activated by the desulfurase enzyme and iron-binding through direct coordination. The third and final step is the transfer of the de novo synthesized Fe-S clusters to the final target protein. In the ISC system, this step is enhanced by co-chaperone HscA and HscB via ATP hydrolysis. In the SUF system, the transfer of nascent Fe-S clusters is less understood, but the system also contains an ATPase, SufC, whose function is required for the functionality of that machinery.

Similarly to prokaryotes, in eukaryotes, the iron-sulfur biogenesis is carried out in mitochondria by a machinery that shares high homology with the bacterial ISC system, as well as in the cytosol by the cytosolic iron-sulfur cluster assembly (CIA) [26]. The mitochondrial ISC encompasses the scaffold protein ISCU, the NFS1-ISD11-ACP complex containing the cysteine desulfurase NFS1, a PLP-dependent enzyme that requires the interaction with both a LYRM superfamily protein (ISD11) and an acyl carrier protein (ACP) for its stability and functionality [27–30]. The CIA system is composed of nine proteins involved in the two major steps for the maturation of cytosolic and nuclear Fe-S proteins. Initially, the cytosolic Fe-S biogenesis comprises the formation of a transient [4Fe-4S] cluster on the scaffolding protein complex formed by the two P-loop NTPases CFD1 and NBP35. The second major step involves the release and delivery of the newly synthesized [4Fe-4S] cluster from CFD1-NBP35 complex to dedicated apoproteins [26].

Characterization of Fe-S cluster proteins is a complex issue because their assembly is really challenging. The most common method applied to produce Fe-S proteins is the plasmid-base expression in *E. coli*. Indeed, this procedure successfully ensures the incorporation of a functional Fe-S cluster synthesized by the housekeeping ISC system. The real challenge lies in the purification protocols. Standard purification methods, carried out in aerobic conditions, generally lead to damage or loss of Fe-S clusters [31]. They are highly labile when exposed to oxygen and are easily oxidized, leading to the isolation of a pure apoprotein (without its cluster) [32]. Although the purification under anaerobic conditions improves the chances of retaining the Fe-S cluster significantly, there are still limitations due to the difficulties of reaching full occupancy or obtaining a nonnative form [31].

While Fe-S cluster biogenesis machineries have been intensively investigated, details at the molecular level are still poorly understood. The lability of the biosynthetic intermediates and the isolation of apoforms of these components makes it challenging to dissect individual steps of this process and characterize how transient intermediates are formed while Fe-S clusters are generated in vivo [33]. Several approaches have been applied to address these issues, from native mass spectrometry [33] to NMR and small-angle light scattering [34–36], resonance Raman and Mossbauer spectroscopies [37].

Within all these different approaches, it is worth to underline that in vitro Fe-S cluster enzymatic reconstitutions are still widely used, for prokaryotes and eukaryotes machineries, respectively [36–44], because they represent still a valuable tool to investigate Fe-S cluster proteins, their biogenesis, and functionalities [45]. Indeed, this method has been extensively used to determine the effect on the kinetics of Fe-S cluster formation in the presence of possible regulators, as frataxin, in their wild type [46, 47] and mutated forms [48, 49] or at different salt conditions [50]. The relevance of this approach is still testified by (1) its wide use to reconstitute iron-sulfur proteins as shown by papers in recent years: ISCU [42], Chromokinesin, KIF4A [51], FNR, and NsrR [52], and (2) the ability to carry out eukaryotic Fe-S cluster reconstitution by the well-characterized prokaryotic systems and/or its essential components such as IscS/IscU, NifS/NifU, and SufS/SufU [53, 54]. This chapter describes a detailed protocol to generate de novo Fe-S clusters on a specific target protein to produce a purified holo-protein. In this example, aconitase has been used, which is a classic Fe-S protein present in the Kreb cycle.

2 Materials

2.1 Supplies and Instrumentation

1. Quartz cuvettes with sealable cup.
2. Syringes and needles of different lengths.
3. Gel filtration column.
4. Anaerobic chamber setup: Anaerobic chamber attached to nitrogen and hydrogen gas tanks and equipped with oxygen meter, cartridges, and cartridge regenerator.
5. Spectrophotometer.

2.2 Stock Solutions Preparation

1. Tris buffer: 1 M Tris–HCl, pH 8.0. Weigh 121.14 g of Tris–HCl base and dissolve into 900 mL of deionized water by mixing thoroughly (*see* **Note 1**). Adjust the pH with HCl (*see* **Note 2**). Transfer the solution into a graduate cylinder or volumetric flask and add deionized water up to 1 L. Filter the solution using ultrafiltration membranes (pore size 0.22 μm) and store it at room temperature or 4 °C.
2. NaCl solution: 5 M NaCl. Dissolve 292.21 g of NaCl in 900 mL of deionized water by mixing thoroughly. Transfer the solution into a graduated cylinder and add deionized water up to 1 L. Filter the solution using ultrafiltration

membranes (pore size 0.22 μm) and store it at room temperature (*see* **Note 3**).

3. Dithiothreitol solution (DTT): 1 M DTT. Weigh 3.085 g of DTT and transfer it to a 20 mL graduate cylinder. Add 19 mL of deionized water and mix it. When all DTT is dissolved, adjust the volume to 20 mL. Filter the solution using ultrafiltration membranes (pore size 0.22 μm), prepare aliquots of 0.2 mL, and store it at −20 °C.
4. Reconstitution buffer: 20 mM Tris–HCl pH 8.0, 150 mM NaCl. Add 20 mL of 1 M Tris–HCl of pH 8.0 to a 1 L graduated cylinder together with 30 mL of 5 M NaCl. Add water to a volume of 1 L and mix it. Filter the solution using ultrafiltration membranes (pore size 0.22 μm) and store it at room temperature (*see* **Note 4**).
5. IscU: IscU scaffolding protein. Use the needed aliquots (200–500 μL) of 1 mM IscU, expressed in *E. coli* and purified aerobically as previously described [55]. Store aliquots at −80 °C.
6. IscS: IscS desulfurase protein. Use the needed aliquots (10–50 μL) of 100 μM IscS, expressed in *E. coli* and purified aerobically as previously described [56]. Store aliquots at −80 °C.
7. AcnB: Aconitase B protein. Use the needed aliquots (200–500 μL) of 800 μM AcnB, expressed in *E. coli* and purified aerobically as previously described [57]. Store aliquots at −80 °C.

2.3 Working Solutions

Solutions need to be prepared fresh each day and stored in a glove box for the duration of the experiment.

1. Fe solution: 40 mM ammonium iron(II) sulfate ($(NH_4)_2Fe(SO_4)_2{\cdot}6H_2O$). Weigh 156.82 mg of ammonium iron(II) sulfate (*see* **Note 5**) and transfer it into a 20 mL graduate cylinder. Transfer the cylinder to the glove box under a controlled atmosphere (*see* **Note 6**). Add 9 mL of degassed deionized water and mix it. When the iron salt is dissolved, adjust the volume to 10 mL (*see* **Note 7**).
2. Fe dilution: 1 mM ammonium iron(II) sulfate ($(NH_4)_2Fe(SO_4)_2{\cdot}6H_2O$). Add 25 μL of 40 mM ammonium iron(II) sulfate to a microfuge tube and add degassed deionized water up to 1 mL. Store it at room temperature in the glove box (*see* **Note 6**).
3. Cysteine: 40 mM cysteine. Weigh 48.64 mg of cysteine in a 20 mL graduate cylinder. Transfer the cylinder to the glove box under a controlled atmosphere. Add 9 mL of degassed reconstitution buffer and mix it. When the cysteine is dissolved,

adjust the volume to 10 mL (*see* **Note 8**). Store it in the glove box at room temperature (*see* **Note 6**).

4. Sulfide: 40 mM Li_2S. Weigh 18.38 mg of Li_2S in a 20 mL graduate cylinder. Transfer the cylinder to the glove box under a controlled atmosphere. Add 9 mL of degassed deionized water and mix it. When the Li_2S is dissolved, adjust the volume to 10 mL. Store it in the glove box at room temperature.

3 Methods

In this section, all the necessary steps are described to produce the Fe-S cluster protein anaerobically in a glove box. Degas all solutions and transfer them into the glove box with all the needed equipment. Carry out reconstitution reactions by adding all components in a precise order. Follow the Fe-S cluster reconstitution spectrophotometrically (Fig. 1).

3.1 Anaerobic Work Setup

All reconstitution assays are carried out in a glovebox to achieve a tightly controlled atmosphere, where all air (oxygen) is removed and replaced with nitrogen or argon. The glovebox is a sealed chamber with an antechamber (separated small box), named portal, on the side. The portal is provided with an independent nozzle for gas purging and is used to place or take out materials/solutions into/from the main chamber. It requires a little purging time (7–10 min) to remove oxygen and avoid contamination. Arm-length nitrile gloves (arm-gloves) in the front side of the

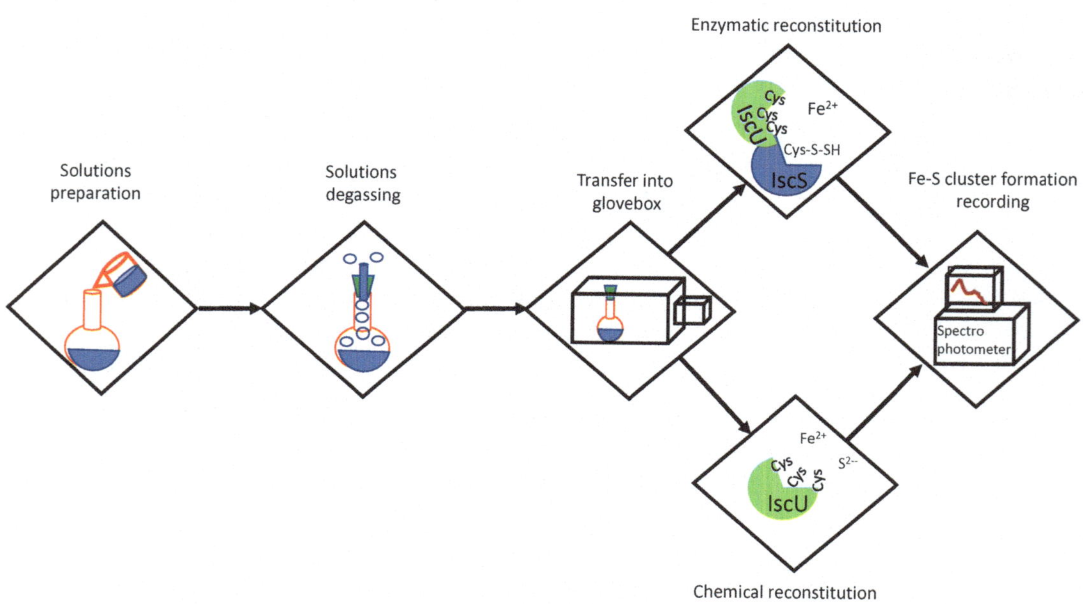

Fig. 1 Fe-S cluster reconstitution. Main steps for chemical and enzymatic in vitro Fe-S cluster reconstitution on bacterial scaffolding IscU

glovebox are arranged in such a way that users can place their hands inside to perform tasks within the box without breaking containment. Manage of the arm-gloves requires extreme care. Indeed, they can easily tear. It is necessary to wear latex gloves and sprinkle them with talcum powder to ensure easy insertion into the arm gloves.

The glove box is made of transparent material (usually plexiglass) to allow visibility inside from every angle. The glove box is connected to a pump, four cartridges, and a nitrogen/argon tank to remove air and fill the glove box compartment with nitrogen or argon. The gas in the chamber is continuously filtered by cartridges to remove any oxygen that can seep through. It is critical to have an oxygen monitor onsite. While the glovebox environment is intended to be closed, small amounts of oxygen can seep through the glove ports. Thus, it is essential that the glove box is purged nightly. Control check of the glovebox must be carried routinely to ensure the correct functioning. Particular attention needs to be focused on the oxygen monitor, the pressure, and the arm-length gloves.

Proper operation of the glove box is essential to perform Fe-S cluster reconstitution in vitro correctly. Indeed, the Fe-S cluster is very sensitive to oxygen, and it is easily oxidized and degraded in aerobic conditions. Acquiring essential skills and the ability to work with restriction and impaired movement is crucial when performing Fe-S reconstitution in a glove box.

3.1.1 Water Degassing

All solutions are prepared with degassed deionized water. This step is critical to ensure all oxygen is removed from water/solutions to be used in Fe-S reconstitution assay.

1. Pour 1 L deionized water out in a glass bottle. The top is sealed with a gas-tight silicon lid. Use a long needle connected to an argon/nitrogen cylinder to puncture the gas-tight lid and deeply immerse it in the water to degas.
2. Purge the solution by bubbling the inert gas. Purging time depends on the amount of water/solution to be degassed (1 L requires a 1-h purging). Care should be taken to prevent water spillage due to the gas bubbling during the degassing procedure. Ensure to leave a considerable gap between the water surface and the sealable lid.

3.1.2 Protein Solutions Degassing

All proteins need to be previously purified and stored at −80 °C. It is essential to degas the protein solutions just before performing the in vitro reconstitution assay. This procedure must be carried out on a fresh protein stock if their stability may be compromised (*see* **Note 9**). If possible, store the degassed proteins in a refrigerator placed in the glove box.

1. Transfer the required aliquots of IscU, AcnB, and IscS into a microfuge tube to be degassed. The top is sealed with a gas-tight silicon lid.
2. Use an insulin-needle connected to an argon/nitrogen cylinder to puncture the gas-tight lid and get as close as possible to the surface of the solution (*see* **Note 10**).
3. Start to inflate gently with nitrogen or argon. It is enough to see the surface riffles. Keep purging for 15 min (*see* **Note 11**).

3.1.3 Cuvettes Preparation

1. Wash the cuvettes by filling them with 50% nitric acid and leave it for 10 min.
2. Carefully remove the acid and wash extensively with water.
3. Finally, rinse with acetone, remove the excess, and let the cuvette air dry or flush nitrogen gas for a quick dry.

3.1.4 Fe-S Cluster Reconstitution Assay Material Preparation

1. Transfer to the glove box and through the portal, sealable cuvettes and degassed protein solutions. All containers must be open (*see* **Note 12**).
2. Purge the portal for the required amount of time (as requested by the manufacturer, usually 7–10 min) to ensure the complete removal of oxygen. Make sure that the external portal door is tightly closed before opening the inside door connected with the glovebox. If this operation is not carried out properly, oxygen will fill the chamber.

3.2 Fe-S Cluster Reconstitution

Carry out all procedures anoxically at room temperature unless otherwise specified.

3.2.1 In Vitro Enzymatic Fe-S Cluster Reconstitution on IscU

This method has been applied successfully to investigate the components involved in the first step of the Fe-S cluster biogenesis (de novo synthesis of Fe-S cluster) and their functions [36, 58]. The reconstitution assay is carried out in a tube, adding all the components in a precise order (*see* **Note 13**).

1. Start the reaction by pipetting 580 μL of reconstitution buffer (always add it first into the cuvette).
2. Add 20 μL of the reducing agent DTT (final concentration 10 mM) (*see* **Note 14**).
3. Add 800 μL of IscU (final concentration 400 μM).
4. Pipette 250 μL of 40 mM ammonium iron(II) sulfate (final concentration 5 mM) and incubate the reaction for 30 min at room temperature.
5. Add 250 μL of cysteine (final concentration 5 mM) (*see* **Note 15**).
6. Finally add 100 μL of IscS (final concentration 5 μM) (*see* **Note 16**) to start the Fe-S enzymatic reconstitution.

7. Mix the reaction gently inverting the tube a few times and start a timer (to define the time 0 of the enzymatic reconstitution). Incubate the reaction for 120 min at room temperature.

3.2.2 In Vitro Chemical Fe-S Reconstitution

This method has been successfully applied to several proteins containing Fe-S clusters and different origins (Archaea, Prokaryotes, and Eukaryotes), but with low occupancy yields [36]. Below, a description for reconstitution of IscU is provided; however, this methodology can be applied to other Fe-S proteins.

1. Add 700 μL of reconstitution buffer (always add it first into the cuvette).
2. Pipette 800 μL of IscU (final concentration 400 μM).
3. Mix 50 μL of 40 mM ammonium iron(II) sulfate.
4. Add 50 μL of 40 mM of Li_2S.
5. Incubate the reaction for 20 min (*see* **Note 17**).
6. Repeat **steps 3–6** an additional four times for a total of 250 μL of each component, and the final concentration of iron and sulfide reaches 5 mM.

3.2.3 In Vitro Enzymatic Fe-S Cluster Reconstitution on Final Acceptors

This method has been successfully applied to investigate the components involved in the second step of the Fe-S cluster biogenesis (transfer of the de novo synthesized Fe-S cluster) and their functions [51]. The method below describes the reconstitution of aconitase (Fig. 2), and it can be adapted to the reconstitution of other targeted proteins. The reconstitution assay is carried out in a

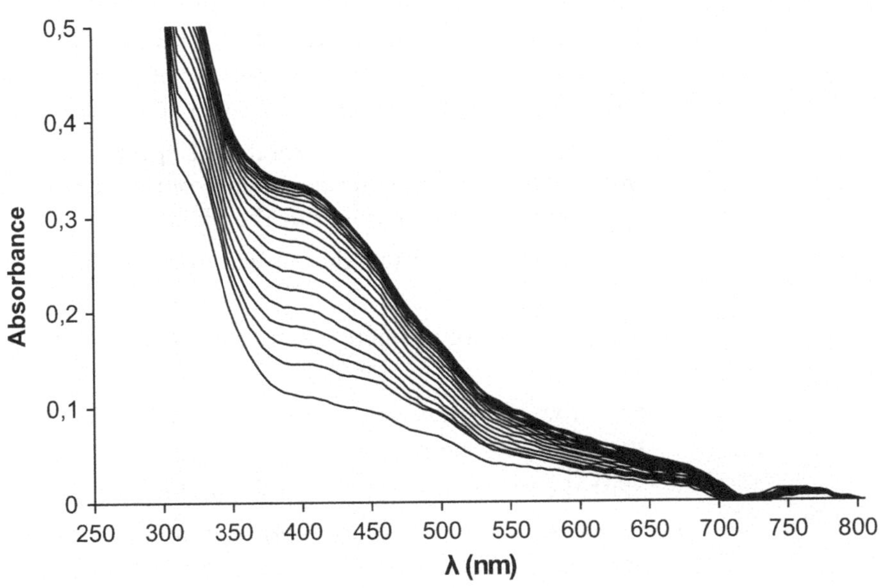

Fig. 2 Kinetics of Fe-S cluster formation on aconitase. UV-Vis absorption spectra of apo-aconitase (50 μM), recorded every 4 min from the addition of substrate

tube by adding all the components in the following precise order (*see* **Note 13**).

1. Start the reaction pipetting 1735 μL of reconstitution buffer (always add it first into the cuvette).
2. Add 10 μL of the reducing agent 1 M DTT (final concentration 5 mM) (*see* **Note 14**).
3. Add 125 μL of AcnB (final concentration 50 μM).
4. Add 10 μL of IscU (final concentration 5 μM).
5. Pipette 50 μL of 10 mM Fe dilution of ammonium iron(II) sulfate (final concentration 250 μM) and incubate for 30 min at room temperature.
6. Add 50 μL of 40 mM cysteine (final concentration 1 mM) (*see* **Note 15**).
7. Finally, add 20 μL of IscS (final concentration 1 μM) (*see* **Note 16**) to start the Fe-S enzymatic reconstitution.
8. Mix the reaction gently inverting the tube a few times and start a timer (to define the time 0 of the enzymatic reconstitution). Incubate the reaction for 120 min at room temperature.

3.3 Recording the Fe-S Cluster Reconstitution Reaction

Fe-S cluster reconstitution can be followed and quantified by UV-Vis absorption spectroscopy. Collect aliquots at different times (e.g., 10, 40, 80, 120 min).

1. Purge the glove box portal for 10 min.
2. Transfer 100 μL aliquots of each time point in a sealable cuvette and mix it gently with 700 μL reconstitution buffer.
3. Prepare a control sample (blank) by transferring all components needed for the reconstitution but the cysteine in a cuvette. This control does not need to be in a sealable cuvette.
4. Use the control blank to calibrate the spectrophotometer. In order to measure only the absorbance of the Fe-S cluster in the solution, we must subtract out the blank.
5. Transfer the cuvettes to the purged portal to get them out of the glove box.
6. Set the spectrophotometer in the 240–700 nm range (*see* **Note 18**).
7. Set the cuvette into the allocated cell within the spectrophotometer.
8. Start the recording of the reconstituted holo-protein sample.
9. Check the presence of the iron-sulfur cluster characteristic maxima. The Fe-S clusters are indeed characterized by specific maxima with different molar extinction coefficients ε_λ. These

molar extinction coefficients can be used to quantify [2Fe-2S] and [4Fe-4S] cluster contents in holo-proteins.

10. The 2Fe-2S clusters are characterized by 2 maxima at 410 nm and 456 nm and a shoulder at 510 nm [43]. Use the $\varepsilon_{452} = 7\text{–}10\ \text{mM}^{-1}\ \text{cm}^{-1}$ to quantify the [2Fe-2S] concentration in holo-proteins [59].
11. The 4Fe-4S clusters have a broad absorption centered around 410 nm. Use the $\varepsilon_{410} = 9\text{–}16\ \text{mM}^{-1}\ \text{cm}^{-1}$ [43] to quantify the [4Fe-4S] in holo-proteins.
12. The absorption ratio Abs 410/280 can also be used to estimate the Fe-S cluster incorporation. For protein containing one 4Fe-4S cluster, this ratio vary between 0.17 and 0.36, depending on the protein sequence [60, 61].
13. The components added to the mix to form the Fe-S cluster could interfere with their concentration measurement. The most accurate measurement will be obtained after the desalting column (Subheading 3.4).

3.4 Purification of Holo-Proteins by Size Exclusion Chromatography

Removing all the unreacted components is essential, with a special focus on iron and sulfur, which could produce undesired species if left unchecked [17]. This can be accomplished by subjecting the reconstituted protein through a size exclusion column.

1. Transfer the desalting column into the glove box. Cut the sealed at the bottom of the column at the notch and remove the top cup. Drain off the column storage solution.
2. To equilibrate the column, fill up the column with the reconstitution buffer and allow the solution to flow completely through the column (repeat it four times) (*see* **Note 19**). Discard the flow-through.
3. Load the reconstituted protein by adding 2.5 mL of sample to the column. Let the sample completely enter the packed bed of the column (*see* **Note 20**). Discard the flow-through.
4. For the elution, prepare several tubes for sample collection and place them under the column. Add 3.5 mL of reconstitution buffer and start collecting the eluate in one test tube or in 0.5 mL fractions (*see* **Note 21**).
5. To quantify the cluster associated with the purified reconstituted sample, transfer 50 μL aliquots of each fraction, obtained from the desalting gel-filtration, in a small sealable cuvette and mix it gently with 150 μL reconstitution buffer. As a blank control, transfer 200 μL reconstitution buffer in a cuvette (it does not need to be a sealable cuvette).
6. Follow the steps from **4–9** described in Subheading 3.3 to quantify the concentration of clusters associated with the purified sample.

In vitro Fe-S cluster chemical reconstitution and subsequent integration into apoproteins can be successfully used for addressing the properties of numerous Fe-S proteins. They may eventually help to elucidate the role of the various components, including some of the pathological relevance, that might be partners in these specialized multiprotein assembly machineries, and clarifying some of the chemistry that ensues from this kind of interactions [42, 56, 62, 63].

4 Notes

1. Add deionized water first helps to dissolve the Tris–HCl relatively easily, allowing the magnetic stir bar to revolve immediately. Tris–HCl can be dissolved faster if the water is warmed at 37 °C. However, it is essential to wait for the solution to reach room temperature before adjusting the pH.
2. Concentrated HCl (12 N) can be used at the beginning to narrow the gap from starting pH to the required one. From then on, it would be better to use a diluted HCl (1 N) to avoid a sudden drop in the pH below the required pH value.
3. A concentration of 5 M is close to the max solubility of NaCl (6 M at 20 °C). Great care is needed to ensure complete dissolution. Warming slightly, the solution is advisable.
4. When concentrated Tris–HCl is diluted, it is advisable to check and eventually adjust the pH after the dilution.
5. Ammonium iron(II) sulfate salt must be stored in a dark glass container and should be pale green (proper of iron(II)).
6. All powders are weighed in sealable containers. When transferred in the glovebox, the lid must be left completely open when using a glove box based on gas exchange pumping nitrogen to replace the air. Powder transfer requires more attention in case a vacuum is created by a mechanical pump prior to gas exchange. In this case, the lid has to be left on the containers but needs to be pierced several times to allow gas exchange while limiting powder dispersion.
7. Ammonium iron(II) sulfate solution must be prepared fresh each day under a controlled atmosphere in the glove box, using degassed deionized water. The controlled atmosphere allows delaying the oxidation of iron(II) to iron(III) that occurs quickly when iron(II) is at pH above 6.0 and exposed to oxygen. This oxidation, producing iron(III), leads to the formation of a yellowish/brown color and precipitation.
8. Cysteine solution must be prepared fresh each day under a controlled atmosphere in the glove box, using degassed reconstitution buffer. The controlled atmosphere delays cysteine

oxidation. It is advisable to use the L-cysteine hydrochloride monohydrate considering its high solubility in water.

9. It is advisable to repeat the degassing of new protein stocks each day, particularly if the protein solutions are stored at room temperature in the glove box.
10. Bubbling inert gas to carry out the purging of a small volume of protein solution is not advisable. The solution will be scattered around and out of the tube. Surface riffles ensure enough gas exchange between oxygen and nitrogen/argon.
11. Time is critical to prevent solvent evaporation caused by gas purging. It is important to measure the volume of the protein solution before and after the purging. It is advisable to take an aliquot of the protein solution after the transfer into the glove box and measure the concentration by UV-Vis spectroscopy.
12. The cuvettes, protein solutions, and tubes with cysteine and iron powders need to be inserted in the small side box for purging with a completely unscrewed lid/cup. This is essential to purge the air trapped in any container.
13. Plan well in advance the experiments to carry out and ensure that all the materials you need are ready and in the glove box.
14. The presence of DTT is necessary to provide reducing equivalents required for cluster generation and reduction of persulfides formed on IscS.
15. The cysteine solution can be injected into the sealed cuvette (the gas-tight lid has a soft silicone area) outside the glove box using a gas-tight syringe filled in the glovebox. The injection can be carried out directly into the spectrophotometer before starting the recording.
16. IscS needs to be used at a low concentration to avoid the minimal contribution of unspecific iron-thiolate polysulfide bound to IscS.
17. It is advisable to add ammonium iron(II) sulfate and cysteine in different steps because these two salts tend to react to form Fe (II)S and/or $Fe(III)_2S_3$, which impairs the Fe-S cluster formation.
18. This broad wavelength range is required to be able to assess the reconstituted protein spectrum both in the UV and visible region.
19. Around 25 mL reconstitution buffer is needed for this step.
20. For sample volume less the 2.5 mL, add the samples, let it enter completely the column packed bed. Add enough reconstitution buffer to adjust the sample volume to 2.5 mL, and once more, let it enter the column packed bed. The reconstitution buffer could be added directly to the sample to reach 2.5 mL

final volume. The former method is more advisable, giving rise to degassed less diluted protein samples.

21. The absorbance of the elution fractions (0.5 mL) can be measured by a spectrophotometer at 280 nm to identify and select only the fractions containing protein. This reduces the sample dilution drastically.

References

1. Fontecave M (2006) Iron–sulfur clusters: ever-expanding roles. Nat Chem Biol 2:171–174. https://doi.org/10.1038/nchembio0406-171
2. Imlay JA (2006) Iron–Sulphur clusters and the problem with oxygen. Mol Microbiol 59:1073–1082. https://doi.org/10.1111/j.1365-2958.2006.05028.x
3. Andreini C, Rosato A, Banci L (2017) The relationship between environmental dioxygen and iron–sulfur proteins explored at the genome level. PLoS One 12:e0171279. https://doi.org/10.1371/journal.pone.0171279
4. Fuss JO, Tsai C-L, Ishida JP, Tainer JA (2015) Emerging critical roles of Fe–S clusters in DNA replication and repair. Biochim Biophys Acta 1853:1253–1271. https://doi.org/10.1016/j.bbamcr.2015.01.018
5. Beinert H (2000) Iron–sulfur proteins: ancient structures, still full of surprises. J Biol Inorg Chem 5:2–15. https://doi.org/10.1007/s007750050002
6. Snyder CH, Merbitz-Zahradnik T, Link TA, Trumpower BL (1999) Role of the Rieske iron–sulfur protein midpoint potential in the protonmotive Q-cycle mechanism of the cytochrome bc1 complex. J Bioenerg Biomembr 31:235–242. https://doi.org/10.1023/A:1005419712731
7. Hunsicker-Wang LM, Heine A, Chen Y et al (2003) High-resolution structure of the soluble, respiratory-type Rieske protein from *Thermus thermophilus*: analysis and comparison. Biochemistry 42:7303–7317. https://doi.org/10.1021/bi0342719
8. Peters JW, Stowell MH, Soltis SM et al (1997) Redox-dependent structural changes in the nitrogenase P-cluster. Biochemistry 36:1181–1187. https://doi.org/10.1021/bi9626665
9. Beinert H, Holm RH, Münck E (1997) Iron-sulfur clusters nature's modular multipurpose structures. Science 277:653–659. https://doi.org/10.1126/science.277.5326.653
10. Iwasaki T (2010) Iron-sulfur world in aerobic and hyperthermoacidophilic archaea *Sulfolobus*. Archaea 2010:842639. https://doi.org/10.1155/2010/842639
11. Rouault TA (2017) Iron-sulfur clusters in chemistry and biology: characterization, properties and applications, 2nd edn. De Gruyter, Berlin. https://doi.org/10.1515/9783110308426
12. Lill R, Broderick JB, Dean DR (2015) Special issue on iron–sulfur proteins structure function biogenesis and diseases. Biochim Biophys Acta Mol 1853:1251–1252. https://doi.org/10.1016/j.bbamcr.2015.03.001
13. Piccioli M (2018) The biogenesis of iron-sulfur proteins: from cellular biology to molecular aspects. J Biol Inorg Chem 23:493–494. https://doi.org/10.1007/s00775-018-1555-7
14. Freibert SA, Weiler BD, Bill E, Pierik AJ, Muhlenhoff U, Lill R (2018) Biochemical reconstitution and spectroscopic analysis of iron-sulfur proteins. Methods Enzymol 599:197–226. https://doi.org/10.1016/bs.mie.2017.11.034
15. Lill R, Freibert S (2020) Mechanisms of mitochondrial iron-sulfur protein biogenesis. Annu Rev Biochem 89:471–499. https://doi.org/10.1146/annurev-biochem-013118-111540
16. Malkin R, Rabinowitz JC (1966) The reconstitution of clostridial ferredoxin. Biochem Biophys Res Commun 23:822–827. https://doi.org/10.1016/0006-291X(66)90561-4
17. Zheng L, White RH, Cash VL, Dean DR (1994) Cysteine desulfurase activity indicates a role for NIFS in metallocluster biosynthesis. Proc Natl Acad Sci U S A 90:2754–2758. https://doi.org/10.1073/pnas.90.7.2754
18. Pala ZR, Saxena V, Saggu G, Garg S (2018) Recent advances in the [FeS] cluster biogenesis (SUF) pathway functional in the apicoplast of *Plasmodium*. Trends Parasitol 34:800–809. https://doi.org/10.1016/j.pt.2018.05.010
19. Xu XM, Moller SG (2010) Iron-sulfur cluster: biogenesis, molecular mechanisms, and their functional significance. Antioxid Redox Sign

15:271–307. https://doi.org/10.1089/ars.2010.3259

20. Rincon-Enriquez G, Crete P, Barras F, Py B (2008) Biogenesis of Fe/S proteins and pathogenicity: IscR plays a key role in allowing *Erwinia chrysanthemi* to adapt to hostile conditions. Mol Microb 67:1257–1273. https://doi.org/10.1111/j.1365-2958.2008.06118.x
21. Outten FW, Wood MJ, Munoz M, Storz G (2003) The SufE protein and SufB complex enhance SufS cysteine desulfurase activity as part of a sulfur transfer pathway for Fe-S cluster assembly in *Escherichia coli*. J Biol Chem 278:45713–45719. https://doi.org/10.1074/jbc.M308004200
22. Huet G, Daffe M, Saves I (2005) Identification of the *Mycobacterium tuberculsosis* SUF machinery as the exclusive mycobacterial system of [Fe-S] cluster assembly: evidence for its implication in the pathogen's survival. J Bacteriol 187:6137–6146. https://doi.org/10.1128/JB.187.17.6137-6146.2005
23. Olson W, Jeffery N, Agar N, Johnson MK, Maier RJ (2000) Characterization of the NifU and NifS Fe-S cluster formation proteins essential for viability in *Helicobacter pylori*. Biochemist 39:16213–16219. https://doi.org/10.1021/bi001744s
24. Zheng L, Cash VL, Flint DH, Dean DR (1998) Assembly of iron-sulfur clusters: identification of an iscSUA-hscBA-fdx gene cluster from *Azotobacter vinelandii*. J Biol Chem 273:13264–13272. https://doi.org/10.1074/jbc.273.21.13264
25. Takahashi Y, Tokumoto U (2002) A third bacterial system for assembly of iron-sulfur clusters with homologs in archea and plastid. J Biol Chem 277:28380–28383. https://doi.org/10.1074/jbc.C200365200
26. Paul VD, Lill R (2015) Biogenesis of cytosolic and nuclear iorn-sulfur proteins and their role in genome stability. Biochim Biophys Acta 1853:1528–1539. https://doi.org/10.1016/j.bbamcr.2014.12.018
27. Adam AC, Bornhövd C, Prokisch H, Neupert W, Hell K (2006) The Nfs1 interacting protein Isd11 has an essential role in Fe/S cluster biogenesis in mitochondria. EMBO J 25:174–183. https://doi.org/10.1038/sj.emboj.7600905
28. Shi Y, Ghosh MC, Tong WH, Rouault TA (2009) Human ISD11 is essential for both ironsulfur cluster assembly and maintenance of normal cellular iron homeostasis. Hum Mol Genet 18:3014–3025. https://doi.org/10.1093/hmg/ddp239
29. Terali K, Beavil RL, Pickersgill RW, van der Giezen M (2003) The effect of the adaptor protein Isd11 on the quaternary structure of the eukaryotic cysteine desulphurase Nfs1. Biochem Biophys Res Commun 440:235–240. https://doi.org/10.1016/j.bbrc.2013.09.039
30. Cory SA, Van Vranken JG, Brignole EJ (2017) Structure of human Fe-S assembly subcomplex reveals unexpected cysteine desulfurase architecture and acyl-ACP-ISD11 interactions. Proc Natl Acad Sci U S A 114:E5325–E5334. https://doi.org/10.1073/pnas.1702849114
31. Tsai CL, Tainer JA (2018) Robust production, crystallization, structure determination, and analysis of [Fe-S] proteins: uncovering control of electron shuttling and gating in the respiratory metabolism of molybdopterin guanine dinucleotide enzymes. Methods Enzymol 599:157–196. https://doi.org/10.1016/bs.mie:2017.11.006
32. Jin Z, Heinnickel M, Krebs C, Shen G, Golbeck JH, Bryant DA (2008) Biogenesis of iron-sulfur clusters in photosystem I. J Biol Chem 283:28426–28435. https://doi.org/10.1074/jbc.M803395200
33. Puglisi R, Boeri Erba E, Pastore A (2020) A guide to native mass spectrometry to determine complex interactomes of molecular machines. FEBS J 287:2428–2439. https://doi.org/10.1111/febs.15281
34. Cai K, Frederick RO, Tonelli M, Markley JL (2018) Interaction of iron-bound frataxin with IscU and ferredoxin on the cysteine desulfurase complex leading to Fe-S cluster assembly. J Inorg Biochem 183:107–116. https://doi.org/10.1016/j.jinorgbio.2018.03.007
35. Parent A, Elduque X, Cornu D, Belot L, Le Caer JP, Grandas A, Toledano MB, D'Autreaux B (2015) Mammalian frataxin directly enhances sulfur transfer of NFS1 persulfide to both ISCU and free thiols. Nat Commun 6:5686. https://doi.org/10.1038/ncomms6686
36. Adinolfi S, Iannuzzi C, Proschi F, Pastore C, Iametti S, Martin SR, Bonomi F, Pastore A (2009) Bacterial frataxin CyaY is the gatekeeper of iron-sulfur cluster formation catalysed by IscS. Nat Struct Mol Biol 16:390–396. https://doi.org/10.1038/nsmb.1579
37. Iannuzzi C, Adinolfi S, Howes BD, Garcia-Serres R, Clemencey M, Latour JM, Smulevich G, Pastore A (2011) The role of CyaY in iron sulfur cluster assembly on the E. coli IscU scaffold protein. PLoS One 6:e21992. https://doi.org/10.1371/journal.pone.0021992

38. Agar JN, Krebs C, Frazzon J, Huynh BH, Dean DR, Johnson MK (2000) IscU as a scaffold for iron-sulfur cluster biosynthesis: sequential assembly of [2Fe-2S] and [4Fe-4S] clusters in IscU. Biochemist 39:7856–7862. https://doi.org/10.1021/bi000931n
39. Chandramouli K, Unciuleac MC, Naik S, Dean DR, Huynh BH, Johnson MK (2007) Formation and properties of [4Fe-4S] clusters on the IscU scaffold protein. Biochemist 46:6804–6811. https://doi.org/10.1021/bi6026659
40. Colin F, Martelli A, Clemencey M, Latour JM, Gambarelli S, Zeppieri L, Birck C, Page A, Puccio H, de Choudens SO (2013) Mammalian frataxin controls sulfur production and iron entry during de novo Fe4S4 cluster assembly. J Am Chem Soc 135:733–740. https://doi.org/10.1021/ja308736e
41. Nuth M, Yoon T, Cowan JA (2002) Iron-sulfur cluster biosynthesis: characterization of iron nucleation sites for assembly of the [2Fe-2S]2+ cluster core in IscU proteins. J Am Chem Soc 124:8774–8775. https://doi.org/10.1021/ja0264596
42. Gervason S, Larkem D, Ben Mansour A, Botzanowski T, Muller CS, Pecqueur L, Le Pavec G, Delaunay-Moisan A, Brun O, Agramunt J, Grandas A, Fontecave M, Schunemann V, Cianferani S, Sizun C, Toledano MB, D'Autreaux B (2010) Physiologically relevant reconstitution of iron-sulfur cluster biosynthesis uncovers persulfide-processing functions of ferredoxin-2 and frataxin. Nat Commun 10:3566. https://doi.org/10.1038/s41467-019-11470-9
43. Sellback BP, Pradhan PK, Dos Santos C (2013) Protected sulfur transfer reaction by Escherichia coli Suf system. Biochemist 52:4089–4096. https://doi.org/10.1021/bi4001479
44. Yan R, Adinolfi S, Iannuzzi C, Kelly G, Oregioni A, Martin S, Pastore A (2013) Cluster and fold stability of E. coli ISC-type ferredoxin. PLoS One 8:e78948. https://doi.org/10.1371/journal.pone:0078948
45. Puglisi R, Yan R, Adinolfi S, Pastore A (2016) A new tessera into the interactome of the isc operon: a novel interaction between HscB and IscS. Front. Mol. Biosci 3:48
46. Adinolfi S, Puglisi R, Crack JC, Iannuzzi C, Dal Piaz F, Konarev PV, Svergun DI, Martin S, Le Brun NE, Pasotre A (2018) The molecular basis of the dual regulation of bacterial Iron Sulfur Cluster Biogenesis by CyaY and IscX. Front Mol Biosci 4:97
47. Rasheed M, Jamshidiha M, Puglisi R, Yan R, Cota E, Pastore A (2019) Structural and FunctionalCharacterization of a Frataxin from thermophilic organism. FEBS J 286:495–506
48. Faggianelli N, Pyglisi R, Venezioano L, Romano S, Frontali M, VAnnocci T, Fortuni S, Testi R, Pastore A (2015) Analyzing the Effects of a G137V Mutation in the FXN gene. Fornt Mol Neurosci 8:66
49. Petrosino M, Pasquo A, Novak L, Toto A, Gianni S, Mantuano E, Venezioano L, Minicozzi V, Pastore A, Puglisi R, Capriotti E, Chiaraluce R, Conslvi V (2019) Characterization of human Frataxin missense variants in cancer tissues. Hum Mutat 40:1400–1413
50. Iannuzzi C, Andover M, Puglisis R, Yan R, Temussi PA, Pastore A (2014) The role of zinc in the stability of the marginally stable IscU scaffolding protein. Protein Sci 23:1208–1219
51. Shimon LB, Paul VD, David-Kadoch G, Volpe M, Stumpfig EB, Muhlenhoff U, Lill R, Ben-Aroya S (2018) Fe-S cluster coordination of the chromokinesin KIF4A alters its subcellular localization during mitosis. J Cell Sci 131: jcs211433. https://doi.org/10.1242/jcs.211433
52. Crack JC, Stewart MY, Le Brun N (2019) Generation of 34S-substituted protein-bound [4Fe-4S] cluster using 34S-L-cysteine. Biol Meth Protoc 4:bpy015. https://doi.org/10.1093/biomethods/bpy015
53. Cai K, Liu G, Xiao R, Montelione GT, Markley JL (2016) Structural/functional properties of human NFU1, an intermediate [4Fe-4S] carrier in human mitochondrial iron-sulfur cluster biogenesis. Structure 24:2080–2091. https://doi.org/10.1016/j.str.2016.08.020
54. Lu J, Bitoun JP, Tan G, Wang G, Min W, Ging H (2010) Iron binding activity of human iron-sulfur assembly protein hIscA-1. Biochem J 428:125–131. https://doi.org/10.1042/BJ20100122
55. Adinolfi S, Rizzo F, Masino L, Nair M, Martin SR, Pastore A, Temussi PA (2004) Bacterial IscU is a well folded and functional single domain protein. Eur J Biochem 271:2093–2100. https://doi.org/10.1111/j.1432-1033.2004.04112.x
56. Prischi F, Pastore C, Carroni M, Iannuzzi C, Adinolfi S, Temussi P, Pastore A (2010) Of the vulnerability of orphan complex proteins: the case study of the E. coli IscU and IscS protein. Prot Expr Purif 73:161–166. https://doi.org/10.1016/j.pep.2010.05.003
57. Jordan PA, Tang Y, Bradbury AJ, Thomson AJ, Guest JR (1999) Biochemical and spectroscopic characterization of *Escherichia coli*

aconitases (AcnA and AcnB). Biochem J 344:739–746. https://doi.org/10.1042/bj3440739

58. Cai K, Frederick RO, Markley LJ (2020) ISCU interacts with NFU1, and ISCU[4Fe-4S] transfers its Fe-S cluster to NFU1 leading to the production of holo-NFU1. J Struct Biol 210:107491. https://doi.org/10.1016/j.jsb.2020.107491

59. Orme-Johnson WH, Orme-Johnson NR (1982) Iron-sulfur proteins: the problem of determining cluster type. In: Spiro TG (ed) The metal ions in biology series, vol 4. Wiley, New York, pp 67–96

60. Morra S, Arizzi M, Valetti F, Gilardi G (2016) Oxygen stability in the new [FeFe]-hydrogenase from *Clostridium beijerinckii* SM10 (CbA5H). Biochemist 55:5897–5900. https://doi.org/10.1021/acs.biochem.6b00780

61. Ollagnier de Choudens S, Barras F (2017) Genetic, biochemical, biophysical methods for studying Fe-S proteins and their assembly. Methods Enzymol 595:1–32. https://doi.org/10.1016/bs.mie.2017.07.015

62. Lill R (2020) From the discovery to molecular understanding of cellular iron-sulfur protein biogenesis. Biol Chem 401:855–876. https://doi.org/10.1515/hsz-2020-0117

63. Fox NG, Yu X, Feng XD, Bailey HJ, Martelli A, Nabhan JF, Strain-Damerell C, Bulawa C, Yue XX, Han S (2019) Structure of the human frataxin-bound iron-sulfur cluster assembly complex provides insight into its activation mechanism. Nat Commun 10:2210. https://doi.org/10.1038/s41467-019-09989-y

Chapter 6

Expression, Isolation, and Characterization of Vanadium Nitrogenase from *Azotobacter vinelandii*

Katharina Parison, Jakob Gies-Elterlein, Christian Trncik, and Oliver Einsle

Abstract

Nitrogenases are the sole enzymes known to mediate biological nitrogen fixation, an essential process for sustaining life on earth. Among the three known variants, molybdenum nitrogenase is the best-studied to date. Recent work on the alternative vanadium nitrogenase provided important insights into the mechanism of nitrogen fixation since this enzyme differs from its molybdenum counterpart in some important aspects. Here, we present a protocol to obtain unmodified vanadium nitrogenase in high yield and purity from the paradigmatic diazotroph *Azotobacter vinelandii*, including procedures for cell cultivation, purification, and protein characterization.

Key words Protein expression, Protein isolation, Protein purification, Anoxic protein biochemistry, Vanadium nitrogenase, Alternative nitrogenase, *Azotobacter vinelandii*, Nitrogen fixation, Biological nitrogen fixation, VFe protein, Fe protein, Vanadium

1 Introduction

1.1 Origin of Nitrogen Fixation by Nitrogenases

Nitrogen is essential for all life on earth, yet the vast majority of it is bound as chemically inert dinitrogen gas (N_2) in the atmosphere [1]. To make this reservoir available for sustaining food crop cultivation, the industrial Haber-Bosch process applies high temperatures and pressures to reduce dinitrogen to ammonium using a syngas [2]. Almost half of the world population today depends on food grown with the help of fertilizers made by this process [3], using as much as 2% of the global energy production [4]. In nature, however, diazotrophic microbes can perform the same reductive chemistry at ambient temperature and pressure [1]. This is realized by nitrogenases, the only enzymes known to be able to reduce the dinitrogen triple bond and produce bioavailable ammonium for the assimilation into amino acids in a pathway called nitrogen fixation [5, 6]. The main system for this is the molybdenum- and iron-dependent enzyme molybdenum nitrogenase. Its presence is

Patricia C. Dos Santos (ed.), *Fe-S Proteins: Methods and Protocols*, Methods in Molecular Biology, vol. 2353,
https://doi.org/10.1007/978-1-0716-1605-5_6,

mandatory in every known diazotrophic organism [7, 8]. The two additional nitrogenase systems, as a consequence, never occur independently from molybdenum nitrogenase and are called alternative nitrogenases. One is dependent on vanadium and iron [9] and the second only uses iron as a metal [10]. Nitrogenases are two-component systems [11] consisting of the main catalytic component and a dedicated reductase, the iron protein [12–14].

Nitrogenase was suggested to have evolved somewhere between 1.5 billion [15] and 3.2 billion years ago [16], when molybdenum was rare and vanadium much more abundant in the oceans [17]. But despite the fact that Earth's crust is generally richer in vanadium [18], vanadium-dependent biological functions are rare in nature [19]. The molybdenum nitrogenase system is thought to have evolved first [16, 20], with the vanadium nitrogenase evolving prior to the iron-only nitrogenase [15, 20]. Nitrogenase function is supposed to have evolved first in archaea and was later transferred to bacteria [21].

1.2 Molybdenum-Independent Nitrogen Fixation

The gammaproteobacterium *Azotobacter vinelandii* is one of the most thoroughly studied diazotrophs [22] and possesses all three nitrogenase systems [13, 23–25]. It is related to *Pseudomonads* [26, 27], and there is some debate about their delineation [28, 29]. The discovery of alternative nitrogenase systems [10, 30] occurred rather late in more than a century of nitrogenase research [5, 22]. Initially, diazotrophic growth was found to be stimulated by the addition of vanadate into the growth medium [31]. It was later discovered that diazotrophic growth is possible without molybdenum altogether, either in the presence [31–33] or absence [9] of vanadium. Diazotrophic growth after deletion of the structural genes for the molybdenum-dependent nitrogenase provided final proof of existence for dedicated alternative nitrogenases [10, 34–36] and their ability to reduce dinitrogen [37].

These alternative systems are negatively regulated at the transcription level [38, 39] by the molybdenum concentration in the growth medium [9, 34, 38]. The second alternative system ("iron only") is also under further negative regulation by vanadate [39]. Interestingly, this second regulation depends on the presence of the structural genes of the vanadium nitrogenase [39]. It is also found to be less tight than the negative molybdenum regulation [23, 40].

The regulation of alternative nitrogenases is furthermore dependent on temperature [41]. At 30 °C, the molybdenum nitrogenase is fully expressed and the alternative nitrogenases are repressed, but at 14 °C, the alternative nitrogenases are no longer inhibited [41]. This could be a physiologically relevant effect, as the vanadium nitrogenase activity is shown to surpass the molybdenum nitrogenase activity at 4 °C [42], although in absolute numbers, all reported activities are very low at these temperatures. Vanadium-dependent diazotrophic growth can be on par with molybdenum-dependent growth if the correct carbon source is available [43]. It was long assumed that alternative nitrogenases work as back-up systems, but a recent study indicates that vanadium nitrogenase might play an important role in nitrogen input into ecological systems under molybdenum-limited conditions [44].

1.3 Structure and Metal Centers of Nitrogenases

The structure of molybdenum nitrogenase was first solved in 1992 by Rees and coworkers [45] with the structure of the reductase NifH [46] solved in the same year. The molybdenum nitrogenase forms a tetramer with a reductase dimer binding to each half [47]. The enzyme contains a [8Fe:7S] cluster, called P-cluster, and the FeMo-cofactor as the catalytic center of substrate reduction [45] (Fig. 1). The FeMo-cofactor is a [Mo:7Fe:9S:C] cluster with *R*-homocitrate binding at the molybdenum atom [45]. The central cavity of the cluster [45] was found to contain a light atom [48] that was later identified as a carbon both in molybdenum [49, 50] and vanadium nitrogenase [51]. The first and so far only structure of an alternative nitrogenase was solved from A. *vinelandii* in 2017 [52], followed by the structure of the cognate reductase VnfH [53]. A new purification method was instrumental for obtaining protein of a quality suitable for crystallization [54] and is described in this chapter.

Vanadium nitrogenase is a $VnfD_2G_2K_2$ heterohexamer [52]. The enzyme contains a P-cluster very similar to the molybdenum nitrogenase [52]. However, while the P-cluster of

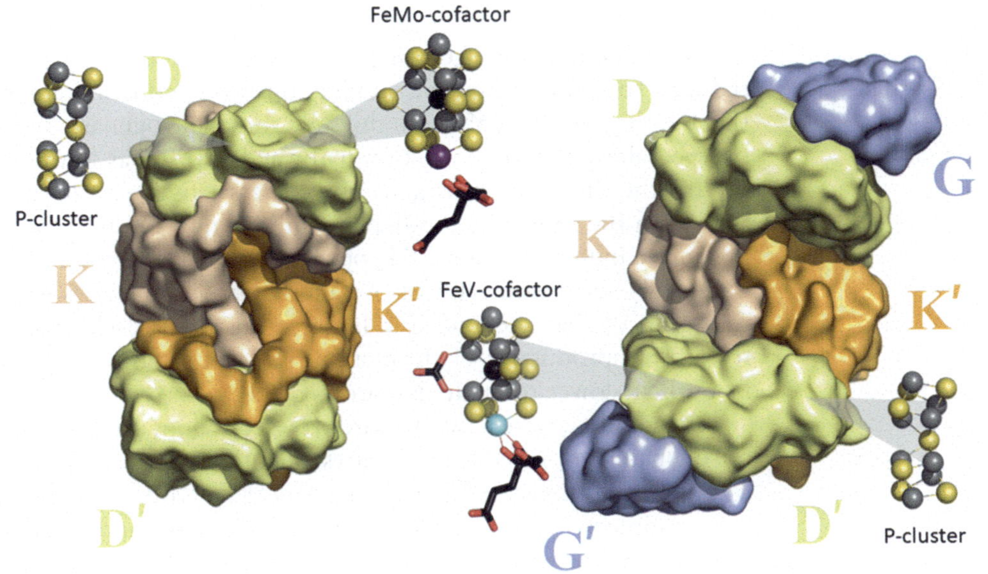

Fig. 1 Surface representation of the molybdenum nitrogenase (PDB 1M1N) (left) and the vanadium nitrogenase (PDB 5N6Y) (right) with the indication of respective subunits. The positions of P-clusters, FeMo-cofactor, and FeV-cofactor are indicated for one catalytic half of the enzyme

molybdenum nitrogenase undergoes a displacement of two iron atoms upon a two-electron oxidation [55], for vanadium nitrogenase the movement of only one iron atom was observed, accordingly in conjunction with a one-electron transfer process [52]. The catalytic center, termed FeV-cofactor in analogy to its Mo-counterpart, was expected to have vanadium replacing molybdenum, but was otherwise hypothesized to be similar to FeMo-cofactor [56]. Surprisingly, the three-dimensional structure in addition revealed the replacement of one belt sulfide with a carbonate ligand [52, 57].

The reductase of vanadium nitrogenase (VnfH) is very similar to its molybdenum counterpart (NifH) [53], and both proteins can functionally replace each other [42, 58, 59]. The reductases form homodimers with a bridging [4Fe:4S] cluster and belong to the family of P-loop NTPases that couple ATP hydrolysis to conformational changes in two "switch" regions [53], mediated via the binding of ATP at a P-loop (or Walker A motifs) [60].

1.4 Reactivity of Alternative Nitrogenases

The basic reaction of nitrogenase consists of eight single-electron transfer steps [1]. For each one of these, the P-cluster first donates an electron to the FeMo-cofactor in what was described as a "deficit-spending mechanism" [61]. Only then, the reductase component (Fe protein) hydrolyzes MgATP [62] to reduce the P-cluster back to the P^N state. The basic reaction scheme for nitrogenase was established as follows [1, 63, 64]:

$$N_2 + 8H^+ + 8e^- + 16MgATP \rightarrow 2NH_3 + H_2 + 16MgADP + 16P_i$$

Molybdenum nitrogenase produces at least one H_2 per N_2 reduced [64], in a reductive elimination of H_2 concomitant with the binding of N_2 [65]. In this, all three nitrogenase systems proceed along the same reaction pathway [66]. It was found early on that the alternative nitrogenases were less active in N_2 reduction [42] than the molybdenum-dependent enzyme [67, 68] and also produce a higher amount of H_2 per N_2 [56, 65]. Interestingly, the higher H_2 formation rates of the alternative nitrogenases are not the result of the diversion of a larger proportion of electron flux toward nonproductive H_2 formation but are thought to originate from their slower reactivity toward N_2 [69].

Although less reactive toward N_2 reduction, the alternative nitrogenases are capable of a series of interesting reactions with carbon-based molecules that yield a variety of products. Acetylene is a nonphysiological substrate for nitrogenase (independently found by Schöllhorn and Burris) [67, 68], and due to the ease of detection of substrate (acetylene) and product (ethylene) by gas chromatography, an acetylene reduction assay has been established as the main method for determining reduction activities in whole cells [70, 71], cell-free extracts [72], and purified protein [73]. Analogous to N_2 reduction, molybdenum nitrogenase also exhibits the highest activity toward acetylene [74]. Vanadium nitrogenase and iron-only nitrogenase produce ethane [23, 75, 76] in addition to ethylene [76] in acetylene reduction, making the detection of ethane a suitable test for the action of alternative nitrogenases [71].

1.5 Purification of Alternative Nitrogenases

All nitrogenases are sensitive toward oxygen. For molybdenum nitrogenase, a dedicated [2Fe:2S] ferredoxin, the FeS II, or Shethna II protein [77, 78], provides protection from oxygen damage by reversibly forming an inactive complex [79, 80]. For vanadium nitrogenase, no such system is known [81]. Nonetheless, the isolation of all nitrogenases requires the strict exclusion of oxygen, but may be carried out at room temperature. Vanadium nitrogenase was first isolated from crude extracts of molybdenum-starved cells grown diazotrophically with supplementation of vanadium [33, 74, 82]. At the time, it was not yet established whether this protein was actually different or simply a molybdenum nitrogenase with an exchanged heteroatom. These early studies established key features of the vanadium-dependent enzyme such as its lower stability as determined by melting temperature, its lower catalytic activity, and also the lack of CO inhibition [33].

Vanadium nitrogenase was first isolated from *Azotobacter chroococcum* in 1986 [30, 81] and from *A. vinelandii* in the same year [58]. The iron-only nitrogenase was also reported from *A. vinelandii* in 1988 [10]. The corresponding Fe proteins VnfH [83] and AnfH [10] were also obtained. Early preparations of vanadium nitrogenase suffered from the instability of the protein complexes [58, 84]. In these preparations, the G-subunit gene products [14, 23, 24, 85] were not recognized as part of the nitrogenase complex [10, 58, 81]. Later purifications showed the presence of the G-subunit for the alternative nitrogenases in the expected [13, 14] 2:2:2 ratio [86, 87], or in a 2:4:2 ratio with 4 G-subunits [88]. The G-subunits are essential for the activity of the alternative nitrogenases in vivo [40, 89] and in vitro [86, 89]. In many cases, preparations were done using deletion strains that lack the *nifDK* genes for molybdenum nitrogenase MoFe protein [58, 84, 88].

A new purification protocol based on unmodified wild-type strains and a long-term culture in the absence of molybdenum yielded a consistent and stable 2:2:2 ratio of subunits [54]. This purification used Q-sepharose ion exchange chromatography followed by an additional high-resolution chromatography on resource Q and a final size exclusion chromatography step [54]. It led to a protein of suitable quality for crystallization of the vanadium nitrogenase from *A. vinelandii* as the first structure of an alternative nitrogenase enzyme [52] and its reductase component,

VnfH [53]. The materials and methods required for vanadium nitrogenase production, isolation, and characterization are described in the following.

2 Materials

2.1 Cell Cultivation

1. Bacterial strain: *A. vinelandii* Type strain Lipman 1903 [5] (DSM-No. 2289) from the Leibniz-Institute DSMZ-German Collection of Microorganisms and Cell Cultures.
2. Modified Burk medium [90–92] (Mo or V, optional addition of N-source): autoclaved sterilized aqueous solution (deionized water) of 58.4 mM sucrose, 0.9 mM $CaCl_2$, 1.67 mM $MgSO_4$, 40 μM $FeSO_4$, 200 μM citric acid, 10 μM Na_2MoO_4 (Mo) or Na_3VO_4 (V), 10 mM or 0 mM NH_4Cl (N-source), 2.5 mM KH_2PO_4 and 10 mM K_2HPO_4. The phosphate components are prepared and autoclaved sterilized separately (*see* **Note 1**). Burk Medium replacing Mo with V is called BV in the following, if a nitrogen source is added to the medium it is labeled BN or BNV.
3. Agar plates (Mo or V) prepared with modified Burk medium (B, BV, BN) and 1.6–5% (w/v) agar. The agar is autoclaved with the phosphate component. When the autoclaved solution reaches approximately 60 °C, the concentrated sterile salts are added prior to plating.
4. Autoclaved 90% (*v/v*) glycerol.

2.2 Protein Purification

All buffers used for protein purification are filtered (0.22 μm) and degassed. Ultrapure water (MilliQ) is used for buffer preparation. Dioxygen is removed by applying 10–15 cycles of alternating evacuation (10^{-4} bar for 10–15 min) and nitrogen flushing using modified Schlenk techniques (*see* **Note 2**). All buffers are supplemented with 2.5 mM $Na_2S_2O_4$ directly before use (*see* **Note 3**). $Na_2S_2O_4$ is used as a reductant, as well as an oxygen scavenger to remove residual dioxygen.

1. IEC A buffer: 50 mM tris(hydroxymethyl)aminomethane (Tris)–HCl, pH 7.4, with 100 mM NaCl and 20 mM $MgCl_2$ (*see* **Note 4**).
2. IEC A_0 buffer: 50 mM Tris–HCl, pH 7.4, with 20 mM $MgCl_2$.
3. IEC B buffer: 50 mM Tris–HCl, pH 7.4, with 1 M NaCl and 20 mM $MgCl_2$.
4. SEC buffer: 20 mM Tris–HCl, pH 7.4, with 100 mM NaCl and 20 mM $MgCl_2$.
5. Centrifugal concentrators for ultrafiltration with a molecular weight cut-off of 30 and 100 kDa.

6. Filters with a pore size of 0.22 and 0.45 μm for buffer filtration and throughout the purification.
7. Anion exchange chromatography column: Q Sepharose HP (50 mL).
8. Anion exchange chromatography column: Resource Q (6 mL).
9. Size exclusion chromatography column: HiLoad 26/600 Superdex 200 pg.

2.3 Protein Characterization

2.3.1 SDS-PAGE

1. Stacking gel (5%): 5% (v/v) 30% acrylamide-bisacrylamide (37.5:1), 125 mM Tris–HCl, pH 6.8, 0.1% (w/v) sodium dodecyle sulfate (SDS), 0.20% (v/v) tetramethylethylenediamine (TEMED), and 0.99% (v/v) 10% ammonium persulfate (APS).
2. Resolving gel (12.5%): 12.5% (v/v) 30% acrylamide-bisacrylamide (37.5:1), 125 mM Tris–HCl, pH 8.8, 0.1% (w/v) SDS, 0.18% (v/v) TEMED, and 0.99% (v/v) 10% APS.
3. SDS running buffer: 25 mM Tris–HCl, pH 8.0, 192 mM glycine, and 0.1% (w/v) SDS.
4. SDS loading dye 5×: 0.5 M Tris–HCl, pH 6.8, 40% (v/v) glycerol, 8% (w/v) SDS, 0.004% (w/v) bromophenol blue, and 9% (v/v) β-mercaptoethanol.
5. Marker size standards: Unstained Protein MW Marker 14.4 kDa to 116 kDa.
6. Staining solution: 10% (v/v) ethanol, 5% (v/v) acetic acid, and 0.01% (w/v) Coomassie Brilliant Blue R250:G250 4:1.
7. Destaining solution: 10% (v/v) ethanol.

2.3.2 BCA Assay

1. BCA protein assay reagent.
2. $CuSO_4$ solution: 4% (w/v) $CuSO_4$.
3. Aqueous solutions of bovine serum albumin with concentrations of 25, 50, 100, 150, 200, and 250 μg/mL (freezer stocks).

2.3.3 Acetylene Reduction Assay

1. Gas chromatography column: 91 cm × 3.2 mm S.S. silica gel column.
2. Activity assay buffer: 20 mM Tris–HCl, pH 7.4, 10 mM $MgCl_2$, 15 mM phosphocreatine, 0.125 mg/mL creatine phosphokinase, and 12.5 mM $Na_2S_2O_4$.
3. ATP solution: 20 mM Tris–HCl, pH 7.4, 10 mM $MgCl_2$, 100 mM Na_2ATP and 1 mM $Na_2S_2O_4$.

   All solutions are prepared fresh.

2.3.4 Dinitrogen Reduction Assay

1. Dinitrogen activity assay buffer: 20 mM 3-(*N*-morpholino) propanesulfonic acid (MOPS)-NaOH, pH 7.4, 10 mM $MgCl_2$, 60 mM phosphocreatine, 0.1 mg/mL creatine phosphokinase, and 10 mM $Na_2S_2O_4$.
2. ATP solution: 20 mM MOPS-NaOH, pH 7.4, 10 mM $MgCl_2$ and 100 mM Na_2ATP.
3. EDTA solution: 500 mM Na_2EDTA-NaOH, pH 8.0.
4. Phenol solution: 140 mM phenol.
5. Nitroprusside solution: 250 mM sodium nitroprusside (*see* **Note 5**).
6. NaOCl solution: 0.12% (v/v) NaOCl and 750 mM NaOH (*see* **Note 6**).

All solutions are prepared fresh.

2.3.5 UV-Vis Absorption Spectroscopy

1. 0.5 M $Na_2S_2O_4$ in 0.5 M Tris base.
2. 0.1 M K_3FeCN_6.
3. Desalting column.

2.3.6 Right-Angle Light Scattering

1. Analytical size-exclusion column: Superdex 200 10/300.
2. SEC buffer (*see* Subheading 2.2).

3 Methods

3.1 Permanent Molybdenum (Mo) Glycerol Culture

For preparing a culture seed stock, a permanent Mo glycerol culture is prepared following an established protocol [54].

1. Resuspend freeze-dried *A. vinelandii* Lipman 1903 cells in sterile-modified Burk medium (B).
2. Use modified Burk medium (BN) to inoculate with these cells and incubate at 30 °C and 180 rpm.
3. Prepare aliquots in Eppendorf tubes as follows. At an OD_{600} of 2.5, mix 0.5 mL of the cell culture with 0.5 mL sterile 90% (v/v) glycerol, shock-freeze it in liquid nitrogen, and store at −80 °C. One aliquot is used for one cultivation.

3.2 Molybdenum Depletion and Permanent Vanadium (V) Glycerol Culture

For a V glycerol culture required for the cultivation of *A. vinelandii* that produces vanadium nitrogenase instead of molybdenum nitrogenase, *A. vinelandii* must be depleted of molybdenum [54].

1. Thaw a permanent Mo glycerol culture stock aliquot on ice and equilibrate to room temperature. Use modified Burk medium (BN) to inoculate with this culture and incubate at 30 °C and 180 rpm.

2. At an OD_{600} of 3.0, streak out the cell culture on agar plates (B) and subsequently incubate at 30 °C. Since no fixed nitrogen is available, *A. vinelandii* initiates nitrogen fixation, as indicated by the formation of brown colonies.
3. Achieve molybdenum depletion by inoculating modified Burk medium (BNV) with a single nitrogen-fixing colony and incubation at 30 °C and 180 rpm.
4. At an OD_{600} of 3.0, streak out the cell culture on agar plates (BV) and then incubate at 30 °C. Brown, nitrogen-fixing colonies grow.
5. Repeat this cycle of a cell culture in modified Burk medium (BNV) and agar plates (BV) (**steps 3** and **4**) at least five times. The procedure leads to cells that produce only vanadium nitrogenase.
6. In order to prepare a permanent V glycerol culture, culture these cells for a final time in modified Burk medium (BNV) at 30 °C and 180 rpm.
7. At an OD_{600} of 2–4, mix 0.75 mL of cell culture with 0.25 mL sterile 90% (v/v) glycerol, shock freeze it in liquid nitrogen, and store at −80 °C in 1 mL aliquots.
8. To ensure that Mo depletion did not result in *nif*-deficient *A. vinelandii*, V glycerol cultures can be used to inoculate modified Burk medium (B) to produce Mo nitrogenase for isolation and assays [49].

3.3 Cultivation of A. vinelandii for Purification of Vanadium Nitrogenase

The absence of a source of fixed nitrogen in the modified Burk medium (BV) leads to a high expression level of vanadium nitrogenase and, therefore, a high protein yield [54]. *A. vinelandii* is an obligate aerobe [90]. Cultures must be grown aerobically. Without the possibility of producing molybdenum nitrogenase, *A. vinelandii* switches to its alternative variant vanadium nitrogenase under N-limiting conditions. Consequently, the medium used for the production of vanadium nitrogenase contains vanadium and lacks molybdenum [39].

1. Grow a preculture in 100 mL of sterile-modified Burk medium (BV) in a baffled flask (*see* **Note 7**). Inoculate with 100–300 μL (*see* **Note 8**) of a thawed aliquot of a permanent V glycerol stock of Mo-depleted *A. vinelandii*.
2. Incubate the preculture for 36–48 h at 30 °C and 180 rpm.
3. Inoculate a preculture with a minimum OD_{600} of 1 to the main culture—average OD_{600} values for inoculation range between 2 and 5.
4. Grow the main culture in 500 mL or 1 L of modified Burk medium (BV) in baffled flasks with a total volume ranging from

10 to 20 L. Inoculate the main culture with 4 mL (500 mL medium) or 8 mL (1 L medium) of the preculture.

5. After incubation at 30 °C and 180 rpm for 18–24 h, an OD_{600} of 2.0–2.5 should be reached. It generally is sufficient to use the optical density to determine the time point for cell harvest. Additionally, cell growth can be monitored by determining the acetylene reduction activity of the cell cultures and harvest cells at the point of maximum activity (*see* Subheading 3.5.3).
6. Harvest cells by centrifugation at 5000 × *g* for 20 min at 4 °C.
7. Cell pellets should be dark brown and can be shock-frozen for storage in liquid nitrogen until purification. On average, a yield of 40 g cells per 10 L of the culture medium is obtained.

3.4 Purification of Vanadium Nitrogenase

Vanadium nitrogenase (with its component proteins VFe and Fe) is isolated following an established protocol [54] with minor modifications (Fig. 2). All steps are carried out under anoxic conditions using modified Schlenk techniques with dinitrogen gas and anoxic chambers with a H_2/N_2 atmosphere (5%/95%, forming gas).

1. Thaw and resuspend cells in a threefold volume of IEC A buffer, and $Na_2S_2O_4$ is added to the cell suspension to a final concentration of 2.5 mM.
2. Lyse cells with a cell homogenizer.
3. Centrifuge the lysate for 30–60 min at 100,000 × *g* at 4 °C.
4. Filter the supernatant containing VFe and Fe proteins (0.45 μm pore size).

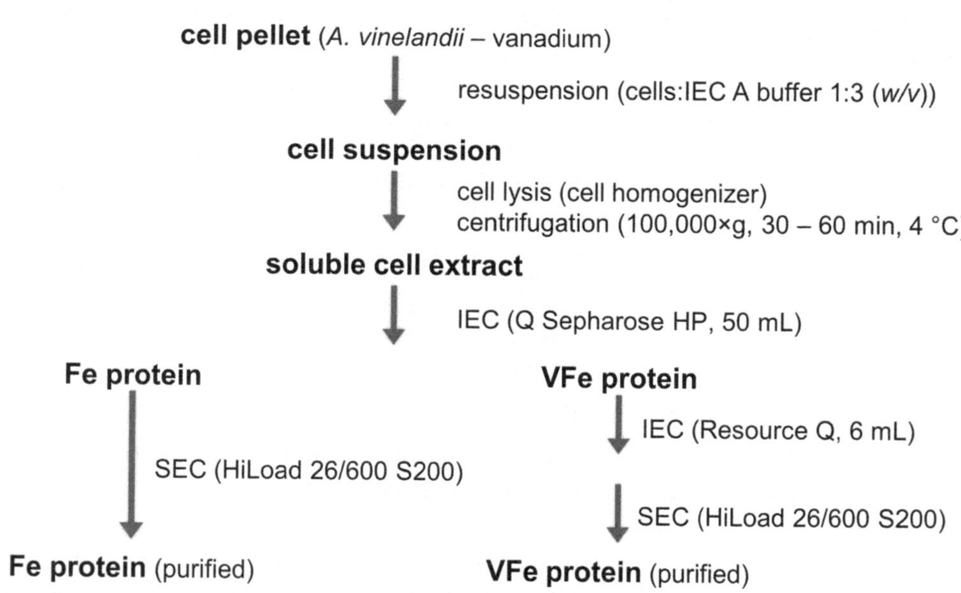

Fig. 2 Protein purification flowchart for VFe and Fe proteins from *A. vinelandii*

5. Perform further purification by several chromatography purification steps at room temperature on an FPLC coupled to a UV detector. Manually collect fractions into sealed N_2-filled vials based on UV absorbance peaks. VFe protein is of dark brown color, whereas Fe protein is light brown in color. Routinely analyze all fractions by SDS-PAGE.
6. Proceed to the first chromatography step using an anion exchange chromatography on Q Sepharose HP (50 mL, flow rate 2–5 mL/min).
7. Equilibrate the column with one column volume (CV) of IEC B buffer followed by one CV of IEC A buffer.
8. Load the sterile filtered supernatant onto the column, followed by a wash step with IEC A buffer until the baseline is reached.
9. Employ a linear gradient of 300 mL from 0 to 50% IEC B buffer. The conductivity at which the proteins elute, as well as the shape of the chromatogram, varies with the amount of $MgCl_2$ and $Na_2S_2O_4$ that was used. With the standard protocol comprising 20 mM $MgCl_2$ and 2.5 mM $Na_2S_2O_4$ in all buffers, VFe protein elutes at a conductivity of approximately 25 mS cm^{-1}, followed by the elution of Fe protein at 28 mS cm^{-1} (Fig. 3a) (*see* **Notes 9** and **10**).
10. After analysis of the fractions via SDS-PAGE and the corresponding chromatography elution profiles, concentrate the Fe protein using ultrafiltration under 3.0 bar nitrogen pressure at 4 °C (centrifugal concentrator, MWCO: 30 kDa). Approximately every 10 min, renew the nitrogen pressure and resuspend the protein solution in the concentrator. Fe protein is sufficiently pure for direct application to a size exclusion chromatography column (*see* **Note 11**).
11. Pool the VFe protein fractions, dilute with IEC A_0 buffer to reach a final NaCl concentration of 100 mM NaCl or lower, and filter before loading onto a second anion exchange

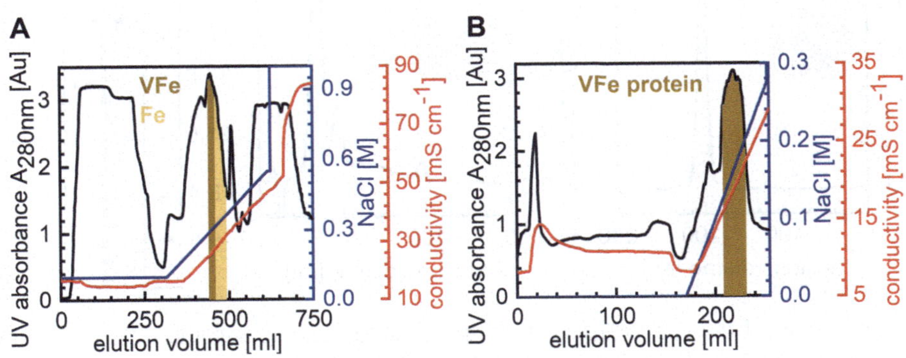

Fig. 3 (**a**) Representative chromatograms of the first chromatography step on a Q Sepharose HP anion exchange column. (**b**) Representative chromatogram of the second chromatography step using a Resource Q anion exchange column. VFe protein fractions are colored in dark brown, Fe protein fractions in light brown

column. The required dilution factor typically ranges between 1:6 and 1:7 (v/v).

12. To obtain VFe protein of maximal purity, use a Resource Q anion exchange column (6 mL, flow rate 2–5 mL/min) as a second chromatography step.
13. Equilibrate the column with 6 CV IEC B buffer, followed by 6 CV IEC A_0 buffer (*see* **Note 12**).
14. Load the filtered, diluted VFe protein solution onto the column and conduct a wash step with IEC A_0 buffer until the baseline is reached.
15. Employ a linear gradient of 120 mL from 0% to 40% IEC B buffer (*see* **Note 13**). VFe protein elutes at a conductivity of approximately 17 mS cm^{-1} (Fig. 3b) (*see* **Note 14**).
16. Concentrate the pooled fractions to 1–5 mL final volume (centrifugal concentrator, MWCO: 100 kDa) using ultrafiltration under 3.5 bar nitrogen pressure at 4 °C. Dilute the protein with SEC buffer and concentrate again by a factor of 2–3 in order to reduce the NaCl concentration in the buffer.
17. Proceed to final size exclusion chromatography using a HiLoad 26/600 Superdex 200 pg size exclusion column.
18. Equilibrate the column with one CV of SEC buffer at a flow rate of 2.6 mL/min.
19. Prior to loading, centrifuge and filter the protein to remove any aggregates.
20. Load the protein and purify on the SEC column. Typical SEC chromatograms are depicted in Fig. 4a for VFe protein and Fig. 4b for Fe protein.

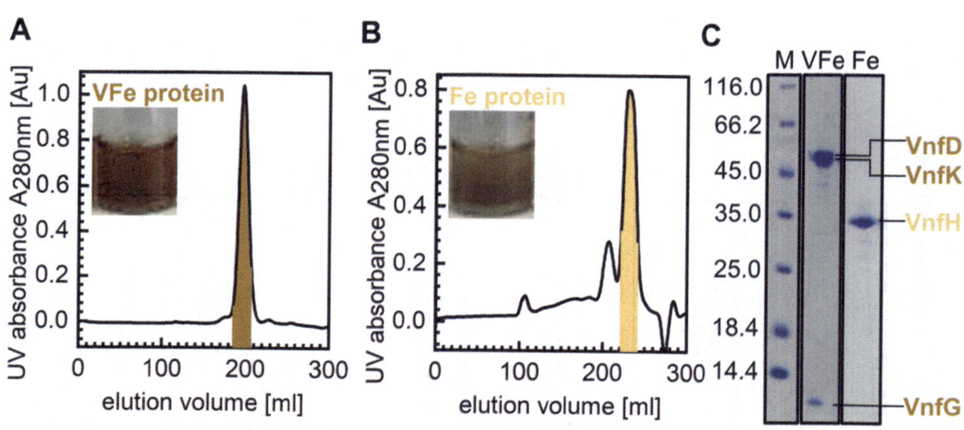

Fig. 4 Chromatograms of the size exclusion step using a HiLoad 26/600 Superdex 200 pg size exclusion column for VFe protein (**a**) and Fe protein (**b**), respectively. Protein fractions are depicted in the photos. (**c**) A typical SDS-PAGE of pure VFe and Fe proteins after size exclusion chromatography with 2μg of protein applied. All subunits are labeled. A molecular weight marker (M) is used as standard ranging from 14.4 kDa to 116.0 kDa

21. Concentrate the pooled protein fractions as described above to 30–70 mg/mL for VFe protein and 50–130 mg/mL for Fe protein, centrifuge, shock-freeze, and store in aliquots in liquid nitrogen until further use. An SDS-PAGE with pure VFe and Fe proteins, respectively, is shown in Fig. 4c. The average protein yield is 1.0–1.5 mg VFe protein and 1.0–1.4 mg Fe protein per gram of cells (wet weight).

3.5 Protein Characterization

To analyze and characterize the purified protein, a selection of available methods is shortly described in the following section. Combined analyses can be used to compare the purified protein to published results and guarantee purity, homogeneity, and activity. Published data [54] confirmed the cultivation and purification protocol described above to provide unmodified, active vanadium nitrogenase from its native organism under diazotrophic growth conditions in high yield and purity.

3.5.1 SDS-PAGE

Collected fractions from chromatography are analyzed by discontinuous sodium dodecyl sulfate polyacrylamide gel electrophoresis (SDS-PAGE) [93] following a modified protocol [94, 95].

1. Mix the protein sample with reducing 5× SDS loading dye, denature at 95 °C for 5 min, and apply it to an SDS-PAGE.
2. The gel consists of a 5% stacking gel and a 12.5% resolving gel. Perform the electrophoresis at 45 mA for 45 min using SDS running buffer.
3. Incubate the gels for 20 min in the staining solution and destain in the destaining solution until a clear background is obtained.

3.5.2 BCA Assay

Protein concentrations are determined with the [2,2′-biquinoline]-4,4′-dicarboxylic acid (bicinchoninic acid, BCA) assay [96]. Bovine serum albumin is used to obtain a calibration curve and serves as a standard. Protein dilutions and standards are prepared at least in triplicates.

1. Mix the BCA Protein Assay Reagent A and $CuSO_4$ solution in a 49:1 (v/v) ratio.
2. Mix an aliquot of 1 mL of solution with 50 μL of sample.
3. Incubate the samples and standard for 30 min at 60 °C and subsequently place on ice.
4. Determine the absorbance at 562 nm.

3.5.3 Acetylene Reduction Assay

The reduction of acetylene, a nonphysiological substrate of nitrogenase, to ethylene is routinely used to measure nitrogenase activity for the ease of detection of both metabolites [97]. The assay is conducted under oxygen exclusion, with VFe protein and Fe protein present in a molar ratio of 1:10 to ensure that the availability of Fe protein does not limit turnover. Phosphocreatine and creatine phosphokinase serve as an ATP-regenerating system, and the formation of ethylene is monitored via gas chromatography.

1. Set up reaction assays of 1 mL volume containing 0.417 nmol VFe protein and 4.17 nmol Fe protein in the activity assay buffer. The reaction is started with the ATP solution that is prepared in a separate vial.
2. Flush with argon the headspace of the sealed, air-tight assay vials (10 mL volume), as well as the ATP solution vial for 3 min.
3. With an air-tight syringe, add 1 mL acetylene to the headspace of all assay vials.
4. Incubate the vials for 3 min at 30 °C and 250 rpm.
5. Add 25 μL ATP solution to each assay vial and further incubate them for 3 min at 30 °C and 250 rpm.
6. Terminate the reaction by the addition of 250 μL glacial acetic acid.
7. Analyze 1 mL of headspace via gas chromatography. A flame ionization detector allows the quantification of the product ethylene, as well as the substrate acetylene. Perform all measurements in triplicates. Our typical specific activities are well in line with literature values, for instance $a_n = (138 \pm 7)$ nmol min^{-1} $nmol(VFe)^{-1}$ [88] (*see* **Note 15**).

Whole-Cell Acetylene Reduction Assay

To determine the activity of cell cultures for assuring the harvesting of cells at the maximum of nitrogenase activity, an acetylene reduction assay can be performed with whole cells instead of pure protein.

1. Fill the vials (10 mL volume) with 2 mL of cell culture.
2. Flush the headspace with argon for 3 min. Replace 1 mL of headspace with 1 mL of dioxygen, since *A. vinelandii* is an obligate aerobe, followed by addition of 1 mL acetylene.
3. Incubate the vials for 30 min at 25 °C and 250 rpm.
4. Monitor the formation of ethylene via gas chromatography as described for the protein acetylene reduction assay.

3.5.4 Dinitrogen Reduction Assay

Dinitrogen is the physiological substrate of vanadium nitrogenase. Isolated protein can therefore be assayed for its dinitrogen reduction activity by quantifying the product ammonia using the Berthelot reaction [98].

The assay is conducted under oxygen exclusion with VFe and Fe proteins present in a 1:10 molar ratio and phosphocreatine and creatine phosphokinase serving as an ATP-regenerating system.

1. Set up assay reactions of 200 μL containing 0.1 mg/mL VFe protein and 0.267 mg/mL Fe protein in dinitrogen activity assay buffer (*see* **Note 16**) with an atmosphere of 95% dinitrogen and 5% hydrogen gas in an Eppendorf tube.
2. Incubate samples for 5 min at 30 °C and 500 rpm.
3. Start the reaction by adding 5 μL of ATP solution to the samples with further incubation for 30 min at 30 °C and 500 rpm.
4. Terminate the reaction by the addition of 50 μL EDTA solution.
5. Vent the samples with air for 10 min to quench residual $Na_2S_2O_4$ (*see* **Note 17**).
6. Prepare a negative control by adding the EDTA solution to the sample before incubation and skipping the addition of ATP.
7. For quantification, mix 30 μL sample with 750 μL phenol solution, 20 μL nitroprusside solution, and 200 μL NaOCl solution and incubate for 30 min at room temperature.
8. Measure the absorbance at 655 nm and use it to quantify the amount of formed ammonia compared to NH_4Cl standards. All measurements are triplicates. Our typical specific activities are comparable to published data with, for instance, $a_m = (230 \pm 28)$ nmol min^{-1} mg(VFe)$^{-1}$, which is in the range of published activities of for example, $a_m = (259 \pm 8)$ nmol min^{-1} mg(VFe)$^{-1}$ for vanadium nitrogenase from *A. chroococcum* [42].

3.5.5 UV-Vis Absorption Spectroscopy

Fe-S clusters can be analyzed by UV-vis absorption spectroscopy as shown for VFe protein [81, 99], since complex Fe-S clusters show typical absorbance features. Multiple individual transitions result in a broad feature at 400 nm. To analyze the metal clusters of VFe protein, UV-vis absorbance spectra of oxidized and reduced protein can be compared [54].

1. Oxidize the isolated VFe protein at a concentration of 15 mg/mL with the addition of 20 mM K_3FeCN_6 to the protein solution. Reduce the isolated protein with the addition of 5 mM $Na_2S_2O_4$.
2. Use a desalting column to exchange the buffer in order to remove excess oxidant or reductant, respectively.
3. Record the UV-vis absorption spectra of $Na_2S_2O_4$-reduced or K_3FeCN_6-oxidized VFe protein, as well as the corresponding difference spectrum as shown in Fig. 5. Upon oxidation, a characteristic feature at 375 nm vanishes.

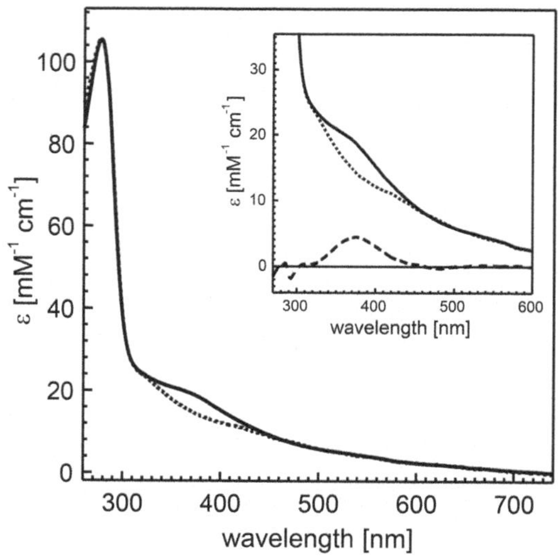

Fig. 5 UV-vis absorption spectra of K_3FeCN_6-oxidized (dotted line) and $Na_2S_2O_4$-reduced VFe protein (solid line). The enlargement shows a difference spectrum (dashed line) of oxidized/reduced VFe protein [54]

3.5.6 Electron Paramagnetic Resonance (EPR) Spectroscopy

EPR spectra allow for the characterization of the metal clusters in vanadium nitrogenase. A simple EPR spectrum can be used as a quality check for the purified VFe protein by comparison with published data [54].

1. To prepare an EPR sample, fill an X-band EPR tube with 300 μL of $Na_2S_2O_4$-reduced VFe protein at a concentration of 20 mg/mL. Then shock-freeze it in liquid nitrogen.
2. An exemplary continuous-wave (cw) X-band spectrum is depicted in Fig. 6 [54]. In this case, a cw X-band EPR-spectrometer with a microwave frequency of 9.339 GHz and a continuous-flow liquid helium cryostat was used [54]. Conduct the measurement at 10 K and a power of 2 mW with a modulation amplitude of 6 G and a receiver gain of 60 dB [54]. The cw EPR spectrum of VFe protein purified according to the protocol described here [54] is comparable to earlier data for VFe protein purified from *A. chroococcum* [81] and *A. vinelandii* [58, 88]. A prominent feature is a $S = ½$ signal with $g_{||} = 2.04$ and $g_{\perp} = 1.93$. The $S = 3/2$ feature at the lower field is more complex [54].

3.5.7 Right-Angle Light Scattering (RALS)

RALS can inform on the molecular mass of a protein, as well as its monodispersity and hence, quality [54].

1. Equilibrate a Superdex 200 10/300 analytical size-exclusion column in SEC buffer containing only 1 mM $Na_2S_2O_4$ using a GPC/SEC system (Viscotek GPCmax) with a triple-detector

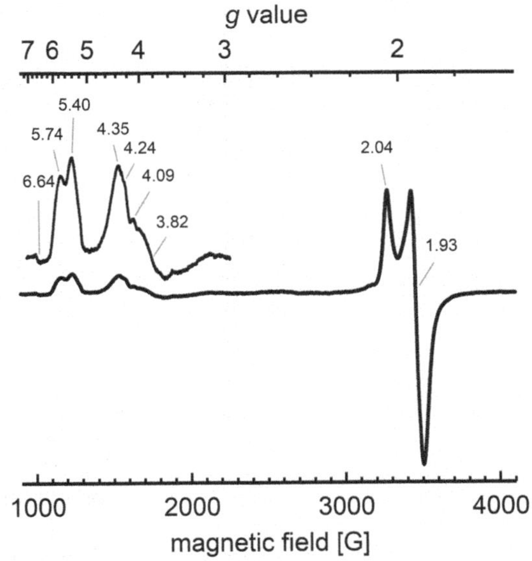

Fig. 6 X-band EPR spectrum of pure VFe protein. A prominent feature is a S = ½ signal with $g_{||} = 2.04$ and $g_{\perp} = 1.93$. The S = 3/2 feature at the lower field is more complex and is described in more detail by Sippel et al. [54]

array and a UV detector. Use a flow rate of 0.5 mL/min with a column and detector temperature set to 27.4 °C.

2. Load 100 μL of pure protein solution with a concentration of 1–4 mg/mL onto the column and UV/vis-, RALS- and refractive index signals are measured.
3. By analyzing the right-angle light scattering signal, calculate the molecular mass of the peak fraction. Exemplary results are shown in Fig. 7. The molecular mass is determined to 235.3 kDa, which corresponds to a heterohexamer $VnfD_2G_2K_2$ [54]. Combined analysis of UV-, RALS-, and refractive index signals demonstrate the monodispersity of the protein sample. The purification protocol described here leads to VFe protein with purity and homogeneity [54] that is comparable to that of the well-established MoFe protein purification.

4 Notes

1. The modified Burk medium is prepared and autoclaved without phosphate. The phosphate solution is prepared as a stock and autoclaved separately. It is added directly prior to usage.
2. The number of cycles and time of evacuation depends on the buffer volume. The remaining dioxygen can be determined using an oxygen electrode or an indicator such as resazurin.

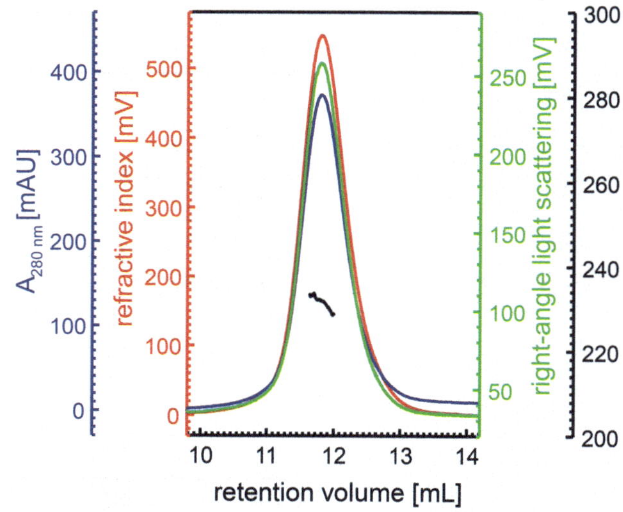

Fig. 7 Chromatograms of an analytical size exclusion chromatography run to confirm monodispersity and molecular mass of the purified VFe protein. The UV absorbance at 280 nm is depicted in blue, the refractive index in red, and the right-angle light scattering signal in green. Calculation of the molecular mass is shown in black [54]

3. Earlier vanadium nitrogenase purification protocols used 2.0 mM sodium dithionite in buffers [54, 100]. We find no distinguishable difference in the isolated protein material when using 2.5 mM sodium dithionite.
4. Earlier vanadium nitrogenase purification protocols omitted $MgCl_2$ from the buffers [54]. For crystallization of the VFe protein of vanadium nitrogenase, $MgCl_2$ is used as an additive [52]. Therefore, the protocol was modified such that all buffers are supplemented with $MgCl_2$. The addition of $MgCl_2$ alters the elution profiles on the anion exchange chromatography columns, but other than that, the isolated protein shows no difference.
5. Sodium nitroprusside in solution is light-sensitive; shielding it from light is obligatory.
6. NaOCl solution is not stable over time and should be prepared fresh when needed.
7. Baffled flasks are used due to the high oxygen demand of *A. vinelandii*.
8. The volume that is used for inoculation depends on the OD_{600} of the glycerol stock.
9. The addition of $MgCl_2$ increases protein purity slightly, but bears the disadvantage of overlapping VFe and Fe protein elution. The protein loss of these mixed fractions is small and, if desired, these fractions can be applied to a size exclusion

column where VFe and Fe proteins separate well due to their difference in size.

10. If Mo depletion did not work as expected, it will become obvious in this step. MoFe and its corresponding Fe protein have different running behavior than vanadium nitrogenase [54], so that the chromatogram will show additional peaks. Furthermore, molybdenum nitrogenase proteins are distinguishable via SDS-PAGE, as shown before [54].
11. This is a good point to interrupt purification and store the protein overnight on ice. VFe protein is relatively stable, and the purification can be stretched over several days. Our standard procedure takes 3 days. The first day comprises cell lysis, centrifugation, and the Q Sepharose HP anion exchange column, as well as the concentration of Fe protein by ultracentrifugation. The second day is used to run the Resource Q anion exchange column and the concentration of VFe protein. Size exclusion chromatography, protein concentration, and shock-freezing is done on the third day. Proteins can be stored on ice overnight. Fe protein is less stable than VFe protein, but since size exclusion is performed on the last day just before protein storage, any formed aggregate can be separated.
12. IEC A_0 buffer can be prepared containing 50 mM NaCl instead of no NaCl. This can increase protein purity further because more impurities are separated on the Resource Q column.
13. This gradient can be varied to reach from 0 to 30% IEC B buffer in 120 mL without any changes in protein quality.
14. For the Resource Q anion exchange column, the shape of the chromatogram may vary significantly depending on the protein amount applied. Rather, collect more fractions and analyze by SDS-PAGE which of these contain VFe protein.
15. Higher activities are obtained when using the proteins instead of ATP to start the reaction, whereas the addition of ATP leads to a higher reproducibility. Note that vanadium nitrogenase reduces acetylene not only to ethylene but also to ethane. Since ethane production is very low compared to ethylene production, it is neglected here with reference to Sippel et al. [54].
16. Tris buffer needs to be avoided in any case when conducting the Berthelot reaction. No buffer (dinitrogen activity assay buffer nor protein buffer) should contain Tris since it is interfering with the Berthelot reaction. We recommend using MOPS buffers instead. It is possible to conduct the last step of the purification (size exclusion chromatography) in MOPS buffer or to exchange the buffer of pure protein using a desalting column.

17. Quenching of residual $Na_2S_2O_4$ is necessary due to its interference with the quantification via Berthelot reaction. $Na_2S_2O_4$ leads to a decrease in absorption.

References

1. Howard JB, Rees DC (1996) Structural basis of biological nitrogen fixation. Chem Rev 96 (7):2965–2982. https://doi.org/10.1021/cr9500545
2. Tamaru K (1991) The history of the development of ammonia synthesis. In: Catalytic ammonia synthesis. Springer, Berlin, pp 1–18
3. Erisman JW, Sutton MA, Galloway J, Klimont Z, Winiwarter W (2008) How a century of ammonia synthesis changed the world. Nat Geosci 1(10):636–639. https://doi.org/10.1038/ngeo325
4. Schrock RR (2006) Reduction of dinitrogen. Proc Natl Acad Sci U S A 103(46):17087. https://doi.org/10.1073/pnas.0603633103
5. Lipman JG (1903) Experiments on the transformation and fixation of nitrogen by bacteria. New Jersey State Agric Exp Sta Ann Rep 24:217–285
6. Rees DC (1993) Dinitrogen reduction by nitrogenase: if N2 isn't broken, it can't be fixed. Curr Opin Struct Biol 3(6):921–928. https://doi.org/10.1016/0959-440x(93)90157-g
7. Dos Santos PC, Fang Z, Mason SW, Setubal JC, Dixon R (2012) Distribution of nitrogen fixation and nitrogenase-like sequences amongst microbial genomes. BMC Genomics 13(1):162. https://doi.org/10.1186/1471-2164-13-162
8. Dos Santos PC, Addo MA (2020) Distribution of nitrogen fixation genes in prokaryotes containing alternative nitrogenases. Chembiochem 21(12):1749–1759. https://doi.org/10.1002/cbic.202000022
9. Bishop PE, Jarlenski DM, Hetherington DR (1980) Evidence for an alternative nitrogen fixation system in *Azotobacter vinelandii*. Proc Natl Acad Sci U S A 77 (12):7342–7346. https://doi.org/10.1073/pnas.77.12.7342
10. Chisnell JR, Premakumar R, Bishop PE (1988) Purification of a second alternative nitrogenase from a nifHDK deletion strain of *Azotobacter vinelandii*. J Bacteriol 170 (1):27–33. https://doi.org/10.1128/jb.170.1.27-33.1988
11. Bulen WA, LeComte JR (1966) The nitrogenase system from Azotobacter: two-enzyme requirement for N2 reduction, ATP-dependent H2 evolution, and ATP hydrolysis. Proc Natl Acad Sci 56 (3):979–986. https://doi.org/10.1073/pnas.56.3.979
12. Brigle KE, Newton WE, Dean DR (1985) Complete nucleotide sequence of the *Azotobacter vinelandii* nitrogenase structural gene cluster. Gene 37(1–3):37–44. https://doi.org/10.1016/0378-1119(85)90255-0
13. Joerger RD, Loveless TM, Pau RN, Mitchenall LA, Simon BH, Bishop PE (1990) Nucleotide sequences and mutational analysis of the structural genes for nitrogenase 2 of *Azotobacter vinelandii*. J Bacteriol 172 (6):3400–3408. https://doi.org/10.1128/jb.172.6.3400-3408.1990
14. Joerger RD, Jacobson MR, Premakumar R, Wolfinger ED, Bishop PE (1989) Nucleotide sequence and mutational analysis of the structural genes (anfHDGK) for the second alternative nitrogenase from *Azotobacter vinelandii*. J Bacteriol 171(2):1075–1086. https://doi.org/10.1128/jb.171.2.1075-1086.1989
15. Boyd E, Hamilton T, Peters J (2011) An alternative path for the evolution of biological nitrogen fixation. Front Microbiol 2:205. https://doi.org/10.3389/fmicb.2011.00205
16. Stueken EE, Buick R, Guy BM, Koehler MC (2015) Isotopic evidence for biological nitrogen fixation by molybdenum-nitrogenase from 3.2 Gyr. Nature 520(7549):666–669. https://doi.org/10.1038/nature14180
17. Anbar AD, Knoll AH (2002) Proterozoic Ocean chemistry and evolution: a bioinorganic bridge? Science 297 (5584):1137–1142. https://doi.org/10.1126/science.1069651
18. Hans Wedepohl K (1995) The composition of the continental crust. Geochim Cosmochim Acta 59(7):1217–1232. https://doi.org/10.1016/0016-7037(95)00038-2
19. Bellenger JP, Wichard T, Kraepiel AM (2008) Vanadium requirements and uptake kinetics in the dinitrogen-fixing bacterium *Azotobacter vinelandii*. Appl Environ Microbiol 74 (5):1478–1484. https://doi.org/10.1128/AEM.02236-07

20. Boyd ES, Anbar AD, Miller S, Hamilton TL, Lavin M, Peters JW (2011) A late methanogen origin for molybdenum-dependent nitrogenase. Geobiology 9(3):221–232. https://doi.org/10.1111/j.1472-4669.2011.00278.x
21. Raymond J, Siefert JL, Staples CR, Blankenship RE (2004) The natural history of nitrogen fixation. Mol Biol Evol 21(3):541–554. https://doi.org/10.1093/molbev/msh047
22. Noar JD, Bruno-Bárcena JM (2018) *Azotobacter vinelandii*: the source of 100 years of discoveries and many more to come. Microbiology 164(4):421–436. https://doi.org/10.1099/mic.0.000643
23. Pau RN, Mitchenall LA, Robson RL (1989) Genetic evidence for an *Azotobacter vinelandii* nitrogenase lacking molybdenum and vanadium. J Bacteriol 171(1):124–129. https://doi.org/10.1128/jb.171.1.124-129.1989
24. Eady RR (1996) Structureminus sign function relationships of alternative Nitrogenases. Chem Rev 96(7):3013–3030. https://doi.org/10.1021/cr950057h
25. Setubal JC, dos Santos P, Goldman BS, Ertesvag H, Espin G, Rubio LM, Valla S, Almeida NF, Balasubramanian D, Cromes L, Curatti L, Du Z, Godsy E, Goodner B, Hellner-Burris K, Hernandez JA, Houmiel K, Imperial J, Kennedy C, Larson TJ, Latreille P, Ligon LS, Lu J, Maerk M, Miller NM, Norton S, O'Carroll IP, Paulsen I, Raulfs EC, Roemer R, Rosser J, Segura D, Slater S, Stricklin SL, Studholme DJ, Sun J, Viana CJ, Wallin E, Wang B, Wheeler C, Zhu H, Dean DR, Dixon R, Wood D (2009) Genome sequence of *Azotobacter vinelandii*, an obligate aerobe specialized to support diverse anaerobic metabolic processes. J Bacteriol 191 (14):4534–4545. https://doi.org/10.1128/JB.00504-09
26. Rediers H, Vanderleyden J, De Mot R (2004) *Azotobacter vinelandii*: a pseudomonas in disguise? Microbiology 150(Pt 5):1117–1119. https://doi.org/10.1099/mic.0.27096-0
27. Setubal JC, Almeida NF (2015) The *Azotobacter vinelandii* genome: an update. In: Biological nitrogen fixation, vol 1 and 2. John Wiley & Sons, Inc., Hoboken, New Jersey, pp 225–234. https://doi.org/10.1002/9781119053095
28. Young JM, Park D-C (2007) Probable synonymy of the nitrogen-fixing genus Azotobacter and the genus pseudomonas. Int J Syst Evol Microbiol 57(12):2894–2901. https://doi.org/10.1099/ijs.0.64969-0
29. Özen AI, Ussery DW (2012) Defining the pseudomonas genus: where do we draw the line with Azotobacter? Microb Ecol 63 (2):239–248. https://doi.org/10.1007/s00248-011-9914-8
30. Robson RL, Eady RR, Richardson TH, Miller RW, Hawkins M, Postgate JR (1986) The alternative nitrogenase of *Azotobacter chroococcum* is a vanadium enzyme. Nature 322 (6077):388–390. https://doi.org/10.1038/322388a0
31. McKenna CE, Benemann JR, Traylor TG (1970) A vanadium containing nitrogenase preparation: implications for the role of molybdenum in nitrogen fixation. Biochem Biophys Res Commun 41(6):1501–1508. https://doi.org/10.1016/0006-291x(70)90557-7
32. Bishop PE, Jarlenski DM, Hetherington DR (1982) Expression of an alternative nitrogen fixation system in *Azotobacter vinelandii.* J Bacteriol 150(3):1244–1251
33. Burns RC, Fuchsman WH, Hardy RWF (1971) Nitrogenase from vanadium-grown : isolation, characteristics, and mechanistic implications. Biochem Biophys Res Commun 42(3):353–358. https://doi.org/10.1016/0006-291x(71)90377-9
34. Premakumar R, Lemos E, Bishop P (1984) Evidence for two dinitrogenase reductases under regulatory control by molybdenum in *Azotobacter vinelandii*☆. Biochim Biophys Acta Gen Subj 797(1):64–70. https://doi.org/10.1016/0304-4165(84)90382-9
35. Bishop PE, Hawkins ME, Eady RR (1986) Nitrogen fixation in molybdenum-deficient continuous culture by a strain of Azotobacter vinelandii carrying a deletion of the structural genes for nitrogenase (nifHDK). Biochem J 238(2):437–442. https://doi.org/10.1042/bj2380437
36. Joerger RD, Premakumar R, Bishop PE (1986) Tn5-induced mutants of *Azotobacter vinelandii* affected in nitrogen fixation under Mo-deficient and Mo-sufficient conditions. J Bacteriol 168(2):673–682. https://doi.org/10.1128/jb.168.2.673-682.1986
37. Bishop PE, Premakumar R, Dean DR, Jacobson MR, Chisnell JR, Rizzo TM, Kopczynski J (1986) Nitrogen fixation by *Azotobacter vinelandii* strains having deletions in structural genes for Nitrogenase. Science 232 (4746):92–94. https://doi.org/10.1126/science.232.4746.92
38. Jacobson MR, Premakumar R, Bishop PE (1986) Transcriptional regulation of nitrogen fixation by molybdenum in *Azotobacter*

vinelandii. J Bacteriol 167(2):480–486. https://doi.org/10.1128/jb.167.2.480-486.1986

39. Luque F, Pau RN (1991) Transcriptional regulation by metals of structural genes for *Azotobacter vinelandii* nitrogenases. Mol Gen Genet MGG 227(3):481–487. https://doi.org/10.1007/BF00273941
40. Waugh SI, Paulsen DM, Mylona PV, Maynard RH, Premakumar R, Bishop PE (1995) The genes encoding the delta subunits of dinitrogenases 2 and 3 are required for mo-independent diazotrophic growth by *Azotobacter vinelandii*. J Bacteriol 177 (6):1505–1510. https://doi.org/10.1128/jb.177.6.1505-1510.1995
41. Walmsley J, Kennedy C (1991) Temperature-dependent regulation by molybdenum and vanadium of expression of the structural genes encoding three Nitrogenases in *Azotobacter vinelandii*. Appl Environ Microbiol 57 (2):622–624
42. Miller RW, Eady RR (1988) Molybdenum and vanadium nitrogenases of *Azotobacter chroococcum*. Low temperature favours N2 reduction by vanadium nitrogenase. Biochem J 256(2):429–432. https://doi.org/10.1042/bj2560429
43. Luxem KE, Kraepiel AML, Zhang L, Waldbauer JR, Zhang X (2020) Carbon substrate re-orders relative growth of a bacterium using Mo-, V-, or Fe-nitrogenase for nitrogen fixation. Environ Microbiol 22(4):1397–1408. https://doi.org/10.1111/1462-2920.14955
44. Darnajoux R, Magain N, Renaudin M, Lutzoni F, Bellenger JP, Zhang X (2019) Molybdenum threshold for ecosystem scale alternative vanadium nitrogenase activity in boreal forests. Proc Natl Acad Sci U S A 116 (49):24682–24688. https://doi.org/10.1073/pnas.1913314116
45. Kim JS, Rees DC (1992) Crystallographic structure and functional implications of the Nitrogenase molybdenum iron protein from Azotobacter-Vinelandii. Nature 360 (6404):553–560
46. Georgiadis MM, Komiya H, Chakrabarti P, Woo D, Kornuc JJ, Rees DC (1992) Crystallographic structure of the nitrogenase iron protein from *Azotobacter vinelandii*. Science 257(5077):1653–1659. https://doi.org/10.1126/science.1529353
47. Schindelin H, Kisker C, Schlessman JL, Howard JB, Rees DC (1997) Structure of ADP·AIF4–stabilized nitrogenase complex and its implications for signal transduction. Nature 387(6631):370–376. https://doi.org/10.1038/387370a0
48. Einsle O, Tezcan FA, Andrade SL, Schmid B, Yoshida M, Howard JB, Rees DC (2002) Nitrogenase MoFe-protein at 1.16 a resolution: a central ligand in the FeMo-cofactor. Science 297(5587):1696–1700. https://doi.org/10.1126/science.1073877
49. Spatzal T, Aksoyoglu M, Zhang L, Andrade SL, Schleicher E, Weber S, Rees DC, Einsle O (2011) Evidence for interstitial carbon in nitrogenase FeMo cofactor. Science 334 (6058):940. https://doi.org/10.1126/science.1214025
50. Lancaster KM, Roemelt M, Ettenhuber P, Hu Y, Ribbe MW, Neese F, Bergmann U, DeBeer S (2011) X-ray emission spectroscopy evidences a central carbon in the nitrogenase iron-molybdenum cofactor. Science 334 (6058):974–977. https://doi.org/10.1126/science.1206445
51. Rees JA, Bjornsson R, Schlesier J, Sippel D, Einsle O, DeBeer S (2015) The Fe-V cofactor of vanadium Nitrogenase contains an interstitial carbon atom. Angew Chem Int Ed Engl 54(45):13249–13252. https://doi.org/10.1002/anie.201505930
52. Sippel D, Einsle O (2017) The structure of vanadium nitrogenase reveals an unusual bridging ligand. Nat Chem Biol 13 (9):956–960. https://doi.org/10.1038/nchembio.2428
53. Rohde M, Trncik C, Sippel D, Gerhardt S, Einsle O (2018) Crystal structure of VnfH, the iron protein component of vanadium nitrogenase. J Biol Inorg Chem 23 (7):1049–1056. https://doi.org/10.1007/s00775-018-1602-4
54. Sippel D, Schlesier J, Rohde M, Trncik C, Decamps L, Djurdjevic I, Spatzal T, Andrade SL, Einsle O (2017) Production and isolation of vanadium nitrogenase from *Azotobacter vinelandii* by molybdenum depletion. J Biol Inorg Chem 22(1):161–168. https://doi.org/10.1007/s00775-016-1423-2
55. Peters JW, Stowell MH, Soltis SM, Finnegan MG, Johnson MK, Rees DC (1997) Redox-dependent structural changes in the nitrogenase P-cluster. Biochemistry 36 (6):1181–1187. https://doi.org/10.1021/bi9626665
56. Eady R (2003) Current status of structure function relationships of vanadium nitrogenase. Coord Chem Rev 237(1–2):23–30. https://doi.org/10.1016/s0010-8545(02)00248-5

57. Benediktsson B, Thorhallsson AT, Bjornsson R (2018) QM/MM calculations reveal a bridging hydroxo group in a vanadium nitrogenase crystal structure. Chem Commun (Camb) 54(53):7310–7313. https://doi.org/10.1039/c8cc03793k
58. Hales BJ, Case EE, Morningstar JE, Dzeda MF, Mauterer LA (1986) Isolation of a new vanadium-containing nitrogenase from *Azotobacter vinelandii.* Biochemistry 25 (23):7251–7255. https://doi.org/10.1021/bi00371a001
59. Chatterjee R, Allen RM, Ludden PW, Shah VK (1997) In vitro synthesis of the iron-molybdenum cofactor and maturation of the nif-encoded apodinitrogenase. Effect of substitution of VNFH for NIFH. J Biol Chem 272(34):21604–21608. https://doi.org/10.1074/jbc.272.34.21604
60. Walker JE, Saraste M, Runswick MJ, Gay NJ (1982) Distantly related sequences in the alpha- and beta-subunits of ATP synthase, myosin, kinases and other ATP-requiring enzymes and a common nucleotide binding fold. EMBO J 1(8):945–951. https://doi.org/10.1002/j.1460-2075.1982.tb01276.x
61. Danyal K, Dean DR, Hoffman BM, Seefeldt LC (2011) Electron transfer within nitrogenase: evidence for a deficit-spending mechanism. Biochemistry 50(43):9255–9263. https://doi.org/10.1021/bi201003a
62. Peters JW, Fisher K, Dean DR (1995) Nitrogenase structure and function: a biochemical-genetic perspective. Annu Rev Microbiol 49 (1):335–366. https://doi.org/10.1146/annurev.mi.49.100195.002003
63. Thorneley RNF, Lowe DJ (1983) Nitrogenase of *Klebsiella pneumoniae.* Kinetics of the dissociation of oxidized iron protein from molybdenum-iron protein: identification of the rate-limiting step for substrate reduction. Biochem J 215(2):393–403. https://doi.org/10.1042/bj2150393
64. Simpson F, Burris R (1984) A nitrogen pressure of 50 atmospheres does not prevent evolution of hydrogen by nitrogenase. Science 224(4653):1095–1097. https://doi.org/10.1126/science.6585956
65. Harris DF, Lukoyanov DA, Shaw S, Compton P, Tokmina-Lukaszewska M, Bothner B, Kelleher N, Dean DR, Hoffman BM, Seefeldt LC (2018) Mechanism of N2 reduction catalyzed by Fe-Nitrogenase involves reductive elimination of H2. Biochemistry 57(5):701–710. https://doi.org/10.1021/acs.biochem.7b01142
66. Harris DF, Lukoyanov DA, Kallas H, Trncik C, Yang ZY, Compton P, Kelleher N, Einsle O, Dean DR, Hoffman BM, Seefeldt LC (2019) Mo-, V-, and Fe-Nitrogenases use a universal eight-electron reductive-elimination mechanism to achieve N2 reduction. Biochemistry 58(30):3293–3301. https://doi.org/10.1021/acs.biochem.9b00468
67. Dilworth MJ (1966) Acetylene reduction by nitrogen-fixing preparations from clostridium pasteurianum. Biochim Biophys Acta Gen Subj 127(2):285–294. https://doi.org/10.1016/0304-4165(66)90383-7
68. R S Burris R Study Intermediates in nitrogen fixation. In: Federation Proceedings, 1966. vol 2 P 1. Federation Amer Soc Exp Biol 9650, Rockville Pike, Bethesda, MD 20814–3998, p. 710
69. Harris DF, Yang ZY, Dean DR, Seefeldt LC, Hoffman BM (2018) Kinetic understanding of N2 reduction versus H2 evolution at the E4(4H) Janus state in the three Nitrogenases. Biochemistry 57(39):5706–5714. https://doi.org/10.1021/acs.biochem.8b00784
70. Riddle GD, Simonson JG, Hales BJ, Braymer HD (1982) Nitrogen fixation system of tungsten-resistant mutants of *Azotobacter vinelandii.* J Bacteriol 152(1):72–80
71. Dilworth MJ, Eady RR, Robson RL, Miller RW (1987) Ethane formation from acetylene as a potential test for vanadium nitrogenase in vivo. Nature 327(6118):167–168. https://doi.org/10.1038/327167a0
72. Schollhorn R, Burris RH (1967) Acetylene as a competitive inhibitor of N-2 fixation. Proc Natl Acad Sci U S A 58(1):213–216. https://doi.org/10.1073/pnas.58.1.213
73. Dilworth MJ, Eady RR (1991) Hydrazine is a product of dinitrogen reduction by the vanadium-nitrogenase from *Azotobacter chroococcum.* Biochem J 277(Pt 2):465–468. https://doi.org/10.1042/bj2770465
74. Benemann J, McKenna C, Lie R, Traylor T, Kamen M (1972) The vanadium effect in nitrogen fixation by azotobacter. Biochim Biophys Acta Gen Subj 264(1):25–38. https://doi.org/10.1016/0304-4165(72)90113-4
75. Dilworth MJ, Eady RR, Eldridge ME (1988) The vanadium nitrogenase of *Azotobacter chroococcum.* Reduction of acetylene and ethylene to ethane. Biochem J 249(3):745–751. https://doi.org/10.1042/bj2490745
76. Smith BE, Eady RR (1993) Metalloclusters of the nitrogenases. In: EJB reviews. Springer, Berlin, Heidelberg, pp 79–93. https://doi.org/10.1007/978-3-642-78046-2_7
77. Scherings G, Haaker H, Veeger C (1977) Regulation of nitrogen fixation by Fe-S

protein II in *Azotobacter vinelandii*. Eur J Biochem 77(3):21–30. https://doi.org/10.1111/j.1432-1033.1977.tb11706.x

78. Robson RL (1979) Characterization of an oxygen-stable nitrogenase complex isolated from *Azotobacter chroococcum*. Biochem J 181(3):569–575. https://doi.org/10.1042/bj1810569
79. Kuhla J, Oelze J (1988) Dependence of nitrogenase switch-off upon oxygen stress on the nitrogenase activity in *Azotobacter vinelandii*. J Bacteriol 170(11):5325–5329. https://doi.org/10.1128/jb.170.11.5325-5329.1988
80. Schlesier J, Rohde M, Gerhardt S, Einsle O (2016) A conformational switch triggers Nitrogenase protection from oxygen damage by Shethna protein II (FeSII). J Am Chem Soc 138(1):239–247. https://doi.org/10.1021/jacs.5b10341
81. Eady RR, Robson RL, Richardson TH, Miller RW, Hawkins M (1987) The vanadium nitrogenase of *Azotobacter chroococcum*. Purification and properties of the VFe protein. Biochem J 244(1):197–207. https://doi.org/10.1042/bj2440197
82. Burns RC, Stasny JT, Hardy RWF (1976). Isolation and characteristics of a modified nitrogenase from *Azotobacter vinelandii* including "vanadium-Fe" protein from cells grown on medium enriched in vanadium. Vol. 1 p.196–207. In W. E. Newton, C. J. Nyman (ed.), Proceedings of the First International Symposium on Nitrogen Fixation, Washington State University Press
83. Hales BJ, Langosch DJ, Case EE (1986) Isolation and characterization of a second nitrogenase Fe-protein from *Azotobacter vinelandii*. J Biol Chem 261 (32):15301–15306
84. Blanchard CZ, Hales BJ (1996) Isolation of two forms of the nitrogenase VFe protein from *Azotobacter vinelandii*. Biochemistry 35(2):472–478. https://doi.org/10.1021/bi951429j
85. Robson RL, Woodley PR, Pau RN, Eady RR (1989) Structural genes for the vanadium nitrogenase from *Azotobacter chroococcum*. EMBO J 8(4):1217–1224. https://doi.org/10.1002/j.1460-2075.1989.tb03495.x
86. Chatterjee R, Allen RM, Ludden PW, Shah VK (1996) Purification and characterization of the vnf-encoded Apodinitrogenase from *Azotobacter vinelandii*. J Biol Chem 271 (12):6819–6826. https://doi.org/10.1074/jbc.271.12.6819
87. Pau RN, Eldridge ME, Lowe DJ, Mitchenall LA, Eady RR (1993) Molybdenum-independent nitrogenases of *Azotobacter vinelandii*: a functional species of alternative nitrogenase-3 isolated from a molybdenum-tolerant strain contains an iron-molybdenum cofactor. Biochem J 293(Pt 1):101–107. https://doi.org/10.1042/bj2930101
88. Lee CC, Hu YL, Ribbe MW (2009) Unique features of the nitrogenase VFe protein from *Azotobacter vinelandii*. Proc Natl Acad Sci U S A 106(23):9209–9214. https://doi.org/10.1073/pnas.0904408106
89. Chatterjee R, Ludden PW, Shah VK (1997) Characterization of VNFG, THE δ subunit of the vnf-encoded Apodinitrogenase from *Azotobacter vinelandii*: implications for its role in the formation of functional dinitrogenase 2. J Biol Chem 272(6):3758–3765. https://doi.org/10.1074/jbc.272.6.3758
90. Burk D, Lineweaver H (1930) The influence of fixed nitrogen on Azotobacter. J Bacteriol 19(6):389–414
91. Strandberg GW, Wilson PW (1968) Formation of the nitrogen-fixing enzyme system in *Azotobacter vinelandii*. Can J Microbiol 14 (1):25–31. https://doi.org/10.1139/m68-005
92. Nagatani HH, Brill WJ (1974) Nitrogenase V. the effect of Mo, W and V on the synthesis of nitrogenase components in Azotobacter vinelandii. Biochim Biophys Acta 362 (1):160–166. https://doi.org/10.1016/0304-4165(74)90037-3
93. Laemmli UK (1970) Cleavage of structural proteins during the assembly of the head of bacteriophage T4. Nature 227 (5259):680–685. https://doi.org/10.1038/227680a0
94. Neuhoff V, Arold N, Taube D, Ehrhardt W (1988) Improved staining of proteins in polyacrylamide gels including isoelectric focusing gels with clear background at nanogram sensitivity using Coomassie brilliant blue G-250 and R-250. Electrophoresis 9(6):255–262. https://doi.org/10.1002/elps.1150090603
95. Kang D, Suh M-K et al (2002) Highly sensitive and fast protein detection with coomassie brilliant blue in sodium dodecyl sulfate-polyacrylamide gel electrophoresis (highly sensitive and fast protein detection with coomassie brilliant blue in sodium dodecyl sulfate-polyacrylamide gel electrophoresis). Bull Korean Chem Soc 23:1511–1512. https://doi.org/10.5012/BKCS.2002.23.11.1511
96. Smith PK, Krohn RI, Hermanson GT, Mallia AK, Gartner FH, Provenzano MD, Fujimoto EK, Goeke NM, Olson BJ, Klenk DC (1985) Measurement of protein using bicinchoninic

acid. Anal Biochem 150(1):76–85. https://doi.org/10.1016/0003-2697(85)90442-7

97. Sippel D (2017) Structure of the vanadium Nitrogenase of *Azotobacter vinelandii* and mechanistic insights into biological nitrogen fixation, https://doi.org/10.6094/UNIFR/13383, Dissertation at the University of Freiburg

98. Searle PL (1984) The berthelot or indophenol reaction and its use in the analytical chemistry of nitrogen. A review. Analyst 109 (5):549–568. https://doi.org/10.1039/AN9840900549

99. Eady RR, Richardson TH, Miller RW, Hawkins M, Lowe DJ (1988) The vanadium nitrogenase of *Azotobacter chroococcum*. Purification and properties of the Fe protein. Biochem J 256(1):189–196. https://doi.org/10.1042/bj2560189

100. Jasniewski AJ, Lee CC, Ribbe MW, Hu Y (2020) Reactivity, mechanism, and assembly of the alternative Nitrogenases. Chem Rev 120(12):5107–5157. https://doi.org/10.1021/acs.chemrev.9b00704

Part II

Biochemical Methods to Probe Steps Involved in the Synthesis of Fe-S Clusters

Chapter 7

Ni-NTA Affinity Chromatography to Characterize Protein–Protein Interactions During Fe-S Cluster Biogenesis

Terrell D. Carter and F. Wayne Outten

Abstract

Affinity chromatography can be repurposed to provide useful information about the specific partner protein (s) to which a protein of interest may bind as well as the relative binding affinity of that partner protein for the protein of interest. Here, we provide a protocol for an Ni-NTA affinity chromatography assay that may be utilized to uncover insightful information about the nature of protein–protein interactions during iron-sulfur (Fe-S) cluster biogenesis reactions.

Key words Iron-sulfur cluster, Affinity chromatography, Anaerobic, IMAC

1 Introduction

Characterizing protein–protein interactions is a common goal among researchers. Characterizing protein–protein interactions during iron-sulfur (Fe-S) cluster trafficking reactions is of paramount importance for the field of Fe-S cluster biogenesis [1]. However, the oxygen sensitivity of Fe-S clusters introduces new challenges for characterizing such interactions [2]. Proteins that contain an Fe-S cluster (holo-proteins) must often be handled under anaerobic conditions to maintain the integrity of the cluster, both prior to and during cluster transfer to an apo-protein acceptor. Immobilized metal ion affinity chromatography (IMAC) is a simple technique that can elucidate protein binding partners and can provide information for the relative binding affinity in protein–protein interactions [3]. Ni-NTA affinity chromatography, a specific type of IMAC, can be carried out under anaerobic conditions using a dedicated FPLC system contained in an anaerobic chamber. Alternatively, the technique can be done manually in disposable anaerobic bags using syringes for sample loading, washing, and elution with HisTrap format columns.

Patricia C. Dos Santos (ed.), *Fe-S Proteins: Methods and Protocols*, Methods in Molecular Biology, vol. 2353, https://doi.org/10.1007/978-1-0716-1605-5_7,

For monitoring protein–protein interactions, a protein with a polyhistidine tag (e.g., His_6 tag) known as the "bait protein" is bound to the Ni-NTA resin. Next, potential interacting protein (s) that do not contain a His_6 motif and do not bind the Ni-NTA resin directly, known as the "prey protein(s)," are loaded and incubated on the column. During this incubation step, the His_6-bait and prey proteins may interact, allowing the prey proteins to be retained on the Ni-NTA column through that interaction. Following the incubation, a wash step is used to remove any prey proteins that do not interact with the bait protein. Then, all proteins are eluted from the column for downstream analysis of load, wash, and elution fractions by SDS-PAGE, UV-visible absorption spectroscopy, or other analytical techniques. The experimental details can be altered to test different aspects of potential Fe-S cluster trafficking interactions (Fig. 1). For example, the His_6-bait protein may be loaded in either the apo- or holo-state to determine how the presence of the Fe-S cluster affects the interaction with the prey protein(s). Similarly, the specific protein used as the His_6-bait can be systematically switched to show the specificity of interactions and to determine the directionality of cluster transfer interactions (Fig. 1).

2 Materials

2.1 Materials for Anaerobic Bag Protocol

1. Anaerobic glove bag (such as Spilfyter Mfr. Model # 690321 or Aldritch Atmosbag glove bag Z530212).
2. Bait protein: His-tagged protein of interest isolated under anaerobic conditions (*see* **Note 1**).
3. Prey protein: untagged protein of interest isolated under anaerobic conditions (*see* **Note 1**).
4. 1 mL Ni-NTA IMAC column with an adapter for Luer tip.
5. 10 mL syringe with Luer tip connector.
6. Ultrapure water (18 MΩ-cm water).
7. Binding Buffer: 20 mM sodium phosphate, 0.5 M NaCl, 10 mM imidazole, pH 7.4. Degas buffer and then sparge the buffer with N_2 gas to limit trace oxygen (*see* **Notes 2** and **3**).
8. Elution Buffer: 20 mM sodium phosphate, 0.5 M NaCl, 500 mM imidazole, pH 7.4. Degas buffer and then sparge the buffer with N_2 gas to limit trace oxygen (*see* **Notes 2** and **3**).
9. Nitrogen gas cylinder with an appropriate regulator or house nitrogen with a line for sparging buffers and filling glove bag.

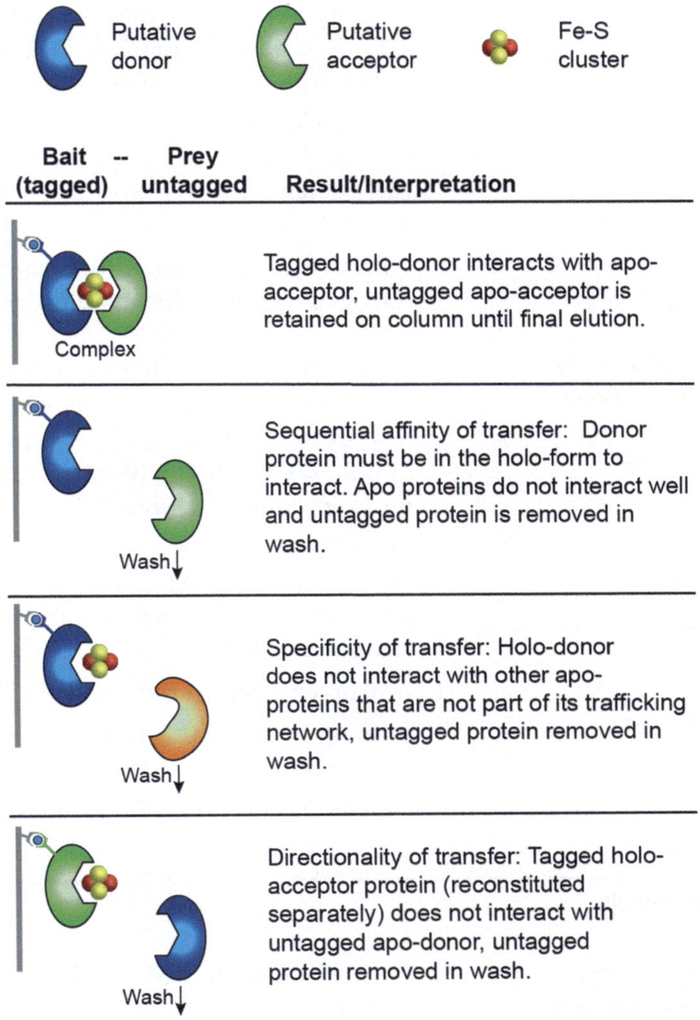

Fig. 1 Diagram illustrating various bait and prey protein combinations that can be used in the affinity chromatography experiment. Several control experiments are also shown (see text for more details)

2.2 Materials for Anaerobic Chamber Protocol

1. Anaerobic chamber: vinyl anaerobic chamber equilibrated with 5% H_2 balanced with N_2 gas (*see* **Note 4**).
2. FPLC (located inside the anaerobic chamber).
3. Bait protein: His-tagged protein of interest isolated under anaerobic conditions (*see* **Note 1**).
4. Prey protein: untagged protein of interest isolated under anaerobic conditions (*see* **Note 1**).
5. 1 mL Ni-NTA IMAC column with adapters for attachment to FPLC lines.
6. Ultrapure water (18 MΩ-cm water).

7. Binding Buffer: 20 mM sodium phosphate, 0.5 M NaCl, 10 mM imidazole, pH 7.4. Degas buffer and then sparge the buffer with N_2 gas to limit trace oxygen (*see* **Notes 2** and **3**).
8. Elution Buffer: 20 mM sodium phosphate, 0.5 M NaCl, 500 mM imidazole, pH 7.4. Degas buffer and then sparge the buffer with N_2 gas to limit trace oxygen (*see* **Notes 2** and **3**).
9. Nitrogen gas cylinder with an appropriate regulator or house nitrogen with lines for sparging buffers.

2.3 Materials for Analysis of Results

1. Concentrator: ultrafiltration concentrators (*see* **Note 5**).
2. SDS-PAGE: Sodium dodecyl sulfate polyacrylamide gel electrophoresis (SDS-PAGE) gel with the appropriate percentage of polyacrylamide (*see* **Note 6**). Use a bis-Tris gel formulation for best results (*see* **Note 7**).
3. SDS-PAGE running buffer for bis-Tris system: 50 mM MOPS, 50 mM Tris, 5 mM EDTA, 0.5% SDS (adjusting pH is not necessary). Add sodium bisulfite to 5 mM as a reducing agent just prior to run.
4. Instrumentation to run standard SDS-PAGE.
5. Staining solution: 0.2–0.3% Coomassie R-250, 10% methanol, 10% acetic acid.
6. Destaining solution: 10% acetic acid.

3 Methods

3.1 Purification of Protein Complexes Using an Anaerobic Bag

1. Place all items needed for the procedure in a glove bag (*see* **Note 8**).
2. Prepare the Ni-NTA column according to the protocol for a 1 mL HisTrap HP column from Cytiva™. A Luer connector is required to manually operate the column using a syringe to administer buffers and samples.
3. Fill the syringe with 5 column volumes (5 mL) of ultrapure water that has been N_2-sparged. Remove the column's top stopper and carefully connect the syringe to the column in a "drop-to-drop" fashion to avoid introducing air into the system. Remove the twist-off end from the bottom of the column.
4. Wash the column with 5 or more column volumes of ultrapure water to remove traces of 20% ethanol (if column was previously stored in that solution). Try to maintain a flow rate of approximately 1 mL/min, as judged by the collection of flow through in a 15 mL conical tube.

5. Equilibrate the column with 5–10 column volumes of Binding Buffer (5–10 mL here).
6. Charge the HisTrap HP column with the His_6-bait protein by loading approximately 500 μg of the His_6-bait protein onto the column. A 1-mL syringe can be used for sample loading at this step. Utilize a buffer flow rate slow enough to allow for efficient binding of the His_6-bait protein to the HisTrap HP column (0.1–0.5 mL/min). After the sample has loaded, wash the column with 3 column volumes to ensure any unbound bait protein has been removed.
7. Load the prey protein(s) suspected to associate with the His-tagged protein onto the column. The exact amount of prey protein is in ten-fold excess over the amount of His_6-bait protein added in **step 6**. The entire amount of prey protein should be loaded in a volume approximately equal to 0.5 column volumes to ensure prey protein is not being forced off the column during loading. After loading, cease or slow the flow rate to allow for a sufficient on-column incubation time of prey and His_6-bait proteins. An incubation time of 5 min can be used for initial experiments; however, shorter or longer incubation times may be desired depending on the nature of the interactions and the specific aims of the research. At this step, the prey protein should be bound to the bait protein if they interact to form a stable complex.
8. Wash the column with 3–5 column volumes of binding buffer to remove prey protein that does not interact with the His_6-bait protein. Retain the wash fractions for further analysis by SDS-PAGE to determine how much prey protein washes off and to ensure that no His_6-bait protein is removed during this step. Depending on the stoichiometry of the interaction and the exact amount of bait/prey proteins used, some of the prey protein may elute if present in excess. Retaining wash fractions allows for mass balance estimates upon completion of the experiment.
9. Elute all proteins directly or indirectly bound to the HisTrap HP column bypassing 3 column volumes of elution buffer through the column and collecting the fractions. Any protein (s) bound to the His_6-bait protein also should be present in the elution fractions. If using an FPLC system, a gradient of binding buffer and elution buffer may be desired. If using a syringe to manually wash and elute, it may be advantageous to collect 0.5 mL fractions in 1.5 mL conical tubes followed by a quick measurement of protein concentration in the elution fractions (*see* **Note 9**). This analysis will allow you to select only those fractions containing protein to carry forward for analysis.

3.2 Purification of Protein Complexes Using an Anaerobic Chamber

1. For the FPLC system, ensure that all tubing is properly connected and that the lines run to appropriate buffer and waste containers. Begin flow manually on the FPLC at 0.5 mL/min flow rate from a line placed in N_2-sparged ultrapure, filtered water (*see* **Note 3**). Remove the column's top stopper and carefully connect the proper line to the column in a "drop-to-drop" fashion to avoid introducing air into the system. After connection, quickly remove the twist-off end from the bottom of the column to allow elution. Connect the waste line to the bottom of the column.
2. Wash the column with 5 or more column volumes of ultrapure water to remove traces of 20% ethanol (if column was previously stored in that solution). If using an FPLC system, wash the lines and column with distilled water until the UV monitor and conductivity readings establish a stable baseline or as recommended by the instrument manual.
3. Equilibrate the column with 5–10 column volumes of binding buffer (5–10 mL here). Pass binding buffer through the column and lines until the UV monitor and conductivity readings stabilize or as recommended by the instrument.
4. Charge the HisTrap HP column with the His_6-bait protein by loading approximately 500 μg of the His_6-bait protein onto the column. Use the appropriately sized sample loading loop (1 mL for 500 μL sample). Reduce flow rate to 0.1–0.5 mL/min to allow for efficient binding of the His_6-bait protein to the HisTrap HP column. After the sample has loaded, wash the column with 3 column volumes to ensure any unbound bait protein has been removed.
5. Load the prey protein(s) suspected to associate with the His-tagged protein onto the column using the sample loop at a flow rate of 0.1–0.5 mL/min. The prey protein should be in ten-fold excess over the amount of His_6-bait protein added in **step 4**. The entire amount of prey protein should be loaded in a volume approximately equal to 0.5 column volumes to ensure prey protein is not being forced off the column during loading. After loading, cease the flow rate to allow for a sufficient on-column incubation time of prey and His_6-bait proteins. An incubation time of 5 min can be used for initial experiments; however, shorter or longer incubation times may be desired depending on the nature of the interactions and the specific aims of the research. At this step, the prey protein should be bound to the bait protein if they interact to form a stable complex.
6. Wash the column with 3–5 column volumes of binding buffer to remove prey protein that does not interact with the His_6-bait protein. The flow rate may be increased to 0.5 mL/min. Retain

the wash fractions for further analysis by SDS-PAGE to determine how much prey protein washes off and to ensure that no His_6-bait protein is removed during this step. Depending on the stoichiometry of the interaction and the exact amount of bait/prey proteins used, some prey protein may elute if present in excess. Retaining wash fractions allows for mass balance estimates upon completion of the experiment.

7. Elute all proteins directly or indirectly bound to the HisTrap HP column bypassing 3 column volumes of elution buffer through the column and collecting the fractions. Any protein (s) bound to the His_6-bait protein also should be present in the elution fractions. A gradient of binding buffer and elution buffer may be used if desired (*see* **Note 10**). It is recommended to collect 0.5 mL fractions if possible. The presence of protein in individual fractions can be determined from the UV-visible absorption peaks on the FPLC elution profile. This analysis will allow you to select only those fractions containing protein to carry forward for analysis.

3.3 Analysis of Results from Either Protocol

1. Since the proteins will likely be too dilute for SDS-PAGE in the initial elution fractions, pool and concentrate the relevant elution fractions using the appropriately sized molecular weight cutoff (MWCO) filtration system. Attempt to concentrate elution fractions from different experiments to the same final volume. After concentrating, analyze equal volumes from each experiment by SDS-PAGE or another desired method. The appearance of proteins in the experimental elution fractions that do not appear in the control elution fractions indicates a potential protein–protein interaction. In addition, the concentration of prey protein retained compared to that initially loaded (estimated by SDS-PAGE) may also provide relative affinity/stoichiometry of a bait protein for a prey protein or a specific conformation of a prey protein compared to other experiments. Fig. 2 shows an example SDS-PAGE result using the apo-form of His_6-SufA Fe-S cluster carrier as the bait protein and untagged apo- or holo-$SufBC_2D$ Fe-S scaffold complex as the prey [4]. In this experiment, the holo-$SufBC_2D$ complex interacts much more strongly with its Fe-S cluster trafficking partner His_6-SufA than apo-$SufBC_2D$ and is, therefore, more highly retained in the elution fractions than apo-$SufBC_2D$. This result provides support for a sequential affinity model of protein–protein interactions between these Fe-S cluster biogenesis proteins [5].
2. As a control, load an identical sample of the untagged prey protein(s) on the HisTrap HP column. Wash and elute sample as directed in **steps 8** and **9** (from Subheadings 3.1 and 3.2), collecting fractions at each step. If the prey protein(s) binds the

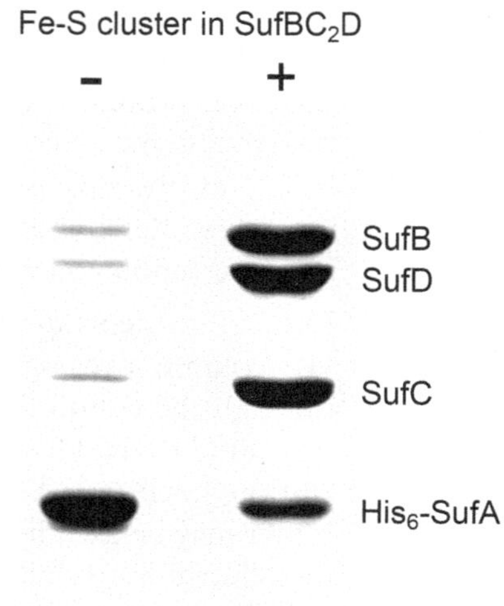

Fig. 2 SDS-PAGE analysis of elution fractions from a protein–protein interaction experiment using this technique. His_6-SufA in the apo-form was used as the bait protein on a 1 mL HisTrap column as described in Methods. Apo- or holo-$SufBC_2D$ scaffold complex (untagged) was used as the prey proteins. The SDS-PAGE analysis shows much greater retention of holo-$SufBC_2D$ than apo-$SufBC_2D$, indicating a stronger interaction between holo-$SufBC_2D$ and apo-SufA than between apo-$SufBC_2D$ and apo-SufA. (Reprinted (adapted) with permission from Chahal HK, Dai Y, Saini A, et al. (2009). The SufBCD Fe-S Scaffold Complex Interacts with SufA for Fe-S Cluster Transfer. Biochemistry 48:10644–10,653. https:/doi.org/10.1021/bi901518y. Copyright 2009 American Chemical Society)

Ni-NTA resin nonspecifically, it will appear in the final elution fractions of the control run rather than the wash fractions. In this case, IMAC may not be appropriate as a technique for analyzing this protein due to its native ability to bind Ni-NTA resin. Alternately, one may increase the concentration of imidazole in the binding buffer to minimize prey protein binding to the Ni-NTA resin while allowing the His_6-bait protein to bind. If apo- and holo-prey proteins are used, this control must be conducted separately on each sample as Fe-S cluster content may influence nonspecific interactions with the column.

3. As a second control, conduct the procedure with His_6-bait protein alone. Retain and analyze fractions at loading, wash, and elution steps by SDS-PAGE to determine if the His_6-bait protein remains stably bound to the column during the loading and wash steps but is efficiently removed from the column at the elution step. If apo- and holo-prey proteins are used, this control must be conducted separately on each sample as Fe-S cluster content may influence nonspecific interactions with the column. One should also analyze the holo-bait protein fractions from this control using UV-visible absorption

spectroscopy to determine if the experimental conditions (especially the imidazole concentrations in binding or elution buffers) may disrupt the Fe-S cluster.

4. Depending on the system, consider additional control experiments. These include testing another apo-Fe-S cluster protein that is not part of the biogenesis pathway as prey to determine the specificity of the interactions or using a tagged version of the holo-prey protein (reconstituted separately) with apo-bait protein to show directionality of cluster transfer (Fig. 1).
5. Even if your bait and prey proteins do not interact on a sufficiently long timescale to allow for their co-purification, this technique may still provide information about the directionality of cluster transfer between them. For example, rapid cluster transfer from the His_6-bait protein to the untagged apo-prey protein that is too fast to allow for co-purification could still be verified by analyzing the cluster content of the holo-prey protein after it elutes off the column in the wash fractions. Time on column and/or ratios of His_6-bait protein/prey protein could be altered to obtain qualitative information on the speed or efficiency of cluster transfer.

4 Notes

1. Preparation of the holo-form of bait or prey protein is often a bottleneck to performing these experiments. Fe-S clusters on scaffold and trafficking proteins can be intrinsically labile to facilitate the movement of the cluster through the trafficking network to target proteins. This property can make it difficult to stably prepare the holo-form of bait or prey proteins. The holo-forms of some proteins may be obtained by purification of the overexpressed protein from a strain that co-expresses a plasmid-borne Fe-S cluster biogenesis pathway [6]. The protein will likely need to be purified under anaerobic conditions to maintain the cluster throughout chromatography steps. In other cases, the protein may need to be purified in its apo-form and reconstituted under anaerobic conditions in vitro, by using iron salts and sulfide salts or by using iron salts with a cysteine desulfurase to provide sulfide from L-cysteine [7]. Detailed protocols for purification or reconstitution of holo-proteins are beyond the scope of this method article. Regardless of how the holo-form is prepared, it must be characterized with UV-visible absorption spectroscopy and other analytical techniques to determine the Fe-S cluster type and content of the protein prior to use in the experiments here [8].
2. The imidazole concentration of binding buffer may vary depending on the optimal binding affinity of your bait

protein(s) of interest. Salt and pH can also be varied to match the conditions required for your specific proteins. Be aware of the negative influence of acidic pH on His_6-tag binding to metal ions due to protonation of imidazole nitrogens in histidine.

3. All buffer solutions should be made with ultrapure water and filtered with a 0.45-micron filter prior to use. Due to the oxygen sensitivity of Fe-S clusters, the buffer solutions should be thoroughly sparged with nitrogen or argon (containing less than 5 ppm O_2) prior to use. Sparging is carried out for at least 60 min at 25 mL/s flow rate using a stoppered bottle with a glass or plastic bubbler submerged in the liquid and a venting hole in the cap to allow gas escape [9]. This method should drop O_2 concentrations below 0.3 ppm in the buffers. Alternatively, buffers can be left uncapped with vigorous stirring in an anaerobic chamber overnight to equilibrate with the N_2 atmosphere.
4. Some chambers may also require an N_2/H_2 gas mix cylinder to supply low levels of H_2 that are needed for oxygen consumption via a catalyst in the chamber.
5. Concentrators may be necessary to concentrate dilute fractions sufficiently to analyze by standard SDS-PAGE. The concentrator should have a molecular weight cutoff (MWCO) less than the size of the smallest bait or prey protein to ensure retention of the protein during concentration. In most cases, a 3 or 10 kDa MWCO is sufficient. Centrifugal concentrators require a centrifuge with a swinging-bucket (preferred) or fixed-angle rotor with wells/carriers that can accommodate 17 mm × 124 mm 15 mL conical-bottomed tubes. Concentration will proceed faster with the higher MWCO values but can take 40–60 min. Examples of concentrators include Amicon® Ultra-4 Centrifugal Filter Unit with 3 kDa MWCO, which will concentrate a maximum fraction volume of 4 mL to a minimum of 50 μL.
6. For proteins of similar molecular weight, use a single percentage gel that results in the separation of your proteins in the lower half of the gel where resolution is highest. For proteins that are 12–45 kDa, an acrylamide percentage of 15% is usually a good starting point. If the molecular weight of the bait and prey proteins differs considerably, use a gradient gel to provide the best separation. If preparing your own "hand cast" gels, be sure to use gloves, lab coat, and eye protection because acrylamide is a potent neurotoxin if exposed repeatedly over time.
7. If you are casting your own bis-Tris gel, use a resolving gel with the appropriate amount of acrylamide for separating bait and prey proteins diluted into concentrated Bis-Tris buffer and

ultrapure water to obtain a final 330 mM Bis-Tris (pH 6.5). The Stacking gel should be prepared similarly, except the acrylamide concentration is always 5%. For simplicity, a 3× Bis-Tris buffer (1.0 M Bis-Tris, pH 6.5 with HCl) can be used as a stock to dilute for gel casting. Polymerization of both gels is initiated with fresh ammonium persulfate and TEMED.

8. The exact procedure for filling and using the glove bag will depend on the manufacturer and method of sealing the bag. It is recommended that an outlet for gas be added to the anaerobic bag if not already present. This can simply be a small hole cut into the bag with a hollow plastic pipette (snapped in half) as the outlet. The outlet can be sealed around the pipet with several layers of vinyl tape or other gas impermeable tape. Having an outlet allows for continuous N_2 flow into the bag without over-inflating or popping the bag. Connect N_2 tank with regulator and line to the deflated glove bag that has been sealed. Slowly fill the bag with N_2; adjust the flow rate, so the bag stays partially inflated even with the outlet open (input flow is slightly greater than output flow). Do not over-inflate the bag, as that will make it difficult to extend your arms through the built-in gloves to reach the reagents.
9. The concentration of total protein in elution fractions can be quickly estimated on a small sample of the fraction using the Bradford assay with Coomassie Brilliant Blue G-250 and bovine serum albumin as the protein standard. Alternatively, the UV-visible absorption at 280 nm can also be used if molar extinction coefficients for the proteins are known.
10. An elution gradient can be utilized to determine if specific subcomplexes might be forming on the column. For example, if a His_6-bait protein is homodimeric, it will contain two His_6-motifs. This homodimer may require more imidazole to elute than a heterodimer of the His_6-bait protein with untagged prey protein, which would have only a single His_6-motif. If a simple step elution from low to high imidazole is used, such subcomplexes would all elute together. In contrast, the use of a gradient may allow them to be individually resolved based on their relative affinity for the Ni-NTA resin.

Acknowledgments

This work was supported by the National Institutes of Health grant GM112919 to F.W.O. and by the National Institutes of Health grant GM076277 to T.D.C.

References

1. Wachnowsky C, Fidai I, Cowan JA (2018) Iron-sulfur cluster biosynthesis and trafficking—impact on human disease conditions. Metallomics 10:9–29. https://doi.org/10.1039/c7mt00180k
2. Imlay JA (2006) Iron-Sulphur clusters and the problem with oxygen. Mol Microbiol 59:1073–1082. https://doi.org/10.1111/j.1365-2958.2006.05028.x
3. Block H, Maertens B, Spriestersbach A et al (2009) Immobilized-metal affinity chromatography (IMAC): a review. Meth Enzymol 463:439–473. https://doi.org/10.1016/S0076-6879(09)63027-5
4. Chahal HK, Dai Y, Saini A et al (2009) The SufBCD Fe–S scaffold complex interacts with SufA for Fe–S cluster transfer. Biochemistry 48:10644–10653. https://doi.org/10.1021/bi901518y
5. Blahut M, Sanchez E, Fisher CE, Outten FW (2020) Fe-S cluster biogenesis by the bacterial Suf pathway. Biochim Biophys Acta, Mol Cell Res 1867:118829. https://doi.org/10.1016/j.bbamcr.2020.118829
6. Corless EI, Mettert EL, Kiley PJ, Antony E (2020) Elevated expression of a functional Suf pathway in *Escherichia coli* BL21(DE3) enhances recombinant production of an iron-sulfur cluster-containing protein. J Bacteriol 202. https://doi.org/10.1128/JB.00496-19
7. Yan A, Kiley PJ (2009) Techniques to isolate O_2-sensitive proteins: [4Fe-4S]-FNR as an example. Meth Enzymol 463:787–805. https://doi.org/10.1016/S0076-6879(09)63042-1
8. Freibert S-A, Weiler BD, Bill E et al (2018) Biochemical reconstitution and spectroscopic analysis of iron-sulfur proteins. Meth Enzymol 599:197–226. https://doi.org/10.1016/bs.mie.2017.11.034
9. Butler IB, Schoonen MAA, Rickard DT (1994) Removal of dissolved oxygen from water: a comparison of four common techniques. Talanta 41:211–215. https://doi.org/10.1016/0039-9140(94)80110-X

Chapter 8

tRNA Modifications as a Readout of S and Fe-S Metabolism

Ashley M. Edwards, Maame A. Addo, and Patricia C. Dos Santos

Abstract

Iron-Sulfur (Fe-S) clusters function as core prosthetic groups known to modulate the activity of metalloenzymes, act as trafficking vehicles for biological iron and sulfur, and participate in several intersecting metabolic pathways. The formation of these clusters is initiated by a class of enzymes called cysteine desulfurases, whose primary function is to shuttle sulfur from the amino acid L-cysteine to a variety of sulfur transfer proteins involved in Fe-S cluster synthesis as well as in the synthesis of other thiocofactors. Thus, sulfur and Fe-S cluster metabolism are connected through shared enzyme intermediates, and defects in their associated pathways cause a myriad of pleiotropic phenotypes, which are difficult to dissect. Post-transcriptionally modified transfer RNA (tRNA) represents a large class of analytes whose synthesis often requires the coordinated participation of sulfur transfer and Fe-S enzymes. Therefore, these molecules can be used as biologically relevant readouts for cellular Fe and S status. Methods employing LC-MS technology provide a valuable experimental tool to determine the relative levels of tRNA modification in biological samples and, consequently, to assess genetic, nutritional, and environmental factors modulating reactions dependent on Fe-S clusters. Herein, we describe a robust method for extracting RNA and analytically evaluating the degree of Fe-S-dependent and -independent tRNA modifications via an LC-MS platform.

Key words tRNA, Bacteria, Iron-sulfur cluster, Sulfur metabolism, LC-MS, RNA extraction, tRNA modification, Thionucleosides

1 Introduction

For nearly six decades, the study of biological Fe-S clusters has captured the interest of the biochemical and bioinorganic scientific community [1]. When associated with proteins, these inorganic cofactors play indispensable roles in central metabolism, gene regulation, DNA repair, respiration, and translation [2, 3]. Due to the essential nature of these molecules, it is therefore not surprising that Fe-S clusters are found throughout all domains of life and remain an active topic of exploration for their functionality and biogenesis [4–6]. Different pathways have been identified for the synthesis of Fe-S clusters. Despite their differences in biosynthetic components, all systems reported thus far include a consortium of proteins dedicated to the recruitment and assembly of individual Fe

Patricia C. Dos Santos (ed.), *Fe-S Proteins: Methods and Protocols*, Methods in Molecular Biology, vol. 2353,
https://doi.org/10.1007/978-1-0716-1605-5_8,

and S components into clusters, followed by transfer to final acceptor proteins. In bacteria, the iron-sulfur cluster (ISC), sulfur mobilization (SUF), and nitrogen fixation (NIF) systems serve as three distinct biosynthetic routes for Fe-S cluster assembly [7–9]. A conserved feature of these systems is the presence of a cysteine desulfurase involved in the abstraction and passage of sulfur from the amino acid L-cysteine to downstream components involved in sulfur and Fe-S cluster mobilization pathways [10, 11]. Thus, defects during the sulfur mobilization step not only affect the activity of Fe-S enzymes, but also impair the synthesis of several S-containing metabolites [12, 13].

The functionality of sulfur and Fe-S cluster pathways affects the efficiency and accuracy of translational processes through modulation of tRNA structure and function [14, 15]. Post-transcriptional modifications are installed at multiple locations on the tRNA structure by specialized tRNA-modifying enzymes. These modifications expand their reactivity while simultaneously promoting proper protein-RNA recognition events and aminoacylation [16, 17]. To date, over 100 modifications to tRNA have been described [18]. The synthesis and functionalities of modified tRNAs are known to be affected by environmental and nutritional factors [19]. Likewise, alterations to sulfur and Fe-S metabolism are reflected in the overall tRNA epitranscriptome through direct and indirect mechanisms [20].

Relevant to the metabolism of sulfur and Fe-S clusters are the modifications listed in Table 1 and Fig. 1. Analysis of these modifications can serve as an in vivo readout of the functionality of their corresponding pathways. Modifications found in the anticodon domain of tRNA are of particular importance due to their role during translation in preventing frameshifting and maintaining accurate codon-anticodon recognition [21]. An example of this type of chemical alteration is the 2-methylthio-N^6-isopentenyladenosine (ms^2i^6A) s-tRNA modification found on A37. Iron starvation or oxidative stress leads to decreased levels of this modification. Loss of ms^2i^6A is attributed to defects in Fe-S cluster synthesis caused by adverse conditions that inactivate MiaB, an Fe-S SAM--dependent methylthiotransferase involved in ms^2i^6A formation [22]. MiaB is a bifunctional enzyme that uses its Fe-S centers along with AdoMet to thiolate and methylate its substrate tRNA, N^6-isopentenyladenosine (i^6A), through a radical-dependent mechanism [23]. In Salmonella species, MiaB performs identical methylthiolation reactions on the same A37 base, but is able to use the hydroxylated form of i^6A (io^6A) as a target substrate [24]. This substrate promiscuity sets it apart from its *E. coli* ortholog in that *E. coli* does not have the hydroxylated precursor nucleoside. The *Bacillus subtilis* MiaB ortholog, MtaB, is also a SAM-dependent Fe-S enzyme and performs similar functions to MiaB [25, 26]. However, *B. subtilis* MtaB uses N^6-threonyl-

Table 1
tRNA modifications whose synthesis is either dependent on FeS proteins or independent of FeS processes in the Gram-positive model organism, *Bacillus subtilis*, and Gram-negative *Escherichia coli*. The latter half of the table entitled "Fe-S-Independent tRNA Modifications" highlights commonly used internal standards comprising highly abundant and stable nucleosides

Fe-S-Dependent tRNA Modifications			
Position	**Modification**	**Observed Molecular Ion (*m/z+*)**	**Biosynthetic Enzyme and Cluster Type**
34	oQ	426.1620	QueE (*B. subtilis* & *E. coli*); [4Fe-4S]
34	Q	410.1670	QueG (*B. subtilis* & *E. coli*); (2) [4Fe-4S]
37	ms^2i^6A	382.1544	MiaB (*B. subtilis* & *E. coli*); (2) [4Fe-4S]
37	m^2A[a]	282.1197	RlmN (*B. subtilis* & *E. coli*); (2) [2Fe-2S]
32	s^2C	260.0699	TtcA (*E. coli*); (2) [4Fe-4S]
37	ms^2t^6A/ms^2ct^6A	459.1293/440.1235	MtaB (*B. subtilis*); (2) [4Fe-4S]

Fe-S-Independent tRNA Modifications			
Position	**Modification**	**Observed Molecular Ion (*m/z+*)**	**Biosynthetic Enzyme and Cluster Type**
55[b]	ψ	245.0767	TruB (*B. subtilis* & *E. coli*)
17[c]	D	247.0925	Dus1 or Dus2/DusA (*B. subtilis* & *E. coli*)
34	I	269.0881	TadA (*B. subtilis* & *E. coli*)
37	t^6A	413.1416	TsaBCDE (*B. subtilis* & *E. coli*)
37	i^6A	336.1667	MiaA (*B. subtilis* & *E. coli*)
37	m^6A	282.1197	TrmN6 (*B. subtilis* & *E. coli*)

[a] m^2A is also found in 23S rRNA and modified by the same biosynthetic enzyme, RlmN

[b] Pseudouridine is also found at positions 32, 38, 39, 40 in *B. subtilis* and *E. coli*, and additionally at positions 13 and 35 in *E. coli*

[c] Dihydrouridine is also found at positions 20 and 20a in *B. subtilis* and *E. coli*, at position 47 in *B. subtilis* only, and at position 16 in *E. coli* only

carbamoyladenosine (t^6A) as a distinct substrate. Interestingly, t^6A's cyclized form, cyclic-N^6-threonyl-carbamoyladenosine (ct^6A), and its hypermodified form, 2-methylthiocyclic-N^6-threonyl-carbamoyladenosine (ms^2ct^6A), have also been reported in this organism [27].

2-thiocytidine (s^2C) found at position C32 of $tRNA^{Arg}$ and $tRNA^{Ser}$ represents another sulfur-containing modification that depends on an Fe-S tRNA- modifying enzyme [28]. The Gram-negative bacterial species, *E. coli* and *Salmonella enterica* serovar

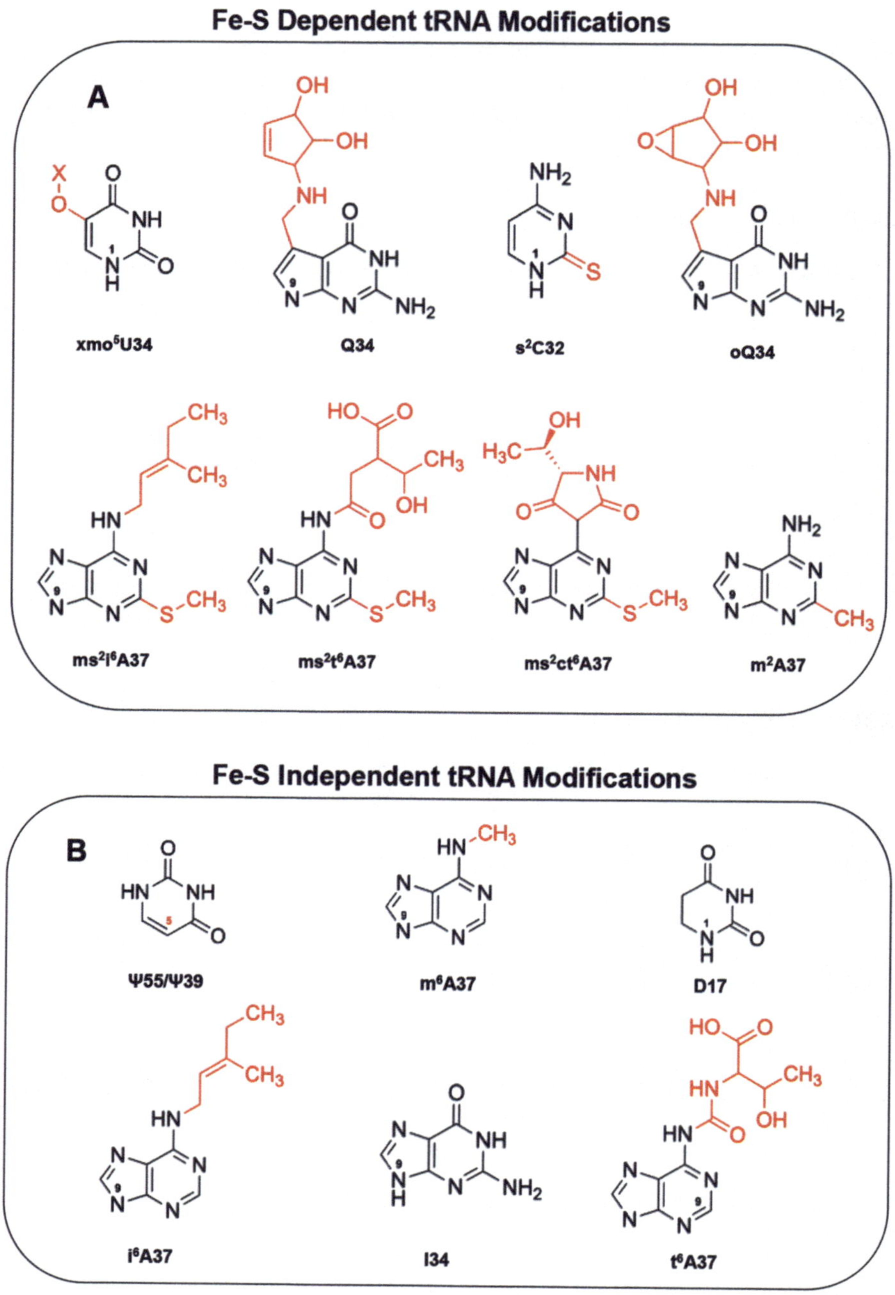

Fig. 1 Structures of modified bacterial tRNA and derivatives. (**a**) Structures of modified tRNAs found at positions 32, 34, and 37 whose synthesis is dependent on Fe-S proteins, and (**b**) structures of modified tRNAs found at positions 17[a], 34, 37, and 55[b] whose synthesis is independent of Fe-S proteins are shown with their corresponding modified/hypermodified moieties indicated in red. The numbers "1" and "9" found within each base denote the β-glycosidic bond orientation found in the pyrimidine and purine bases, with their adjacent riboses in tRNA. (**a**) Pseudouridine is also found at positions 32, 38, 39, 40 in *B. subtilis* and *E. coli*, and additionally at positions 13 and 35 in *E. coli*. (**b**) Dihydrouridine is also found at positions 20 and 20a in *B. subtilis* and *E. coli*, at position 47 in *B. subtilis* only, and at position 16 in *E. coli* only

typhimurium, use the [4Fe-4S] tRNA thioltransferase TtcA for the synthesis of s^2C . This unique enzyme contains a PP-loop known for its ATPase activity and six highly conserved cysteine residues, four of which are bundled into CysXXCys motifs. Possibly one of the most intriguing characteristics of TtcA is that it was the first Fe-S enzyme of its kind demonstrated to thiolate its tRNA substrates in a nonradical catalyzed mechanism [29, 30]. Other thiolating enzymes that share common PP-loop motif characteristics with TtcA include the 4-thiouridine (s^4U) modifying enzyme, ThiI, and the 2-thiouridine (s^2U) thiouridylase, MnmA [31–35]. In sharp contrast, ThiI's and MnmA's functionality is not contingent on the presence of an Fe-S cluster biosynthetic apparatus for thiolation and catalysis and additionally contain no CysXXCys motifs in the majority of their sequences. However, it was recently shown that a few archaeal and extreme thermophilic bacterial species possess an MnmA or a ThiI that requires an Fe-S cluster for their tRNA-modifying activities. *Thermus thermophilus* MnmA bears conserved cysteines, which are canonically found in Fe-S proteins. These cysteines coordinate an oxygen-sensitive [4Fe-4S] cluster that is required for catalytic activity [36]. Similarly, in *Methanococcus maripaludis*, ThiI has a redox-active [3Fe-4S] cluster that is also required for the tRNA thiouridylation reaction [37]. Surprisingly, like several methanogenic species, *M. maripaludis*' genome does not code for a cysteine desulfurase, and the formation of thionucleosides is thought to use sulfide as a sulfur source [37, 38].

A distinct subset of tRNA modifications are dependent on Fe-S tRNA-modifying enzymes but do not require sulfur to be incorporated into their structures. For instance, in *E. coli*, the Fe-S SAM-dependent enzyme RlmN implements the 2-methyladenosine (m^2A) base modification onto messenger RNA (mRNA) as well as A37 in tRNA, making this enzyme unique for its dual substrate specificity [39]. At the neighboring G34 position of tRNAs, the 7-deaza-guanosine hypermodification known as queuosine (Q) possesses an elaborate and highly coordinated biosynthetic pathway. In both *B. subtilis* and *E. coli*, the 7-carboxy-7-deazaguanine synthase, QueE, uses a [4Fe-4S] cluster along with AdoMet to catalyze the heterocyclic, radical-mediated conversion of 6-carboxy-5,6,7,8-tetrahydropterin (CPH4) to 7-carboxy-7-deazaguanine (CDG), a highly conserved biosynthetic step intrinsic to all compounds containing 7-deazapurine [40]. In the same queuosinilation pathway, these organisms also employ the epoxide reductase, QueG, to complete the last Q synthesis step. In this scheme, QueG uses a redox chain of two [4Fe-4S] clusters to reduce the epoxide found on the Q-tRNA precursor, epoxyqueuosine (oQ), and form the cyclopentediol product Q [41, 42]. Lastly, the inventory of Fe-S-dependent modifications has been recently expanded to include the hydrouridine (hoU) tRNA modification in a pathway that depends on a specialized ferredoxin in *B. subtilis* and *E. coli* [43].

The type and number of Fe-S clusters vary substantially in these tRNA-modifying Fe-S proteins. Characterization of these proteins has been extensively undertaken using a combination of genetic, biochemical, and spectroscopic techniques, in vitro Fe-S cluster assays, and Fe/sulfide analyses. However, integrating this information into Fe-S protein pathway functionality also requires a broad, in vivo assessment. Analysis of tRNA modifications provides a valuable method for identifying factors that affect Fe-S and/or S metabolism at the organismal level. LC-MS-mediated analysis of total tRNA isolated from an organism of interest allows for the determination of the composition and identity of modified tRNA nucleosides. The quantification of tRNA modifications that depend on Fe-S modifying enzymes provides a unique avenue for monitoring Fe-S cluster metabolism in the cell.

Epitranscriptomic analyses can also provide a readout on the functionality of pathways involving sulfur transfer reactions, specifically those that do not employ Fe-S enzymes in their biosynthetic pathways. Thus, this approach permits deconvolution of thio-modifications that are dependent and independent of Fe-S clusters. The abundance of these modifications can then be normalized to the levels of modifications that are independent of Fe-S and S metabolism, such as dihydrouridine (D), inosine (I), pseudouridine (ψ), and many others [14] (Table 1). Dihydrouridine is particularly useful as a normalizing tool due to its consistency, high abundance, and near exclusivity in bacterial tRNA [44]. The necessity of this modified nucleotide is further highlighted when attempting to quantify low abundance tRNA modifications. For instance, in most organisms, the C34 lysidine (k^2C) modification is present in a single $tRNA^{Ile}$ isoacceptor bearing the CUA anticodon sequence [45]. Since this modification is exceptionally rare, normalizing k^2C abundance to its highly abundant unmodified cytidine precursor base would grossly underrepresent lysidine's physiological prevalency in tRNA. Additionally, cytidine is also present in ribosomal RNA (rRNA) and mRNA, further diluting the significance of the tRNA-specific k^2C when using total RNA during these types of analyses. Factors like these reiterate the importance of using moderately abundant normalizing factors such as dihydrouridine to get an accurate representation of modification levels in the overall pool of cellular tRNA.

Cultures exposed to varying stressors known to affect Fe-S protein competency impact the cellular tRNA pool composition and give insight into interlinked metabolic processes that affect translation and protein synthesis [13, 19, 46, 47]. The methods described in this chapter enable the use of tRNA modifications as a readout of Fe-S protein activity as well as proteins involved in sulfur metabolism. This methodology allows for a comprehensive understanding of how Fe-S proteins, specifically those that modify tRNAs, function at any given time under a range of environmental and nutritional conditions.

2 Materials

2.1 Growth of Cells for Total RNA Extraction

1. LB broth: 12.5 g of Luria Bertani (LB) medium and 500 mL of water autoclaved. Stored at room temperature (*see* **Note 1**).
2. LB agar: 12.5 g of LB medium and add 7.5 g agar; add 500 mL of water and autoclave. Aseptically, pour roughly 20 mL of autoclaved LB agar into sterile Petri dishes and store them at 4 °C.
3. *Bacillus subtilis* PS832 strain (*see* **Note 2**).

2.2 Saturation of Phenol with NaOAc pH 4.3

1. Phenol, 95.5%, extra pure, solid (*see* **Note 3**).
2. NaOAc buffer: 1 M sodium acetate (NaOAc), 0.26 M sodium chloride (NaCl) (*see* **Note 4**), pH 4.3. Stock solutions should be filtered with 0.45 μM mixed cellulose ester (MCE) membrane filter.
3. 500 mL separatory funnel.
4. Chemical fume hood.
5. MilliQ water.

2.3 Total RNA Extraction

1. Cell resuspension buffer: 300 mM sodium acetate, 190 mM NaCl, pH 4.5, 10 mM ethylenediaminetetraacetic acid (EDTA) (*see* **Notes 4** and **5**). NaOAc and EDTA stock solutions should be filtered with 0.45 μM MCE membrane filter before being combined for the final buffer solution.
2. NaOAc-saturated phenol, pH 4.3.
3. Chloroform-isoamyl alcohol mix (IAA) (24:1).
4. Pure ethanol (EtOH), 200 proof, chilled in −20 °C freezer.
5. 70% EtOH, chilled in −20 °C freezer.
6. High-speed polypropylene copolymer (PPCO) 50 mL centrifuge tubes.
7. 6 M ammonium acetate (NH_4OAc), pH 5.3.
8. Water, Optima LC-MS grade.

2.4 Digestion of Total RNA and LC-MS Sample Preparation

1. Nuclease P1: 250 U/mL Nuclease P1 from *Penicillium citrinum*, resuspended from the lyophilized form in Water, Optima LC-MS grade; stored as 50 μL aliquots in −80 °C freezer.
2. rSAP: 1000 U/mL Shrimp Alkaline phosphatase.

3. Nuclease P1 digestion buffer: 400 mM NH_4OAc (HPLC grade); 200 mM $ZnCl_2$, pH 5.3.
4. CutSmart Digestion Buffer: 50 mM Potassium Acetate, 20 mM tris (hydroxymethyl) aminomethane (Tris)-acetate, pH 7.9; 10 mM Magnesium Acetate; 100 μg/mL Bovine Serum Albumin (BSA).
5. 100 mM NH_4HCO_3 (*see* **Note 6**).
6. Formic acid 99.0 + %, Optima LC-MS grade.
7. Water, Optima LC-MS grade.
8. Blue screw-top LC-MS sample vials with 200 μL glass inserts.
9. 80:4 MeOH:FA: 80% mixture of methanol (MeOH), Optima, LC-MS grade containing 4% formic acid (FA) 99.0 + % Optima LC-MS grade and 16% water, Optima, LC-MS grade.

2.5 LC-MS Method Run, Mobile Phases, and Data Analysis

1. Solvent A: Water with 0.1% formic acid, Optima LC-MS grade.
2. Solvent B: Methanol and 0.1% formic acid, Optima LC-MS grade.
3. Formic acid 99.0 + %, Optima LC-MS grade.
4. LC-18 analytical column, 3 μm particle size, 100 mm column length, 4.6 mm internal diameter.
5. LC-MS: HPLC instrument coupled to a linear ion trap mass spectrometer.
6. Data analysis software (*see* **Note 7**).

3 Methods

Assessing the function of Fe-S proteins in vivo gives an insightful look into cellular redox status, degree of oxidative stress, and bioavailability of core elements such as iron and sulfur at the organismal level [4, 48]. In the past few decades, LC-MS technology has been used to analyze the tRNAome and closely monitor changes in its composition and degree of modifications. These changes can act as signals of depleted nutritional status or markers of oxidative stress and cellular competency. Fluctuating modification levels also serve as direct/indirect readouts of upstream Fe-S protein activity, as many tRNA-modifying enzymes utilize Fe-S clusters for their activities. Challenging growth cultures with varying stressors such as nutrient limitation or peroxide stress allows for the evaluation of tRNA modification levels and the linking of those patterns to Fe-S protein expression levels and activity profiles. These analyses also provide information about Fe-S cluster independent modifications, which serve as internal controls in these biological samples.

The following methods outline the procedures used for cellular growth, total RNA extraction, sample preparation, and LC-MS sample processing/analysis. The levels and types of modification vary across species and growth conditions and are growth phase-dependent. Therefore, it is important to maintain the same experimental conditions and harvest cells at the same cell density to obtain meaningful results. Below we describe the workflow for isolation and analysis of tRNA modified nucleosides from *B. subtilis* cultures. We have successfully implemented these methods using bacterial species such as *B. subtilis, E. coli, Azotobacter vinelandii,* and *Borrelia burgdorferi.* This methodology may be successfully extended to other organisms.

3.1 Growth of Cells for Total RNA Extraction

1. Using a sterile technique, streak the desired bacterial strain onto an LB agar plate and incubate at 37 °C to obtain isolated colonies.
2. Inoculate 250 mL of sterile LB liquid medium (*see* **Note 8**) with a single colony from overnight LB agar plate and place it in a 37 °C shaking incubator until OD_{600} ~ 1.0 is reached (mid-exponential growth phase).
3. Dilute the starting culture in 500 mL fresh LB liquid medium to OD_{600} 0.01, and place in 37 °C shaking incubator until OD_{600} 0.4–0.6 is reached (early exponential growth phase).
4. Centrifuge 500 mL of culture at 6744 × *g* for 10 min at 4 °C, decanting supernatant in between spins; store pellet in −80 °C freezer for up to 1 year.

3.2 Saturation of Phenol with NaOAc pH 4.3

1. Melt 100 g bottle of phenol by loosening cap and placing the bottle in a 65 °C water bath. Phenol should be completely melted in 20–30 min (*see* **Notes 9** and **10**).
2. Transfer melted phenol to separatory funnel in a fume hood while phenol is still hot. Make sure the valve on the funnel is in the horizontal closed position.
3. Add 100 mL of NaOAc buffer to liquid phenol, close the funnel with the appropriate ground glass stopper, carefully invert the separatory funnel, and shake vigorously to mix phases. Stop and vent every four to five inversions.
4. Allow phases to separate in a ring clamp. This step may take 6–8 h at room temperature. At that point, the bottom organic phase (phenol) should be clear if full separation is achieved.
5. Place a clean glass beaker under the opening of the separatory funnel, then open the valve to the vertical open position, and collect the bottom organic layer. Stop collecting when the interphase, or the murky layer found between two phases, is reached. Discard the upper aqueous layer.

6. Return the bottom organic layer to the funnel, add 100 mL of NaOAc buffer, and repeat **steps 4** and **5** for two additional times.
7. Transfer the bottom organic layer to a brown, light-resistant glass container and store at 4 °C for up to 6 months (*see* **Note 11**).

3.3 Total RNA Extraction

The following protocol uses manually presaturated acidic phenol with chloroform/IAA for cell disruption and subsequent RNA extraction and EtOH precipitation (*see* **Note 12**). When using any phenol/chloroform/IAA-containing reagents, all experiments must be performed in a well-ventilated fume hood. The appropriate gloves, eye protection, and lab coats should be used.

1. Thaw a bacterial cell pellet and add 10 mL of cell resuspension buffer. Mix until resuspension is homogenous and transfer to a 50-mL high-speed PPCO centrifuge tube.
2. Add 3.5 mL of NaOAc-saturated phenol (bottom layer) to the 10 mL resuspension, and vortex vigorously for 60 s, followed by a 30-s rest and resume with a final 90 s of vortexing.
3. Centrifuge the mixture at 33,745 × *g* for 15 min at 25 °C.
4. Carefully remove the spun tubes without disturbing extracted layers. Remove the top layer (aqueous layer) and transfer it to a clean 50-mL high-speed PPCO tube.
5. Extract again by adding 3.5 mL of NaOAc-saturated phenol to the aqueous layer and vortex for 60 s.
6. Centrifuge the mixture at 33,745 × *g* for 15 min at 25 °C.
7. Carefully remove the spun tubes from the centrifuge, remove the top aqueous layer, and transfer to a clean 50-mL high-speed PPCO tube.
8. Add 0.5 g of NaCl and 10 mL of 24:1 chloroform/IAA mixture to aqueous layer and vortex for 10 s.
9. Centrifuge the mixture at 33,745 × *g* for 3 min at 25 °C.
10. Remove the top layer (take care to not disturb the grainy, white interphase) and transfer to a clean 50-mL high-speed PPCO tube.
11. Fill the 50 mL tube containing aqueous layer with ice-cold, 200 proof, pure EtOH.
12. Allow the RNA to precipitate for 4 h to overnight in a −20 °C freezer.
13. After precipitation, centrifuge tubes at 33,745 × *g* for 25 min at 4 °C.
14. Carefully remove the tubes from the centrifuge and gently decant the liquid EtOH from the tubes. At this point, there

should be a distinct pellet at the bottom that can be easily dislodged during decanting, so care should be taken to prevent this occurrence.

15. Add 10 mL of chilled 70% EtOH to the pellet along with 166 μL of 6 M NH_4OAc, pH 5.3 (final concentration of 100 mM), and vortex for 15 s.
16. Centrifuge tubes at 33,745 × *g* for 30 min at 4 °C.
17. Carefully remove the tubes from the centrifuge and gently decant the liquid without disturbing the pellet.
18. Repeat **steps 15** and **16** *without* the addition of the 100 mM NH_4OAc, pH 5.3.
19. Carefully remove the tubes from the centrifuge and gently decant the liquid without disturbing the pellet.
20. Invert tubes in a rack over a dry surface and allow them to air-dry for 30–45 min, or until pellets are dry and no residual EtOH exists on the sides of the tubes (*see* **Note 13**).
21. Resuspend RNA pellet in 750 μL of Optima, LC-MS grade water, and store in −80 °C freezer for further use.

3.4 Digestion of Total RNA and LC-MS Sample Preparation

1. Perform a 1:10 RNA/water dilution and measure the absorbance of the RNA sample at 260 nm. Calculate the concentration of RNA using an extinction coefficient of 0.025 (μg/mL)$^{-1}$ cm^{-1}.
2. Aliquot 40 μg of total RNA into a 0.6-mL high-binding microfuge tubes.
3. Heat denature for 5 min in a 95 °C heating block.
4. Perform a quick spin to pull denatured RNA to the bottom of the tube and allow it to come to room temperature before adding any digestion components.
5. Add Optima, LC-MS grade water to cooled RNA up to 150 μL.
6. Add 4.1 μL of nuclease P1 digestion buffer and 5 μL of 0.25 U/ μL nuclease P1 and incubate at 50 °C for 4 h.
7. Perform a quick spin to collect evaporated digestion products in the tube.
8. Add 20 μL of CutSmart buffer, 20 μL of 100 mM NH_4HCO_3, and 10 μL of the rSAP enzyme, and incubate at 37 °C for 2 h.
9. Spin digested RNA for 10 min at 16,873 × *g* at room temperature (*see* **Note 14**).
10. Take 196 μL of RNA supernatant and transfer to the new 0.6 mL microfuge tube.
11. Add 5 μL of freshly prepared 80:4 MeOH/FA mixture to the supernatant.

12. Transfer MeOH/FA-spiked mixture to glass 200 μL LC-MS vial insert and place inside LC-MS sample vial for subsequent analysis.

3.5 LC-MS Method Run and Tune Parameters

1. Perform all nucleoside analyses on an LC-MS using the following column equilibration prior to experimental sample run: 98% Solvent A and 2% Solvent B for 15 column volumes (45 min) at 300 μL min^{-1} flow rate.
2. Adjust the tune file to the following settings: Voltage (kV): 4.01; Sheath Gas Flow Rate (arb): 47.00; Auxiliary Gas Flow Rate (arb): 30.00; Sweep Gas Flow Rate (arb): 0.00; Capillary Voltage (V): 2.00; Capillary Temperature (C): 350.00; Tube Lens Voltage (V): 49.89.
3. Over the course of a 39-min run, set up the gradient of solvent B as follows: 0–4 min, hold at 2% B (v/v); 4–25 min, increase B from 2% to 100%; 25–33 min, hold at 100% B; 33–39 min switch from 100% B to 2% B to allow column reequilibration at 2% B.

3.6 LC-MS Data Analysis Using MestReNova

1. The Fourier transformed m/z data from each experimental run will be saved in a *.raw* file format; transfer these files from the computer that controls the OrbiTrap to your personal/shared device with a copy of MestReNova data analysis software along with the appropriate license files.
2. Open the MestReNova application, then open your *.raw* file. This command will open with two screens; one is a total ion chromatogram (TIC), and the other is a mass spectrum (MS+ Spectrum).
3. In the top panel of the MestReNova application, click on the "Show Mass Browser" icon (Fig. 2a arrow), and a mass browser panel will be displayed on the left side of the screen. Uncheck the "TIC" panel (Fig. 2b arrow).
4. In the mass browser, click in the white space next to the dropdown boxes, and then select the "open new chromatograph" icon, shown as a red chromatogram (Fig. 2c, d arrows).
5. In the new window, select the "Mass" option. In the dropdown box, add the mass over charge ratio (m/z) calculated from the monoisotopic mass of the targeted modification of interest with four decimal places. Lastly, change the "tolerance" from "Da" to "PPM" in the dropdown menu, and set the threshold to five parts per million (PPM) (Fig. 3a, b arrows).
6. The new panel will display the *extracted ion chromatogram* (EIC) (Fig. 4). This chromatogram shows the profile of species containing the selected m/z within five PPMs with associated retention times.

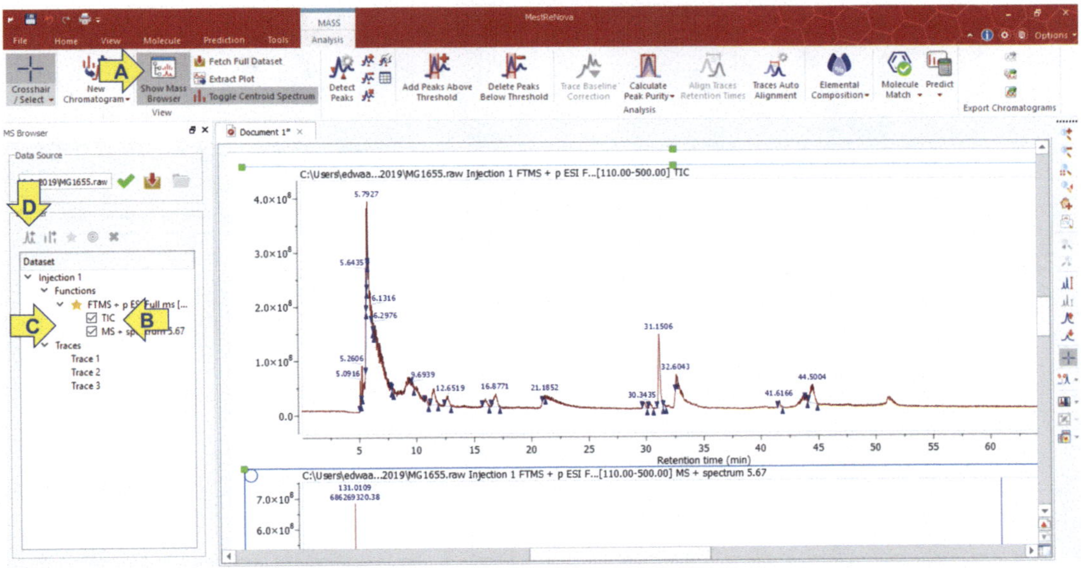

Fig. 2 The layout of MestreNova application displaying a total ion chromatograph of a 70-min LC-MS run. Arrows indicate the application commands for selecting the mass browser (**a**), TIC panel (**b**), new chromatogram (**c**), and a new window (**d**)

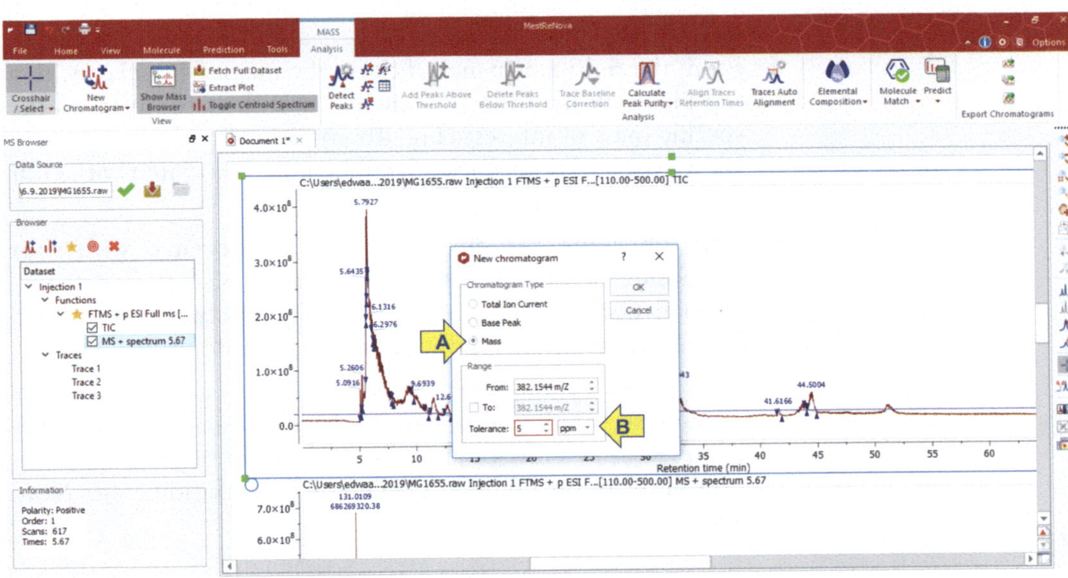

Fig. 3 Layout of MestreNova application when mass browsing to produce an extracted ion chromatogram (EIC). Arrows indicate commands for selecting type of chromatogram (**a**) and tolerance range (**b**)

7. Select the blue crosshair icon located on the right panel of icons and drag the crosshair from one side of the peak of interest to the other side, indicated by up/down faced blue triangles on EIC (Fig. 4a arrow). The "MS + Spectrum" of the highlighted

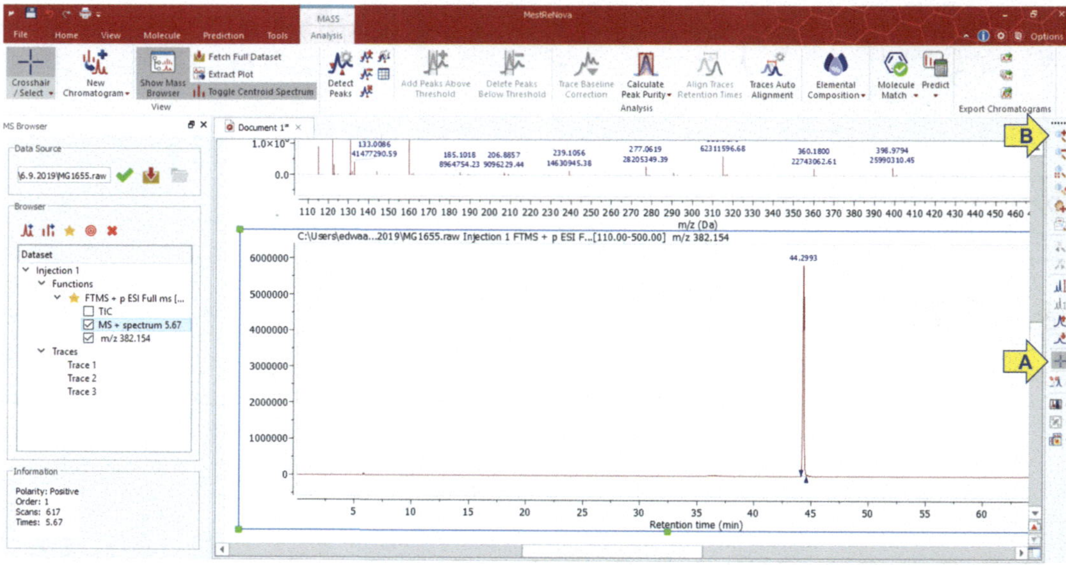

Fig. 4 Extracted ion chromatogram of 382.1544 m/z and the corresponding retention times of all analytes that fall within 5 PPM of this m/z of interest. Arrows indicate commands for selecting the range (**a**) and zoom in (**b**)

peak will be displayed in the top panel with the corresponding range of masses.

8. From the right-side panel, click on the magnifying glass with a red plus sign icon for "Zoom In," and then zoom in on the specific mass of interest (Fig. 4b arrow). The zoom in feature allows visualization of the spectral counts associated with your m/z of interest (Fig. 5).

9. After obtaining the quantitative value associated with your nucleoside of interest, **steps 3–8** should be repeated for an internal normalizing factor that is highly abundant, robust, and chemically stable (dihydrouridine, pseudouridine, etc.).

10. After gathering the values corresponding to all other analytes of interest, use the value associated with your chosen normalizing factor as a divisor and obtain the ratio of your modification in relation to dihydrouridine or pseudouridine (*see* **Note 15**).

4 Notes

1. Our lab uses 500 mL cultures for these analyses, but successful extraction and LC-MS experiments have been performed using culture volumes as small as 5 mL.

2. Our lab chose to use *B. subtilis* PS832 to perform the majority of our analyses. However, this method can be applied to other bacterial species and produce comparable results.

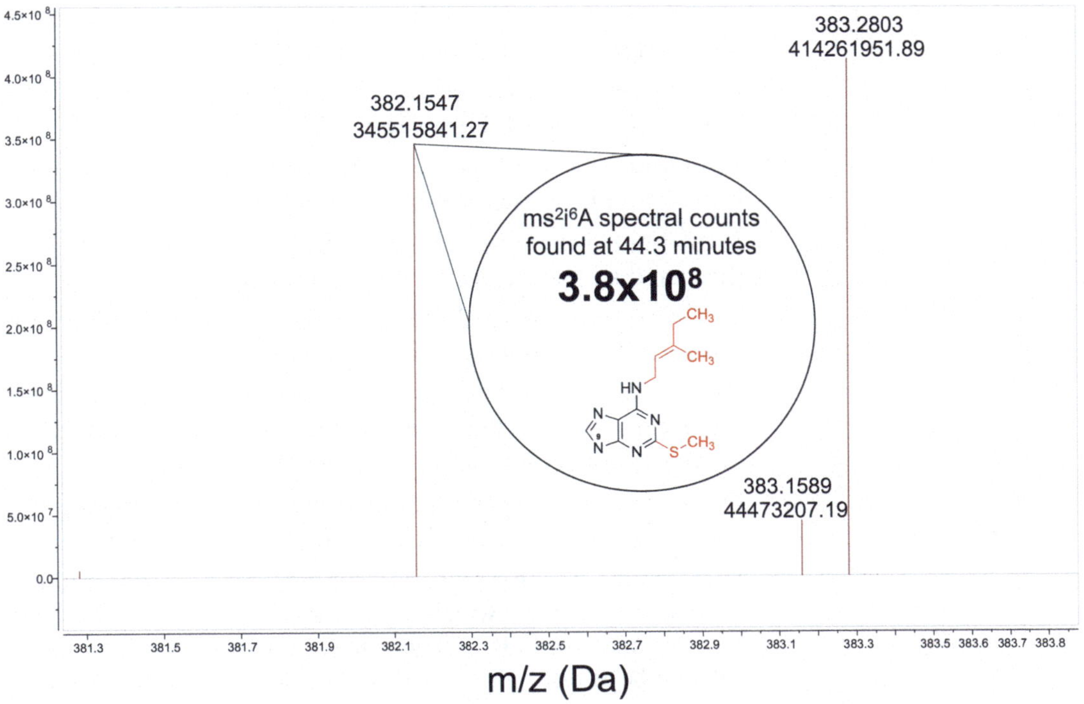

Fig. 5 Mass spectrum of the 43.7- to 44.9-min retention time, the m/z values found within this time range, and the spectral count associated with the ms^{2}i^{6}A37 modified nucleoside

3. Solid phenol is the preferred state in this method. Liquid phenol and crystallized phenol yield different extraction efficiencies in our hands.
4. A final concentration of 1 M NaOAc salt dissolved in water and pH is adjusted with concentrated HCl. At equilibrium, the solution will contain NaOAc and NaCl as indicated.
5. The NaOAc portion of the buffer should be made using NaOAc salt. The EDTA should be made as a liquid stock solution at 0.5 M EDTA, pH 8.0. The stock EDTA can then be added to the NaOAc buffer, and finally, pH adjusted to pH 4.3 using glacial acetic acid.
6. All enzymatic digestion buffers should use compounds that are at least 99% assay purity, preferably Optima or HPLC-grade purity. If not, the impurities lead to a background signal and increased signal/noise ratio of MS analyses.
7. Our lab uses MestReNova analytical software, but there are multiple different data analysis software available.
8. The growth medium can be modified to test the effects of different stressors/nutritional conditions on the relative abundance of tRNA modifications. For example, when *Saccharomyces cerevisiae* is grown in sulfur amino acid limited-standard

glucose minimal media, a sharp decrease in tRNA thiolation is observed in a dose-dependent manner [49].

9. If 100 g phenol is not in the bottle, saturate the phenol with NaOAc buffer at a 1:1 v/v ratio.
10. Saturation of phenol should be performed in a chemical fume hood with gloves, safety eyewear, and lab coat.
11. If the phenol turns brown/pink, the phenol has become oxidized and should be disposed appropriately.
12. The extraction of total RNA can be performed using any standard RNA extraction kit (i.e., Trizol).
13. Do not over-dry pellets and do not use speed-vac to speed up the drying process. This prevents the resolubilization of RNA in your resuspending buffer of choice or water.
14. Do not disturb the pellet of digest products; this can clog the column and produce a high background signal on MS.
15. Each sample within an LC-MS run will have its own internal normalizing factors, which are the levels of modification that are not fluctuating with experimental conditions. These internal standards normalize the desired analyte quantification and account for variations on extraction, digestion, and MS quantification across samples.

Acknowledgments

The research and methodology developed in the PDS laboratory was supported by the National Science Foundation (1716535). A. M.E was a recipient of a T32 NIH predoctoral fellowship (GM 95440-9).

References

1. Nicholas DJ, Wilson PW, Heinen W et al (1962) Use of electron paramagnetic resonance spectroscopy in investigations of functional metal components in micro-organisms. Nature 196:433–436
2. Kiley PJ, Beinert H (2003) The role of Fe-S proteins in sensing and regulation in bacteria. Curr Opin Microbiol 6:181–185
3. Py B, Barras F (2010) Building Fe-S proteins: bacterial strategies. Nat Rev Microbiol 8:436–446
4. Lill R, Freibert SA (2020) Mechanisms of mitochondrial iron-sulfur protein biogenesis. Annu Rev Biochem 89:471–499
5. Blahut M, Sanchez E, Fisher CE et al (2020) Fe-S cluster biogenesis by the bacterial Suf pathway. Biochim Biophys Acta, Mol Cell Res 1867:118829
6. Zheng C, Dos Santos PC (2018) Metallocluster transactions: dynamic protein interactions guide the biosynthesis of Fe-S clusters in bacteria. Biochem Soc Trans 46:1593–1603
7. Takahashi Y, Tokumoto U (2002) A third bacterial system for the assembly of iron-sulfur clusters with homologs in archaea and plastids. J Biol Chem 277:28380–28383
8. Zheng L, Cash VL, Flint DH et al (1998) Assembly of iron-sulfur clusters. Identification

of an iscSUA-hscBA-fdx gene cluster from *Azotobacter vinelandii.* J Biol Chem 273:13264–13272

9. Dos Santos PC, Dean DR (2014) A retrospective on the discovery of [Fe-S] cluster biosynthetic machineries in *Azotobacter vinelandii.* In: Rouault T (ed) Iron-sulfur clusters in chemistry and biology. Verlag Walter de Gruyter, Berlin, pp 267–296
10. Dos Santos PC (2017) *B. subtilis* as a model for studying the assembly of Fe-S clusters in gram-positive bacteria. Methods Enzymol 595:185–212
11. Mihara H, Kurihara T, Yoshimura T et al (2000) Kinetic and mutational studies of three NifS homologs from *Escherichia coli*: mechanistic difference between L-cysteine desulfurase and L- selenocysteine lyase reactions. J Biochem (Tokyo) 127:559–567
12. Lauhon CT, Kambampati R (2000) The iscS gene in *Escherichia coli* is required for the biosynthesis of 4-thiouridine, thiamin, and NAD. J Biol Chem 275:20096–20103
13. Lauhon CT (2002) Requirement for IscS in biosynthesis of all thionucleosides in *Escherichia coli.* J Bacteriol 184:6820–6829
14. Zheng C, Black KA, Dos Santos PC (2017) Diverse mechanisms of sulfur decoration in bacterial tRNA and their cellular functions. Biomol Ther 7:33
15. Cavuzic M, Liu Y (2017) Biosynthesis of sulfur-containing tRNA modifications: a comparison of bacterial, archaeal, and eukaryotic pathways. Biomol Ther 7:27
16. Agris PF, Vendeix FA, Graham WD (2007) tRNA's wobble decoding of the genome: 40 years of modification. J Mol Biol 366:1–13
17. Motorin Y, Helm M (2010) tRNA stabilization by modified nucleotides. Biochemistry 49:4934–4944
18. Boccaletto P, Machnicka MA, Purta E et al (2018) MODOMICS: a database of RNA modification pathways. 2017 update. Nucleic Acids Res 46:D303–D307
19. Edwards AM, Addo MA, Dos Santos PC (2020) Extracurricular functions of tRNA modifications in microorganisms. Genes (Basel) 11(8):907
20. Alings F, Sarin LP, Fufezan C et al (2015) An evolutionary approach uncovers a diverse response of tRNA 2-thiolation to elevated temperatures in yeast. RNA 21:202–212
21. Yarian C, Townsend H, Czestkowski W et al (2002) Accurate translation of the genetic code depends on tRNA modified nucleosides. J Biol Chem 277:16391–16395
22. Buck M, Griffiths E (1982) Iron mediated methylthiolation of tRNA as a regulator of operon expression in *Escherichia coli.* Nucleic Acids Res 10:2609–2624
23. Zhang B, Arcinas AJ, Radle MI et al (2020) First step in catalysis of the radical S-Adenosylmethionine Methylthiotransferase MiaB yields an intermediate with a [3Fe-4S](0)-like auxiliary cluster. J Am Chem Soc 142:1911–1924
24. Esberg B, Bjork GR (1995) The methylthio group (ms2) of N6-(4-hydroxyisopentenyl)-2-methylthioadenosine (ms2io6A) present next to the anticodon contributes to the decoding efficiency of the tRNA. J Bacteriol 177:1967–1975
25. Arragain S, Handelman SK, Forouhar F et al (2010) Identification of eukaryotic and prokaryotic methylthiotransferase for biosynthesis of 2-methylthio-N6-threonylcarbamoyladenosine in tRNA. J Biol Chem 285:28425–28433
26. Anton BP, Russell SP, Vertrees J et al (2010) Functional characterization of the YmcB and YqeV tRNA methylthiotransferases of *Bacillus subtilis.* Nucleic Acids Res 45:2124–2136
27. Kang BI, Miyauchi K, Matuszewski M et al (2017) Identification of 2-methylthio cyclic N6-threonylcarbamoyladenosine (ms2ct6A) as a novel RNA modification at position 37 of tRNAs. Nucleic Acids Res 45:2124–2136
28. Reichle VF, Petrov DP, Weber V et al (2019) NAIL-MS reveals the repair of 2-methylthiocytidine by AlkB in E. coli. Nat Commun 10:5600
29. Jager G, Leipuviene R, Pollard MG et al (2004) The conserved Cys-X1-X2-Cys motif present in the TtcA protein is required for the thiolation of cytidine in position 32 of tRNA from Salmonella enterica serovar typhimurium. J Bacteriol 186:750–757
30. Bouvier D, Labessan N, Clemancey M et al (2014) TtcA a new tRNA-thioltransferase with an Fe-S cluster. Nucleic Acids Res 42:7960–7970
31. Rajakovich LJ, Tomlinson J, Dos Santos PC (2012) Functional analysis of *Bacillus subtilis* genes involved in the biosynthesis of 4-Thiouridine in tRNA. J Bacteriol 194:4933–4940
32. Black KA, Dos Santos PC (2015) Abbreviated pathway for biosynthesis of 2-Thiouridine in *Bacillus subtilis.* J Bacteriol 197:1952–1962
33. Palenchar PM, Buck CJ, Cheng H et al (2000) Evidence that ThiI, an enzyme shared between thiamin and 4-thiouridine biosynthesis, may be a sulfurtransferase that proceeds through a

persulfide intermediate. J Biol Chem 275:8283–8286

34. Numata T, Ikeuchi Y, Fukai S et al (2006) Snapshots of tRNA sulphuration via an adenylated intermediate. Nature 442:419–424
35. Bimai O, Arragain S, Golinelli-Pimpaneau B (2020) Structure-based mechanistic insights into catalysis by tRNA thiolation enzymes. Curr Opin Struct Biol 65:69–78
36. Shigi N, Horitani M, Miyauchi K et al (2020) An ancient type of MnmA protein is an iron-sulfur cluster-dependent sulfurtransferase for tRNA anticodons. RNA 26:240–250
37. Liu Y, Zhu X, Nakamura A et al (2012) Biosynthesis of 4-thiouridine in tRNA in the methanogenic archaeon Methanococcus maripaludis. J Biol Chem 287:36683–36692
38. Liu Y, Vinyard DJ, Reesbeck ME et al (2016) A [3Fe-4S] cluster is required for tRNA thiolation in archaea and eukaryotes. Proc Natl Acad Sci U S A 113:12703–12708
39. Schwalm EL, Grove TL, Booker SJ et al (2016) Crystallographic capture of a radical S-adenosylmethionine enzyme in the act of modifying tRNA. Science 352:309–312
40. Mccarty RM, Krebs C, Bandarian V (2013) Spectroscopic, steady-state kinetic, and mechanistic characterization of the radical SAM enzyme QueE, which catalyzes a complex cyclization reaction in the biosynthesis of 7-deazapurines. Biochemistry 52:188–198
41. Phillipson DW, Edmonds CG, Crain PF et al (1987) Isolation and structure elucidation of an epoxide derivative of the hypermodified nucleoside queuosine from *Escherichia coli* transfer RNA. J Biol Chem 262:3462–3471
42. Payne KA, Fisher K, Sjuts H et al (2015) Epoxyqueuosine reductase structure suggests a mechanism for cobalamin-dependent tRNA modification. J Biol Chem 290:27572–27581
43. Lauhon CT (2019) Identification and characterization of genes required for 5-Hydroxyuridine synthesis in *Bacillus subtilis* and *Escherichia coli* tRNA. J Bacteriol 201
44. Dalluge JJ, Hashizume T, Mccloskey JA (1996) Quantitative measurement of dihydrouridine in RNA using isotope dilution liquid chromatography-mass spectrometry (LC/MS). Nucleic Acids Res 24:3242–3245
45. Soma A, Ikeuchi Y, Kanemasa S et al (2003) An RNA-modifying enzyme that governs both the codon and amino acid specificities of isoleucine tRNA. Mol Cell 12:689–698
46. Schwartz CJ, Djaman O, Imlay JA et al (2000) The cysteine desulfurase, IscS, has a major role in in vivo Fe-S cluster formation in *Escherichia coli*. Proc Natl Acad Sci U S A 97:9009–9014
47. Skovran E, Lauhon CT, Downs DM (2004) Lack of YggX results in chronic oxidative stress and uncovers subtle defects in Fe-S cluster metabolism in *Salmonella enterica*. J Bacteriol 186:7626–7634
48. Imlay JA (2006) Iron-Sulphur clusters and the problem with oxygen. Mol Microbiol 59:1073–1082
49. Gupta R, Walvekar AS, Liang S et al (2019) A tRNA modification balances carbon and nitrogen metabolism by regulating phosphate homeostasis. Elife 8:e44795

Chapter 9

Methods to Unravel the Roles of ATPases in Fe-S Cluster Biosynthesis

Christa N. Molé, Kinjal Dave, and Deborah L. Perlstein

Abstract

Complex biosynthetic pathways are required for the assembly and insertion of iron-sulfur (Fe-S) cluster cofactors. Each of the four cluster biogenesis systems that have been discovered requires at least one ATPase. Generally, the function of nucleotide hydrolysis in Fe-S cluster biogenesis is understudied. For example, the cytosolic Fe-S cluster assembly (CIA) pathway is proposed to begin with a scaffold, which assembles nascent Fe-S clusters destined for cytosolic and nuclear enzymes. This scaffold, comprised of Nbp35 and Cfd1 in yeast, possesses an ATPase site that is necessary for CIA function, but the role of nucleotide hydrolysis is poorly understood. Herein, we describe the in vitro methods that have been developed to uncover how the ATPase site of the scaffold regulates interaction with one of its partner proteins, Dre2. We describe a qualitative affinity copurification assay and a quantitative assay for evaluating the dissociation constant for the scaffold–partner protein complex. Finally, we describe kinetic methods to measure the k_{cat} and K_M values for ATP hydrolysis by the scaffold–partner protein complex and the execution of the ATPase assays in an anaerobic environment. These methods could be applied to study other ATPases to advance our mechanistic understanding of nucleotide hydrolases involved in metallocluster biogenesis.

Key words Iron-sulfur protein, Iron-sulfur cluster biosynthesis, Cytosolic iron-sulfur cluster assembly, ATPases, Enzyme kinetics, Deviant Walker A ATPase, Anaerobic techniques

1 Introduction

Nucleotide hydrolases (NTPases) are required for many metal trafficking and metallocluster biosynthetic pathways. For instance, the complex metalloclusters of urease, nitrogenase, carbon monoxide dehydrogenase, [Ni-Fe] and [Fe-Fe] hydrogenases, and acetyl CoA synthase are quite divergent in their structures. However, they all require an ATPase or GTPase for the biogenesis of their metallocofactors [1–6]. In many cases, the roles these NTPases play during metallocluster biogenesis are not well defined. For example, iron-sulfur (Fe-S) clusters are biosynthesized by one of four major biosynthetic machineries, the SUF (*su*lfur *f*ixation), ISC (iron-*s*ul-fur *c*luster), NIF (*ni*trogen *f*ixation), and CIA (*c*ytosolic iron-sulfur

Patricia C. Dos Santos (ed.), *Fe-S Proteins: Methods and Protocols*, Methods in Molecular Biology, vol. 2353, https://doi.org/10.1007/978-1-0716-1605-5_9, © Springer Science+Business Media, LLC, part of Springer Nature 2021

cluster *a*ssembly) systems [7–10]. Each of these cluster biogenesis systems requires an ATPase, but the role nucleotide hydrolysis plays during cluster biogenesis is generally poorly understood.

The Hsp70/J-protein chaperone/cochaperone complexes are the best studied of the Fe-S cluster biogenesis ATPases [11]. They are required for ISC cluster biogenesis system found in some bacteria and within the mitochondria of eukaryotic organisms. These chaperone/co-chaperone proteins, called HscA/HscB in bacteria and Ssq1/Jac1 in *Saccharomyces cerevisiae*, exploit nucleotide binding and hydrolysis to orchestrate the transfer of a nascent Fe-S cluster from the Isu scaffold to an apo-recipient [12, 13].

In comparison to ISC, the roles of the nucleotide hydrolases found in the SUF, NIF, and CIA systems are less well defined. In the SUF pathway, the SufC ATPase is required for the formation of the nascent cluster [14]. Although recent structural studies of the $SufBC_2D$ scaffold complex suggest that ATP binding to SufC promotes conformational changes, which is required to form the cluster assembly site, the mechanistic details of the ATPase cycle and how it is integrated with the scaffold's function remain unclear [14, 15]. In addition to the chaperone/cochaperone system analogous to those found in ISC, the NIF system also requires the NifH ATPase for the delivery of a homocitrate-bound molybdenum during the biogenesis of nitrogenase's FeMoCo cofactor [7, 16]. But many of the biochemical details of how metal transfer is coordinated with ATP hydrolysis by NifH remain challenging to decipher. In the CIA pathway, the scaffold which assembles nascent Fe-S clusters destined for cytosolic and nuclear Fe-S proteins is an ATPase [10, 17, 18]. In fungi and animals, the CIA scaffold comprises two homologous proteins, called Nbp35 and Cfd1 in *Saccharomyces cerevisiae*. Site-directed mutagenesis studies have revealed that the ATPase site is required for both cluster acquisition by the scaffold and transfer of the cluster to an apo-recipient [19, 20]. However, the specifics related to how ATP binding and hydrolysis coordinates cluster assembly and transfer remain unclear.

Recently, we have begun investigating how the ATPase sites of Nbp35 and Cfd1 regulate interaction with the scaffold's partner proteins, Nar1 and Dre2. These partner proteins are CIA factors that bind to the scaffold in a nucleotide-dependent fashion and alter its ATP hydrolysis kinetics [21]. Here, we describe a qualitative in vitro affinity copurification assay and a quantitative assay based on the kinetics of ATP hydrolysis, which can be applied to study the affinity of the scaffold–partner protein complex. We also describe our method to understand how the interaction of the scaffold with its partner protein affects the k_{cat} and K_m for ATP hydrolysis. Finally, we describe how the ATPase assays can be completed in an anaerobic environment so that the oxygen-sensitive, Fe-S-bound forms of these proteins can be evaluated. We expect the approaches could be applied to reveal more information about the mechanism of cluster biogenesis during CIA as well as uncover new information about the roles of the ATPases and GTPase required for the biogenesis of other complex metallocofactors.

2 Materials

Prepare all buffers using 18 M Ω/cm water unless otherwise stated. For experiments executed within the anaerobic chamber, place all required items (microcentrifuge tubes, pipet tips, cuvettes, etc.) and solutions (water, buffers, etc.) into the anaerobic glove box for a minimum of 72 h before use unless it is noted that the reagent needs to be freshly prepared.

2.1 Common Proteins and Solutions

1. Nbp35-Cfd1 scaffold: 20 μM purified His-tagged Nbp35-Cfd1 scaffold dimer in Buffer B [22].
2. Dre2 partner protein: 100 μM purified untagged Dre2 partner protein in Buffer B [23].
3. ATP stock: 200 mM adenosine 5′-triphosphate disodium salt trihydrate (ATP) in 1 M Tris pH 8 (freshly prepared).
4. β-Mercaptoethanol (β-ME).
5. DTT stock: 1 M dithiothreitol (DTT) (freshly prepared).
6. Buffer A: 50 mM Tris–HCl pH 8, 100 mM NaCl, 10% glycerol.
7. Buffer B: 50 mM Tris–HCl pH 8, 100 mM NaCl, 10% glycerol, 5 mM β-ME (freshly added β-ME).
8. 1× Bradford protein assay reagent (freshly prepared).

2.2 Affinity Copurification Assay

1. ADP stock: 200 mM adenosine 5′-diphosphate sodium salt (ADP) in 1 M Tris pH 8 (freshly prepared).
2. Imidazole stock: 3 M imidazole pH 8 in Buffer A (*see* **Note 1**).
3. $MgCl_2$ stock: 100 mM $MgCl_2$.
4. Buffer C: 50 mM Tris–HCl pH 8, 100 mM NaCl, 10% glycerol, 5 mM β-ME, 5 mM imidazole, 10 mM $MgCl_2$ (freshly added β-ME).
5. Nickel-NTA resin, or other affinity resin corresponding to the affinity tag of bait protein.
6. Microchromatography polypropylene disposable spin columns, 0.8 mL bed volume.
7. Sodium dodecyl sulfate-polyacrylamide gel electrophoresis (SDS-PAGE) gel, 10% acrylamide.
8. 5× SDS-PAGE protein loading buffer: 10% SDS, 0.5 M DTT, 0.25 M Tris pH 6.8, 0.02% bromophenol blue, 30% glycerol.

2.3 Measurement of the Apparent Dissociation Constant for the Scaffold–Partner Protein Complex/Steady-State Kinetic Analysis of the Scaffold–Partner Protein Complex

1. 5× ATPase assay buffer: 250 mM Tris–HCl pH 8, 50 mM $MgCl_2$, 1 M KCl.
2. PEP stock: 100 mM phosphoenolpyruvate (PEP) (freshly prepared).
3. NADH stock: 10 mM nicotinamide adenine dinucleotide hydrate (NADH) in 100 mM Tris pH 8 (freshly prepared).
4. Coupling enzyme mixture: 600–1000 units/mL pyruvate kinase (PK) and 900–1400 units/mL lactate dehydrogenase (LDH) from rabbit muscle.
5. UV-Vis spectrophotometer with temperature control accessory.
6. Masked quartz microcuvette, 1 cm path length.

2.4 Anaerobic ATP Hydrolysis Assays Using the Fe-S Cluster-Bound Form of the Scaffold

1. *See* materials described in Subheading 2.3 for ATP hydrolysis, with the substitution of a quartz cuvette with airtight screw cap.
2. Iron solution: 10 mM iron(II) ammonium sulfate hexahydrate stock in Buffer D (anaerobic, freshly prepared).
3. Sulfide solution: 10 mM sodium sulfide stock in Buffer D (anaerobic, freshly prepared).
4. Sephadex G-25 column (anaerobic).
5. Amicon centrifugal filter unit.
6. Anaerobic glove box (kept at <3 ppm O_2).
7. Coolbox or lab armor beads.
8. 18 M Ω/cm water (anaerobic).
9. Buffer D: 50 mM Tris–HCl pH 8, 150 mM NaCl, 4% glycerol (anaerobic).
10. Stock: 1 M Tris–HCl pH 8 (anaerobic).
11. Masked quartz microcuvette with airtight screw cap, 1 cm pathlength.

3 Methods

3.1 Affinity Copurification Assay

Here, we present our qualitative assay for monitoring the interaction of the Nbp35-Cfd1 scaffold with the Dre2 partner protein and how this interaction is affected by ATP or ADP (*see* Fig. 1). In the example below, Nbp35-Cfd1 is the His-tagged bait protein. Other affinity-tagged bait proteins (Strep-tag, maltose binding protein, glutathione-S-transferase, etc.) could be also used with appropriate changes to the affinity resins and buffers.

1. Calculate the concentrations of Nbp35-Cfd1 and Dre2 via the Bradford assay using the 1× Bradford protein assay dye

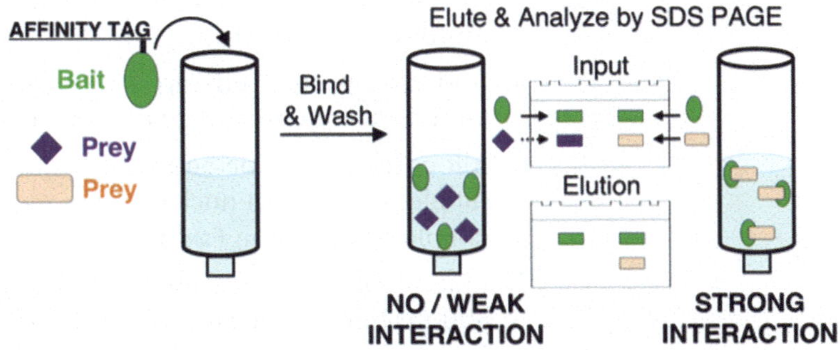

Fig. 1 Diagram of the affinity copurification experiment. An affinity-tagged bait protein (green oval) is incubated with one or more prey proteins (purple diamond and orange rectangle). The protein mixture is incubated with resin specific for the bait protein's affinity tag. After washing the column, the bound bait and any associated prey proteins are eluted. Both the input and elution fractions are analyzed by SDS-PAGE. If a prey protein (purple diamond) has no or a weak interaction with the bait protein, it does not appear in the elution samples. If the prey protein (orange rectangle) has a strong interaction with the bait protein, it will be observed in the elution fraction

Table 1
Sample components for a typical affinity copurification assay

Tube #			1	2	3	4
Component	**[Stock]**	**[Final]**	**Volume (μL)**			
Nbp35-Cfd1	20 μM	4.5 μM	90	90	90	0
Dre2	50 μM	9 μM	72	72	72	72
ATP	200 mM	10 mM	0	20	0	0
ADP	200 mM	10 mM	0	0	20	0
$MgCl_2$	100 mM	10 mM	40	40	40	40
Buffer B			198	178	178	288
Total volume (μL)		400				

reagent. If required, dilute Nbp35-Cfd1 to 20 μM and Dre2 to 50 μM using Buffer B.

2. Set up four experimental samples according to Table 1: (1) Nbp35-Cfd1 and Dre2; (2) Nbp35-Cfd1 and Dre2 with ATP; (2) Nbp35-Cfd1 and Dre2 with ADP; and (4) a Dre2 only negative control (*see* **Notes 2–4**).
3. Incubate samples for 1 h at 4 °C. Reserve 20 μL of each input sample for SDS-PAGE analysis.
4. Prepare four 200 μL Ni-NTA columns by equilibrating the resin with 10 column volumes (CV) of Buffer C supplemented

with nucleotide, as required, to match the input sample composition (*see* Table 1).

5. Add each protein mixture to its corresponding Ni-NTA column. Seal column and gently rock for 1 h at 4 °C.
6. Allow resin to settle and wash column with 10 CV Buffer C supplemented with nucleotide, as required, to match the input sample composition (*see* Table 1).
7. Elute column by adding 5 CV Buffer B supplemented with 300 mM imidazole and collect 80–100 μL fractions (*see* **Notes 5** and **6**).
8. Determine which elution fractions contain the highest protein concentration by visual inspection of a Bradford spot assay. Mix 3 μL of each fraction with 40 μL of 1× Bradford reagent. Identify the fraction with the most intense blue color (*see* **Note 7**).
9. Make 8 samples for SDS-PAGE analysis (4 input samples and 4 elution samples) by mixing 20 μL of each input sample with 5 μL 5× SDS-PAGE protein loading buffer, and 40 μL of each elution sample with 10 μL of 5× SDS-PAGE protein loading buffer. Boil samples for 5 min and then cool to room temperature (*see* **Note 8**).
10. Analyze input and elution samples via 10% SDS-PAGE. With the described setup, we have found that analyzing 5 μL of the input sample and 30 μL of the elution samples allows for adequate visualization of protein bands (*see* **Note 9**).
11. Analyze the results of the experiment. All proteins (bait and prey) should be visible, as appropriate, in the input samples. In the elution fractions, the bait protein should be visible in all samples, except in the negative control (Tube 4, Table 1), but the prey protein will only be present if it binds to the bait with sufficiently high affinity. Elution samples should be run side by side so that the relative band intensities can be directly compared using visual inspection for qualitative analysis or via densitometry for more quantitative analysis (*see* Fig. 1). A stoichiometric amount of the prey protein(s) in the elution fraction is indicative of a high affinity, stable interaction. Whereas, sub-stoichiometric amount of the prey indicates a weaker, more dynamic interaction (*see* **Note 10**).

3.2 Measurement of the Apparent Dissociation Constant for the Scaffold–Partner Protein Complex

The dissociation constant for the scaffold–partner protein complex can be determined by titrating Dre2 into a solution of Nbp35-Cfd1. The Dre2-dependent increase in Nbp35-Cfd1's ATPase activity, as assessed via a kinetic assay, is used as a readout for complex formation. The ATPase activity is monitored spectroscopically by a continuous assay using a pyruvate kinase (PK) and lactate dehydrogenase (LDH) coupling system. The PK and LDH are used to convert one molecule of NADH into NAD^+ for every molecule

Table 2
Preparing the 1× master mix

Component	Stock concentration	Final concentration	Volume for 1 reaction (μL)	Volume for 20 reactions (μL)
ATPase assay buffer	5×	1×	100	2000
DTT	1 M	1 mM	0.5	10
PEP	100 mM	4 mM	20	400
NADH	10 mM	0.2 mM	10	200
Water	–	–	259.5	7785

of ATP that is hydrolyzed by the scaffold to produce ADP. The consumption of NADH is monitored spectrophotometrically by the decrease of absorbance at 340 nm.

1. Calculate the concentrations of Nbp35-Cfd1 and Dre2 via the Bradford assay using the 1× Bradford reagent. If required, dilute Nbp35-Cfd1 to 2 μM and Dre2 to 20 μM using Buffer B.
2. Prepare 5× ATPase assay buffer, 100 mM PEP, 1 M DTT, 10 mM NADH, and 100 mM ATP. All solutions except for the 5× ATPase assay buffer must be freshly prepared for best results (*see* **Note 11**).
3. Prepare 1× master mix according to Table 2 (*see* **Note 12**).
4. Mix 390 μL 1× master mix, 50 μL of 100 mM ATP, and 10 μL of PK-LDH and incubate for 5 min at 37 °C. Use the UV-Vis spectrophotometer to monitor for the presence of any contaminating ADP in the ATP stock by examining the change in absorbance at 340 nm ($\Delta A_{340\text{nm}}$, *see* **Notes 13–15**).
5. Add 25 μL of 20 μM Dre2 and gently invert the cuvette to mix (*see* **Notes 16** and **17**). For activity measurements in the absence of Dre2, add 25 μL of Buffer B instead of Dre2. Incubate for ≤10 min at 37 °C while monitoring the $\Delta A_{340\text{nm}}$ (Fig. 2). Fit the linear portion of the data to determine the slope. This value will be the $\Delta A_{340nm^{background}}$ in Eq. (1).
6. Add Nbp35-Cfd1 to a concentration of 0.1 μM by adding 25 μL of a 2 μM stock. The volume of the assay will be 500 μL. Gently invert the cuvette to mix and then monitor the $\Delta A_{340\text{nm}}$ for ≤10 min (*see* **Notes 16** and **17**). Fit the linear portion of the data to determine the slope. This value will be the $\Delta A_{340nm^{ATPase}}$ in Eq. (1).
7. Repeat **steps 5** and **6** to measure the ATPase activity of Nbp35-Cfd1 at each concentration of Dre2. We have found that

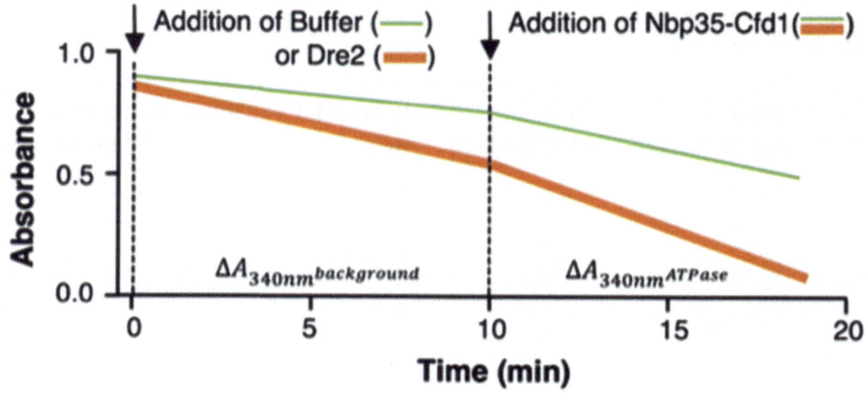

Fig. 2 Typical assay results for Nbp35-Cfd1's ATPase activity in the presence or the absence of Dre2. The background rate of ATP hydrolysis is determined by measuring the rate of change in absorbance at 340 nm ($\Delta A_{340nm^{background}}$) of all assay components except the Nbp35-Cfd1 ATPase. In the case of assays with Dre2 (thick orange line), the background activity reflects ATP consumption due to contaminating ATPases in the Dre2 preparation. For assays without Dre2 (thin green line), little to no background activity is expected. After the background ATPase activity has been determined, the Nbp35-Cfd1 ATPase is added, and the rate of change in absorbance at 340 nm ($\Delta A_{340nm^{ATPase}}$) is determined. The specific activity for ATP hydrolysis is determined according to Eq. (1)

measurements at 20, 10, 7.5, 5, 2.5, 1.25, and 0 μM Dre2 are sufficient.

8. Calculate the ATPase specific activity at each Dre2 concentration by using Eq. (1).

$$SA = \frac{\frac{\left(\Delta A_{340nm^{ATPase}} - \Delta A_{340nm^{background}}\right)}{\left(6220\ M^{-1} cm^{-1} \times cuvette\ pathlength\right)} \times \frac{10^9\ nmol}{1\ mol} \times 0.0005\ L}{nmol\ enzyme\ active\ site} \tag{1}$$

Equation 1 describes the calculation of specific activity, where: *SA* is the specific activity (nmol ATP hydrolyzed min^{-1} nmol ATPase active sites^{-1}); $\Delta A_{340nm^{ATPase}}$ is the slope of the line after Nbp35-Cfd1 addition (*see* Fig. 2); $\Delta A_{340nm^{background}}$ is the slope of the line before Nbp35-Cfd1 addition (*see* Fig. 2); 6220 M^{-1} cm^{-1} is the extinction coefficient of NADH; cuvette pathlength is 1 cm in this example; 0.0005 L is the volume of the assay; and nmol enzyme active site is the amount of ATPase active sites, which is 0.05 nmol in this example as there are two ATPase active sites per Nbp35-Cfd1 heterodimer.

9. To calculate the apparent binding affinity, plot the specific activity versus the concentration of Dre2. Fit the data to the quadratic binding equation (Eq. 2) to determine the dissociation constant (*see* **Note 18**).

$$SA = (SA_{max} - SA_{min})^* \frac{(R_T + L_T + K_D) - \sqrt{(R_T + L_T + K_D)^2 - 4(R_T{}^*L_T)}}{(2R_T)} + SA_{min} \quad (2)$$

Equation 2 describes the quadratic binding model, where: SA is the specific activity, SA_{max} is the maximum observed specific activity, SA_{min} is the minimum observed specific activity, R_T is the concentration of Nbp35-Cfd1 (in μM), L_T is the total concentration of the Dre2 (in μM), and K_D is the apparent dissociation constant for the scaffold–partner protein complex.

3.3 Steady-State Kinetic Analysis of the Scaffold–Partner Protein Complex

Here, the method for determination of the steady-state kinetic constants (k_{cat} and K_M) for the scaffold–partner protein complex is described. The approach is similar to that in Subheading 3.2, except for this application, the concentrations of Nbp35-Cfd1 and Dre2 are held constant while the concentration of ATP is varied. Also, the dissociation constant (K_D), which was determined using Eq. (2) in Subheading 3.2, is used to estimate the concentration of Dre2–scaffold complex under the assay conditions. This assay is performed under saturating conditions, with the concentration of Dre2 at least 3× the K_D value (*see* **Note 19**). This value is critical for determining the k_{cat} value of the partner protein–scaffold complex.

1. Calculate the concentrations of Nbp35-Cfd1 and Dre2 via the Bradford assay using the 1× Bradford reagent. If required, dilute Nbp35-Cfd1 to 2 μM and Dre2 to 100 μM using Buffer B.
2. Prepare 5× ATPase assay buffer, 100 mM PEP, 1 M DTT, 10 mM NADH, and 200 mM ATP. All, but the 5× ATPase assay buffer, must be freshly prepared for best results (*see* **Note 11**).
3. Prepare 1× master mix according to Table 2 (*see* **Note 12**).
4. Use the 200 mM ATP stock to prepare 60 μL of ATP dilutions between 1 and 150 mM in 1 M Tris pH 8.0.
5. Mix 390 μL 1× master mix, 50 μL ATP (1–150 mM, *see* **step 4**), and 10 μL of PK-LDH and incubate for 5 min at 37 °C. Use the UV-Vis spectrophotometer to monitor for the presence of any contaminating ADP in the ATP stock by examining $\Delta A_{340\mathrm{nm}}$ (*see* **Notes 13–15**).
6. Add 25 μL of 100 μM Dre2 and gently invert the cuvette to mix (*see* **Notes 16** and **17**). Incubate for 10 min at 37 °C while monitoring the change in absorption at 340 nm ($\Delta A_{340nm^{background}}$, Fig. 2).
7. Add 25 μL of the 2 μM Nbp35-Cfd1 solution. At this point, the assay volume should be 500 μL. Gently invert the cuvette

to mix and then monitor the change in absorbance at 340 nm ($\Delta A_{340nm^{ATPase}}$, Fig. 2, *see* **Notes 16** and **17**).

8. Calculate the initial velocity at each concentration of ATP using Eq. (3).

$$v_o = \frac{(\Delta A_{340nm^{ATPase}} - \Delta A_{340nm^{background}})}{(6220\ M^{-1}cm^{-1} \times cuvette\ pathlength)} \times \frac{10^9\ nmol}{1\ mol} \times 0.0005\ L \quad (3)$$

Equation 3 describes the calculation of initial velocity, where: v_o is the initial velocity (nmol ATP hydrolyzed min^{-1}); $\Delta A_{340nm^{ATPase}}$ is the slope after Nbp35-Cfd1 addition (Fig. 2); $\Delta A_{340nm^{background}}$ is the slope before Nbp35-Cfd1 addition (Fig. 2); 6220 M^{-1} cm^{-1} is the extinction coefficient of NADH; cuvette pathlength is the width of the cuvette used, which is 1 cm in this example; and 0.0005 L is the volume of the assay.

9. Plot initial velocity (nmol ATP hydrolyzed min^{-1}) versus the ATP concentration (mM) and fit the data to a Michaelis–Menten equation (Eq. 4).

$$v_o = \frac{V_{max}\,[S]}{K_m + [S]} \quad (4)$$

Equation 4 is the Michaelis-Menten equation, where: v_o is the initial velocity (nmol min^{-1}), V_{max} is the maximum observed velocity (nmol min^{-1}), $[S]$ is the ATP concentration (mM), and K_m is the Michaelis constant.

10. Calculate k_{cat} by taking the V_{max} value obtained in **step 9** and dividing by E_t, which is the amount of ATPase active sites in the Dre2-Nbp35-Cfd1 complex (in nmol). The K_D value obtained in Subheading 3.2 is used to estimate the concentration of the Dre2-Nbp35-Cfd1 complex under the assay conditions ($RL_{complex}$, *see* Eq. 5). After obtaining $RL_{complex}$ from Eq. 5, E_t is obtained by converting the concentration (in μM) to the amount of ATPase sites (in nmol) using volume of the ATPase assay (0.0005 L) and multiplying by two to account for the fact that there are two ATPase sites per Nbp35-Cfd1 complex.

$$K_D = \frac{[R_T - RL_{complex}]\,[L_T - RL_{complex}]}{[RL_{complex}]} \quad (5)$$

Equation 5 is used to estimate the concentration of the Dre2-Nbp25-Cfd1 complex, where: K_D is the apparent dissociation constant obtained in Subheading 3.2, R_T is the total concentration of Nbp35-Cfd1 (both bound and free, in μM), L_T is the total concentration of Dre2 (both bound and free, in μM), and $RL_{complex}$ is the concentration of Dre2-Nbp35-Cfd1 complex (in μM).

11. Compare the k_{cat} and K_M values of Nbp35-Cfd1 in the absence and presence of the Dre2 (*see* **Note 20**).

3.4 Anaerobic ATP Hydrolysis Assays Using the Fe-S Cluster-Bound Form of the Scaffold

Since the scaffold's function is to assemble a nascent cluster that is subsequently transferred to an apo-recipient protein, the scaffold's Fe-S clusters are bound close to the protein surface and are prone to oxidative degradation [24]. Therefore, ATPase assays using Fe-S cluster-bound scaffold must be carried out in an anaerobic environment. The method below describes the protocol required for examination of the scaffold's ATPase activity under anaerobic conditions. The assay samples are set up inside an anaerobic chamber, then they are then placed in a cuvette fitted with an airtight screw cap. The cuvette is sealed before removal from the chamber to avoid oxidative degradation of Fe-S cluster while monitoring ATPase activity in a spectrophotometer on the benchtop.

1. Reconstitute the Fe-S clusters of Nbp35-Cfd1 according to Fig. 3a. Briefly, bring an aliquot of Nbp35-Cfd1 into an anaerobic chamber and add 5 mM DTT from an anaerobic 1 M stock solution. Degas protein at room temperature for 2 h while gently stirring. Iron is added by the dropwise addition of 10 mM iron(II) ammonium sulfate hexahydrate. Sulfide is added dropwise from a 10 mM sodium sulfide stock solution. Both iron and sulfide are added to an eightfold molar excess over the concentration of the Nbp35-Cfd1 heterodimer, which

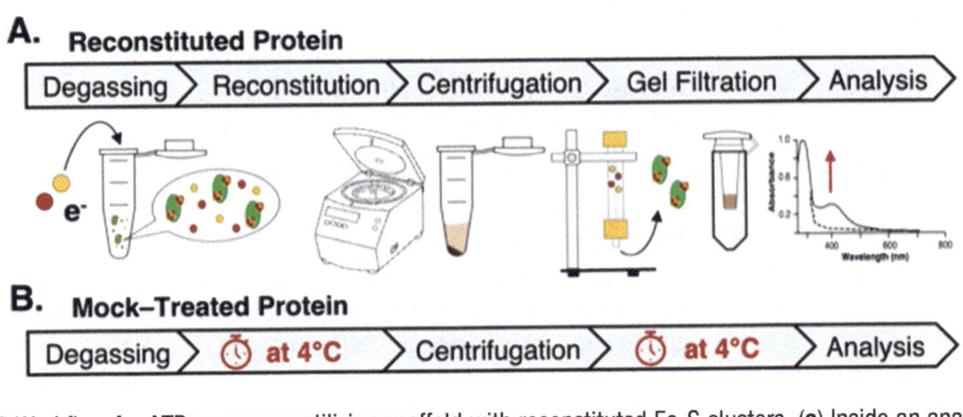

Fig. 3 Workflow for ATPase assays utilizing scaffold with reconstituted Fe-S clusters. (**a**) Inside an anaerobic chamber, the scaffold is degassed, and its Fe-S clusters are reconstituted by incubation with iron and sulfide in the presence of a reductant. Precipitated proteins are removed by centrifugation, and excess iron or sulfide is removed via gel filtration. The scaffold is concentrated and assayed for Fe-S cluster content and ATPase activity. (**b**) To account for any ATPase activity loss due to the reconstitution, a mock-treated control is evaluated in parallel. This scaffold sample is treated similarly as part (**a**), except no iron or sulfide is added, and the gel filtration step is omitted. Instead, the protein is incubated at 4 °C for the duration of the reconstitution and gel filtration steps

binds two [4Fe-4S] clusters. Stir the mixture at 4 °C for 2 h to reconstitute the Fe-S clusters. Remove precipitated protein by centrifugation. Remove any unbound iron and sulfur by gel filtration using a Sephadex G-25 column (*see* **Notes 21–23**).

2. To control for any changes in ATPase activity due to the lengthy reconstitution protocol or instability of the scaffold, a mock-treated control sample is evaluated in parallel. For this control, bring an aliquot of Nbp35-Cfd1 into the anaerobic chamber along with the sample that is to be reconstituted. Treat this control sample identically to the reconstituted sample except the iron and sulfide addition steps, and the gel filtration steps are omitted. Instead, incubate the protein at 4 °C (*see* Fig. 3b).
3. Approximately 2 h before beginning the ATPase assays, bring ATP, NADH, $MgCl_2$, DTT, KCl, NADH, and PEP into the anaerobic chamber as solids. Degas at room temperature for ≥2 h. Also, bring the Dre2 solution into the box at this time and add DTT to a concentration of 5 mM. Gently stir Dre2 at room temperature for 2 h to degas (*see* **Note 22**).
4. After Fe-S cluster reconstitution steps are completed, centrifuge all protein samples (Dre2, reconstituted Nbp35-Cfd1, and the mock-treated control) for 20 min at 12.1 krcf at 4 °C to remove any precipitated protein.
5. Calculate the protein concentrations via the Bradford assay using the 1× Bradford reagent. If required, dilute Nbp35-Cfd1, the reconstituted sample or the mock-treated sample, to 6 μM using Buffer B. Dre2 should be diluted to 60 μM using Buffer B.
6. Prepare the ATPase assay components, including the 5× ATPase assay buffer, 100 mM PEP, 1 M DTT, 10 mM NADH, and 100 mM ATP (*see* **Note 11**).
7. Prepare 1× master mix according to Table 2 (*see* **Note 12**).
8. Combine 390 μL 1× master mix, 50 μL of 100 mM ATP, and 10 μL of the PK-LDH coupling enzyme mixture in the cuvette and gently invert to mix. Cap cuvette with screwcap and remove the cuvette from the anaerobic chamber. Monitor the change in absorbance at 340 nm ($\Delta A_{340\text{nm}}$) for ≥5 min at 37 °C to monitor the turnover of any contaminating ADP and bring the assay mixture up to the appropriate temperature (*see* **Notes 13–15** and **24**).
9. Return the cuvette to the anaerobic chamber, add 25 μL of 60 μM Dre2 into the assay mixture in the cuvette, and gently invert to mix. For assays evaluating the scaffold's ATPase activity in the absence of Dre2, add 25 μL of Buffer B in place of the Dre2 solution. Cap the cuvette, gently invert it to mix, and

remove it from the anaerobic chamber (*see* **Notes 16**, **17** and **24**).

10. Place the cuvette in the UV-Vis spectrophotometer and monitor the background change in absorbance at 340 nm ($\Delta A_{340nm^{background}}$) for ≤10 min at 37 °C (*see* Fig. 2).
11. Return the cuvette to the anaerobic chamber and add the 25 μL of 6 μM Nbp35-Cfd1 (either reconstituted or the mock-treated control sample). At this point, the total volume of the sample should be 500 μL. Gently invert the cuvette to mix and remove the cuvette from the anaerobic chamber (*see* **Notes 16**, **17** and **24**).
12. Place the cuvette in the UV-Vis spectrophotometer and monitor the change in absorbance at 340 nm ($\Delta A_{340nm^{ATPase}}$) for ≤10 min (*see* Fig. 2).
13. Calculate the specific activity of each sample (reconstituted and mock-treated Nbp35-Cfd1, both in the presence and absence of Dre2) via equation Eq. (1) (*see* **Note 25**).
14. Determine the fold change activity by comparing the change in ATP hydrolysis in the absence and presence of Dre2 for both the mock-treated control and reconstituted Nbp35-Cfd1.

4 Notes

1. Imidazole, when added to a buffer at high concentrations, will raise the pH of a neutral solution. Therefore, the pH of the imidazole containing buffers must be readjusted to a pH of 8.0 if the imidazole powder is added into a buffered solution.
2. The bait protein should be the only protein with an affinity tag that binds to the affinity column. In many cases, it is convenient to design the constructs with a protease cleavable affinity tag, such as those that can be removed with the TEV or SUMO protease.
3. For each set of experiments, bait-only and prey-only controls are typically performed in parallel. A bait-only control is a positive control to ensure that the bait efficiently binds to the resin under the assay conditions. A prey-only control is a negative control useful for identifying any nonspecific interaction of the prey proteins with the resin.
4. The prey protein will be found in the elution fraction if it forms a high affinity, stable complex with the bait protein. We typically use a 1.5–2× molar excess of the prey protein over the bait protein. The ideal bait-to-prey protein ratio is dependent on both the stoichiometry and affinity of the bait-prey protein interaction and may require optimization.

5. The conditions for the highest efficiency elution, including pH and ionic strength of the elution buffer, must be optimized for each type of affinity resin.
6. While performing multiple affinity copurification experiments, the timing of the steps for each experimental sample is important. We perform each step (binding, washing, and eluting) in tandem for all samples (for 4 total samples in the presented example). All samples are allowed to equilibrate with the affinity resin for the same amount of time. In addition, care is taken to ensure that the time spent washing and eluting each column is similar and minimized as much as possible.
7. Elution fractions are obtained by collecting approximately 4–5 drops per fraction. In our experience, elution fractions 3–5 typically have the highest concentration, but this could vary depending on the fraction size and must be optimized for each application. For the SDS-PAGE analysis of the elution samples, it is important that the same fraction be selected and analyzed for all trials. For example, it is not desirable to analyze "elution fraction 4" of one sample and "elution fraction 2" of another.
8. The volume of sample required for SDS-PAGE analysis may need to be optimized depending on the concentrations of proteins used, the efficiency of binding and release of the bait protein from the column, and the affinity and stoichiometry of the bait-prey protein complex.
9. The molecular weights of all proteins should be calculated prior to the experiment to ensure that the proteins can be effectively separated by SDS-PAGE. In addition, the percentage of acrylamide in the separating gel can be optimized to ensure sufficient separation.
10. For lower affinity complexes, the protocol must be optimized. Care should be taken to minimize the time required for the washing and elution steps (*see* **Note 6**). Also, the no-bait, negative control is used to identify the smallest amount of wash buffer required to remove any nonspecific interaction of the prey protein with the resin.
11. The concentration of NADH and ATP should be determined spectroscopically using the absorbance at 340 nm and 260 nm, respectively, and the molar extinction coefficient (6220 M^{-1} cm^{-1} and 15,400 M^{-1} cm^{-1}, respectively). It is also important that these solutions are at neutral pH, which can be achieved by dissolving them in an appropriate buffer or by dissolving the powders in water and titrating the solution to neutral pH.
12. The volume of master mix one prepares should be a little larger than required for the planned experiments. The extra reagents will be useful if a point needs to be repeated or if, while the

experiment is in progress, the decision is made to collect data with a higher or a lower Dre2 concentration than originally planned.

13. The purpose of this step is to ensure the assay mix is at 37 °C prior to protein addition. It is possible that the absorbance at 340 nm may decrease during this incubation as any contaminating ADP is consumed by the coupling enzyme mixture. After a rapid decrease, the absorbance value should stabilize. If the absorbance drop corresponds to greater than 5% of ATP, then a higher purity ATP should be obtained.
14. The temperature of the assay might require optimization. Lowering of temperature to 25 °C, for example, can be useful for proteins that are unstable and precipitate at higher temperatures.
15. The coupling enzyme mixture must be present in excess to ensure that the rate of NADH consumption directly corresponds to the rate of ATP hydrolysis. To verify that the coupling system is not rate limiting, perform assays under conditions expected to produce the highest ATP hydrolysis rate, such as those with the highest Dre2 concentration, and vary the amount coupling enzymes added. If the rate of NADH consumption increases when more coupling enzymes are added, then the amount of PK-LDH enzymes should be increased. Similarly, if a decrease in coupling enzymes does not affect the rate of NADH consumption, then the amount of coupling enzymes could be decreased to conserve reagents.
16. The optimal protein concentrations depend on both the background activity and the rate of ATP hydrolysis in the presence and absence of Dre2. Background ATP hydrolysis activity may occur if the partner protein contains any contaminating ATPases. In our experience, $\Delta A_{340nm^{background}}$ should be $\leq 0.002 \text{ min}^{-1}$, and $\Delta A_{340nm^{ATPase}}$ values between 0.02 and 0.2 min^{-1} are optimal.
17. A few minutes prior to the addition of proteins, an aliquot of the protein stock solution should be preincubated to bring the solutions up to the temperature to be used for the assay. For example, if 25 μL of Dre2 is being used, typically 28 μL of Dre2 is preincubated at 37 °C for a few minutes.
18. Each individual system will determine the appropriate binding model (e.g. hyperbolic, sigmoidal, two-site binding, etc.). In our system, the K_D value is close to the concentration of the fixed component in the assay, Nbp35-Cfd1. Therefore, the data are fit to the quadratic binding model.
19. Concentrations of the scaffold and partner protein should be selected using the K_D value determined in Subheading 3.2 to ensure the maximum amount of complex formation while also

conserving protein reagents. As described, our assays have between 70% and 90% of Nbp35-Cfd1 in the complex and have an $\Delta A_{340nm^{ATPase}}$ value around 0.1–0.2 min^{-1} at the higher ATP concentrations.

20. If the k_{cat} and K_m values of the ATPase are not known, then repeat this experiment in the absence of the partner protein by substituting Buffer B for Dre2 in **step 6**.

21. The anaerobic chamber must be kept at <3 ppm O_2 at all times. We used Coy chamber filled with 95% nitrogen and 5% hydrogen at hydrogen levels ≥2.5% H_2.

22. If the partner protein also binds an Fe-S cluster, there is a possibility of Fe-S cluster transfer between it and the scaffold. In this case, it might be necessary to reconstitute the partner protein's Fe-S clusters in parallel with the scaffold or modify the partner protein, such as via site-directed mutagenesis, to prevent Fe-S cluster transfer.

23. Optimization of the reconstitution may be needed. For additional information about Fe-S cluster reconstitution, *see* Ref. 22 or Chapter 5 of this volume.

24. To ensure the reaction mixture remains anaerobic when removed from the chamber, the cuvette is capped and sealed. It is then removed from the anaerobic chamber by opening the internal door to the airlock and placing the cuvette inside. The internal door is closed, and the external door of the airlock is opened. The cuvette can then be removed and placed in the holder of the spectrometer. After the measurement is complete, the cuvette is returned through the airlock. The cuvette is placed inside the airlock, and the purging cycle run two times. When the chamber indicator reads "anaerobic," the internal door is opened, and the cuvette can be moved into the anaerobic chamber.

25. As an alternative to the single-point assay described, one could measure the ATPase activity of varying amounts of Nbp35-Cfd1 (for example, from 0.1 to 15 μM). Plot the initial velocity (nmol ATP hydrolyzed min^{-1}) versus the amount of ATPase active sites (in nmol, noting also that each Nbp35-Cfd1 dimer has two ATPase sites). These data should form a line, the slope of which is equivalent to the specific activity (nmol ATP hydrolyzed min^{-1} nmol active $site^{-1}$).

References

1. Yuen MH, Fong YH, Nim YS, Lau PH, Wong KB (2017) Structural insights into how GTP-dependent conformational changes in a metallochaperone UreG facilitate urease maturation. Proc Natl Acad Sci U S A 114:10890–10898

2. Jeoung J-H, Giese T, Grünwald M, Dobbek H (2009) CooC1 from Carboxydothermus hydrogenoformans is a nickel-binding ATPase. Biochemistry 48:11505–11513

3. Wittenborn EC, Cohen SE, Merrouch M, Léger C, Fourmond V, Dementin S, Drennan

CL (2019) Structural insight into metallocofactor maturation in carbon monoxide dehydrogenase. J Biol Chem 294:13017–13026

4. Lacasse MJ, Summers KL, Khorasani-Motlagh-M, George GN, Zamble DB (2019) Bimodal nickel–binding site on *Escherichia coli* [NiFe]--hydrogenase metallochaperone HypA. Inorg Chem 58:13604–13618
5. Gregg CM, Goetzl S, Jeoung J-H, Dobbek H (2016) AcsF catalyzes the ATP-dependent insertion of nickel into the Ni, Ni-[4Fe-4S] cluster of acetyl-CoA synthase. J Biol Chem 291:18129–18138
6. Shepard EM, McGlynn SE, Bueling AL, Grady-Smith CS, George SJ, Winslow MA, Cramer SP, Peters JW, Broderick JB (2010) Synthesis of the 2Fe subcluster of the [FeFe]--hydrogenase H cluster on the HydF scaffold. Proc Natl Acad Sci U S A 107:10448–10453
7. Burén S, Jiménez-Vicente E, Echavarri-Erasun C, Rubio LM (2020) Biosynthesis of nitrogenase cofactors. Chem Rev 120:4921–4968
8. Lill R, Freiber S-A (2020) Mechanisms of mitochondrial iron–sulfur protein biogenesis. Annu Rev Biochem 89:471–499
9. Pérand J, de Choudens SO (2017) Iron-sulfur clusters biogenesis by the SUF machinery: close to the molecular mechanism understanding. J Biol Inorg Chem 23:581–596
10. Lill R, Dutkiewicz R, Freibert SA, Heidenreich T, Mascarenhas J, Netz DJ, Paul VD, Pierik AJ, Richter N, Stümpfig M, Srinivasan V, Stehling O, Mühlenhoff U (2015) The role of mitochondria and the CIA machinery in the maturation of cytosolic and nuclear iron-sulfur proteins. Eur J Cell Biol 94:280–291
11. Dutkiewicz R, Nowak M, Craig EA, Marszalek J (2017) Fe–S cluster Hsp70 chaperones: the ATPase cycle and protein interactions. Methods Enzymol 595:161–184
12. Silberg JJ, Tapley TL, Hoff KG, Vickery LE (2004) Regulation of the HscA ATPase reaction cycle by the co–chaperone HscB and the iron–sulfur cluster assembly protein IscU. J Biol Chem 279:53924–53931
13. Chandramouli K, Johnson MK (2006) HscA and HscB stimulate [2Fe–2S] cluster transfer from IscU to Apoferredoxin in an ATP–dependent reaction. Biochemistry 45:11087–11095
14. Saini A, Mapolelo DT, Chahal HK, Johnson MK, Outten FW (2010) SufD and SufC ATPase activity are required for iron acquisition during in vivo Fe-S cluster formation on SufB. Biochemistry 49:9402–9412
15. Hirabayashi K, Yuda E, Tanaka N, Katayama S, Iwasaki K, Matsumoto T, Kurisu G, Outten FW, Fukuyana K, Takahashi Y, Wada K (2015) Functional dynamics revealed by the structure of the SufBCD complex, a novel ATP-binding cassette (ABC) protein that serves as a scaffold for iron-sulfur cluster biogenesis. J Biol Chem 290:29717–29731
16. Hu Y, Corbett MC, Fay AW, Webber JA, Hodgson KO, Hedman B, Ribbe MW (2006) Nitrogenase Fe protein: a molybdate/homocitrate insertase. Proc Natl Acad Sci U S A 103:17125–17130
17. Netz DJ, Pierik AJ, Stümpfig M, Mühlenhoff U, Lill R (2007) The Cfd1–Nbp35 complex acts as a scaffold for iron–sulfur protein assembly in the yeast cytosol. Nat Chem Biol 3:278–286
18. Camire EJ, Grossman JD, Thole GJ, Fleischman NM, Perlstein DL (2015) The yeast Nbp35–Cfd1 cytosolic iron–sulfur cluster scaffold is an ATPase. J Biol Chem 290:23793–23802
19. Netz DJA, Pierik AJ, Stümpfig M, Bill E, Sharma AK, Pallesen LJ, Walden WE, Lill R (2012) A bridging [4Fe–4S] cluster and nucleotide binding are essential for function of the Cfd1–Nbp35 complex as a scaffold in iron–sulfur protein maturation. J Biol Chem 287:12365–12378
20. Grossman JD, Gay KA, Camire EJ, Walden WE, Perlstein DL (2019) Coupling nucleotide binding and hydrolysis to iron–sulfur cluster acquisition and transfer revealed through genetic dissection of the Nbp35 ATPase site. Biochemistry 58:2017–2027
21. Molé CN, Browder LK, Grossman JD, Seguinot BO, Vo A, Perlstein DL (2020Nar1 and Dre2 are the partner proteins of the Nbp35–Cfd1 cytosolic iron–sulfur cluster assembly scaffold's ATPase domain. In preparation
22. Grossman JD, Camire EJ, Perlstein DL (2018) Approaches to interrogate the role of nucleotide hydrolysis by metal trafficking NTPases: the Nbp35–Cfd1 iron–sulfur cluster scaffold as a case study. Methods Enzymol 599:293–325
23. Netz DJ, Stümpfig M, Doré C, Mühlenhoff U, Pierik AJ, Lill R (2010) Tah18 transfers electrons to Dre2 in cytosolic iron-sulfur protein biogenesis. Nat Chem Biol 6:758–765
24. Stehling O, Jeoung J-H, Freibert SA, Paul VD, Banfer S, Niggemeyer B, Rosser R, Dobbek H, Lill R (2018) Function and crystal structure of the dimeric P–loop ATPase Cfd1 coordinating an exposed [4Fe–4S] cluster for transfer to apoproteins. Proc Natl Acad Sci U S A 115:E9085–E9094

Chapter 10

Methods to Investigate the Kinetic Profile of Cysteine Desulfurases

Maame A. Addo, Ashley M. Edwards, and Patricia C. Dos Santos

Abstract

Biological iron-sulfur (Fe-S) clusters are essential protein prosthetic groups that promote a range of biochemical reactions. In vivo, these clusters are synthesized by specialized protein machineries involved in sulfur mobilization, cluster assembly, and cluster transfer to their target proteins. Cysteine desulfurases initiate the first step of sulfur activation and mobilization in cluster biosynthetic pathways. The reaction catalyzed by these enzymes involves the abstraction of sulfur from the amino acid L-cysteine, with concomitant formation of alanine. The presence and availability of a sulfur acceptor modulate the sulfur-transferase activity of this class of enzymes by altering their reaction profile and catalytic turnover rate. Herein, we describe two methods used to probe the reaction profile of cysteine desulfurases through quantification of alanine and sulfide production in these reactions.

Key words Fe-S clusters, Cysteine desulfurase, Alanine, Sulfide, Sulfur acceptor, Methylene blue, NDA

1 Introduction

Iron-sulfur clusters are among the most ubiquitous metal cofactors and are found in metalloenzymes that possess important redox, catalytic, and regulatory functions [1, 2]. These clusters mediate processes such as gene expression, electron transfer, and catalysis while also playing key roles in central metabolic pathways [3]. The most prevalent clusters exist as [2Fe-2S], [4Fe-4S], and [3Fe-4S] assemblies of iron ions and inorganic sulfide, and these cofactors are commonly bound to their protein partners via cysteine-thiolate ligands [1]. Each iron of a [2Fe-2S] cluster is typically coordinated by two cysteine thiolate ligands, while the [4Fe-4S] and [3Fe-4S] clusters are coordinated by one cysteine ligand. Thus, in all cases, the iron ion is in a tetrahedral coordination environment, enabling facile one-electron redox reactions [4].

It is well known that certain apo forms of cluster-containing proteins can be activated in vitro by simply adding sulfide and iron

Patricia C. Dos Santos (ed.), *Fe-S Proteins: Methods and Protocols*, Methods in Molecular Biology, vol. 2353,
https://doi.org/10.1007/978-1-0716-1605-5_10,

ions [5, 6]. Although this observation led to the initial assumption that Fe-S clusters can be spontaneously fused into their protein partners in vivo, this view was not compatible with the physiological toxicity of Fe and S at concentrations needed for in vitro cluster maturation. A more realistic scenario entails the delivery of Fe and S to apo-[Fe-S] proteins via specific carrier proteins known to shuttle Fe and S in nonhazardous forms [7]. The *Azotobacter vinelandii* Nif system, comprising NifS and NifU, was the first biosynthetic system to be identified, and is involved in the initial assembly and shuttling of Fe-S clusters destined for nitrogenase [8]. The identification of the Nif system led to the subsequent discovery of the Isc system, which is involved in the housekeeping synthesis of Fe-S clusters in *A. vinelandii* and many other bacterial species [9–11].

Fe-S cluster biosynthetic pathways have been identified in a wide range of species belonging to various ecological and phylogenetic groups [12–17]. All Fe-S cluster biosynthetic pathways studied thus far utilize a general three-step scheme that includes: a sulfur activation reaction, the preformation of Fe-S clusters, and the incorporation of these clusters into their final acceptor proteins. Hence, a model for the biological formation of Fe-S clusters minimally requires a sulfur-activating enzyme and a protein that serves as the assembly site of Fe-S clusters, more commonly known as the scaffold protein. Once synthesized, clusters are then transferred to their final targets, where they perform their biological functions [1, 18].

The first step of sulfur mobilization in the synthesis of Fe-S clusters is catalyzed by cysteine desulfurases (EC 2.8.1.7). These enzymes are also required for the initiation of several biosynthetic pathways in which sulfur-containing cofactors are synthesized [19, 20]. So far, all characterized cysteine desulfurases adopt homodimeric structures containing a pyridoxal 5′-phosphate (PLP) active site cofactor coordinated by a highly conserved lysine via a Schiff base [21]. These enzymes are known to catalyze the Cys:sulfur acceptor sulfurtransfer reactions via a double displacement mechanism [22, 23] (Fig. 1a).

The first half of a cysteine desulfurase-catalyzed reaction results in the formation of alanine and a persulfide intermediate in the active site of the enzyme, defined here as CD-SSH [20, 21]. In vitro, the rate of this half-reaction can be determined in the presence of reducing agents that reduce the enzyme's persulfide intermediate, releasing free sulfide and regenerating the enzyme for the next catalytic cycle (Fig. 1b). In the absence of a reducing agent and/or a sulfur acceptor, the enzyme reaction leads to the formation of polysulfides and the release of colloidal sulfur (S_8) (Fig. 1c). The reactivity of the persulfurated and polysulfurated (CD-SS-$S_{(1-7)}$H) forms is reduced, as evidenced by slower turnover rates of Ala formation under nonreducing conditions [21]. In vivo, the persulfide intermediate initiates the second half of the catalytic

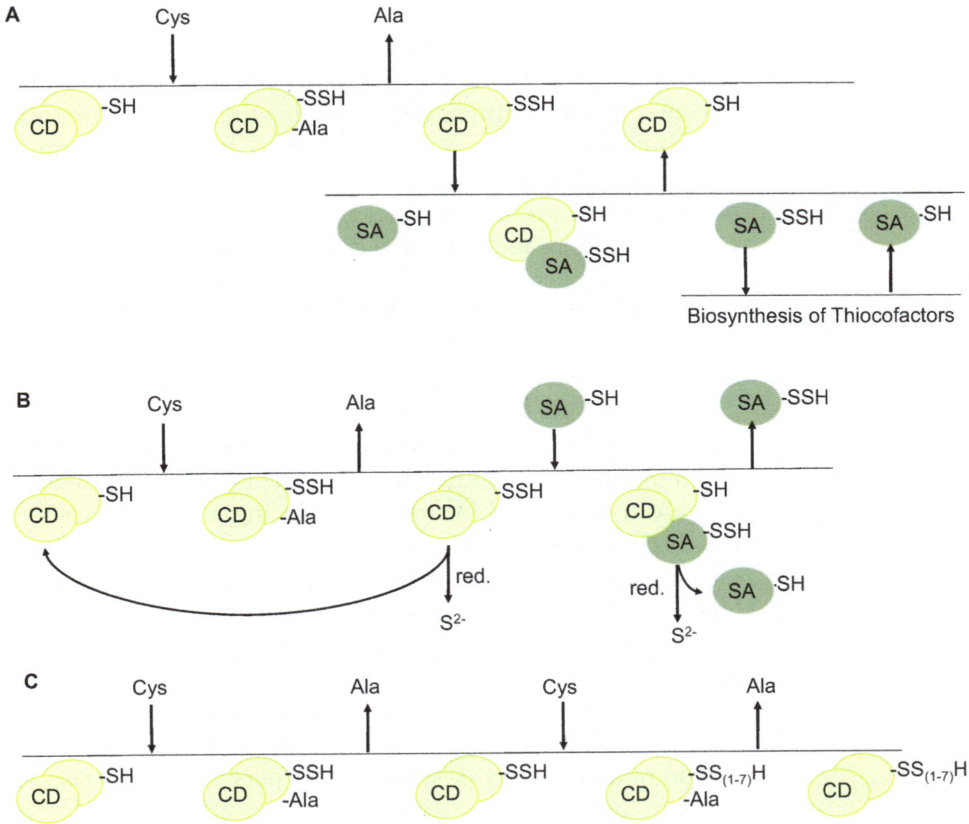

Fig. 1 Double displacement mechanism of Cys:sulfur acceptor sulfurtransferase reactions. (**a**) In vivo reaction. Persulfide sulfur is transferred from the cysteine desulfurase to the sulfur acceptor for the biosynthesis of thiocofactors. (**b**) Reducing/sulfur acceptor reaction. Persulfide sulfur is either transferred from the cysteine desulfurase to sulfur acceptor or directly reduced by a reductant, releasing sulfide. If sulfur transfer is achieved, the persulfide on the sulfur acceptor is also susceptible to reduction. (**c**) Nonreducing/no sulfur acceptor reaction. The cysteine desulfurase undergoes multiple turnovers leading to the formation of colloidal sulfur (S_8)

cycle, which includes the transfer of the persulfide sulfur to an acceptor molecule for the synthesis of Fe-S clusters or other S-containing cofactors [19, 24–26] (Fig. 1a). Thus, kinetic assays probing the complete reaction cycle include the cysteine desulfurase and corresponding sulfur acceptors.

Cysteine desulfurases are classified into two groups based on sequence similarity and tertiary structural motifs [27]. Members of the group I desulfurases, including *Escherichia coli* IscS and *A. vinelandii* NifS, contain a characteristic 12-amino acid residue sequence insertion following the active site cysteine [28]. These conserved moieties form a flexible, structural loop proposed to adopt distinct conformations during catalysis. Conversely, group II cysteine desulfurases, such as *E. coli* SufS and CsdA, lack this structural element and react through a conserved cysteine residue located within a shorter and more structured loop [29]. This distinction is thought to result in greater steric hindrance and consequently require the involvement of specific sulfur acceptor for effective sulfur transfer. For example, the *Bacillus subtilis* SufS class II enzyme activity is enhanced up to 200-fold in the presence of its dedicated sulfur acceptor, SufU [23, 30]. Therefore, the distinct kinetic profiles of class I and class II enzymes are attributed to the functional and structural differences among members of this general class of enzymes.

In addition to sulfur acceptors modulating cysteine desulfurase activity, the presence and type of reducing agents can also affect the activity of cysteine desulfurases. The synthetic reducing agent dithiothreitol (DTT) can directly interact with the PLP cofactor, thereby decreasing the enzymes' effectiveness in sulfur abstraction and transfer. Furthermore, the persulfide bond on a cysteine desulfurase can be cleaved by artificial reducing agents. In some cases, this direct reaction interferes with the sulfur transfer step by excluding the involvement of sulfur acceptors, ultimately skewing the kinetic behavior of the cysteine desulfurase and distorting the assessment of their in vivo functionality [18, 23, 28, 31–43]. For instance, the enhancement of *B. subtilis* NifZ activity by its sulfur acceptor ThiI is only observed in the absence of a reducing agent [44]. Physiological reducing agents, such as the thioredoxin/thioredoxin reductase (Trx) system, display persulfide reductase activity and can participate in coupled reaction schemes with cysteine desulfurases and sulfur acceptors. For the *B. subtilis* SufS and YrvO cysteine desulfurases, the Trx system is effective in reducing persulfides that are formed within sulfur acceptors but not cysteine desulfurases [45]. The distinct reactivity of physiological and artificial reducing agents must be considered when analyzing the kinetic profile of reactions promoted by these enzymes.

The characterization of the mechanistic steps involved in the synthesis of Fe-S clusters can be quite challenging and generally requires multiple validation methods. This chapter describes experimental approaches used for the kinetic investigation of the first step of Fe-S cluster assembly reactions, which involves cysteine desulfurases. These methods include the detection of alanine, a byproduct in these reactions, and sulfide, a constituent of nascent clusters, and a product released upon reductive cleavage.

2 Materials

2.1 Assay Components

1. Cysteine desulfurase: 20 μM cysteine desulfurase purified in MOPS buffer (*see* **Notes 1** and **2**). Protein stock is stored at −80 °C in 100 μL aliquots.

2. Sulfur acceptor: 20 μM sulfur acceptor purified in MOPS buffer (*see* **Notes 1** and **2**). Protein stock is stored at −80 °C in 100 μL aliquots.
3. MOPS buffer: 50 mM 3-(*N*-morpholino)propanesulfonic acid, pH 8.0. The buffer is prepared in deionized (DI) water, and the pH is adjusted with 10 M NaOH. This buffer should be degassed before use in assays that are coupled to the synthesis Fe-S clusters (*see* **Notes 1** and **2**).
4. L-Cysteine solution: 10 mM L-cysteine hydrochloride prepared fresh in MOPS buffer (*see* **Note 2**).
5. Fe solution: 20 mM ferrous ammonium sulfate. This solution is freshly prepared in DI water (*see* **Note 2**).
6. DTT solution: 0.5 M dithiothreitol in MOPS buffer, pH 8. This solution can be prepared and stored in 500 μL aliquots at −20 °C for up to a year.

2.2 Sulfide Quantification

1. NaOH solution: 12% sodium hydroxide in DI water.
2. ZnOAc solution: 10% zinc acetate in DI water.
3. DMPD solution: 20 mM *N*,*N*-dimethyl-*p*-phenylenediamine dihydrochloride in 7.2 M HCl.
4. Fe solution: 30 mM iron(III) chloride in 1.2 M HCl.
5. Sulfide standard solution: 1 mM sodium sulfide nonahydrate prepared fresh with DI water in a volumetric flask.
6. 25 μL and 200 μL gas-tight syringes.
7. Vials with rubber septa (for Subheading 3.2.1).
8. Plastic cuvette (for Subheading 3.2.1).
9. Spectrophotometer (for Subheading 3.2.1).
10. Parafilm (for Subheading 3.2.2).
11. 96-Well plate (for Subheading 3.2.2).
12. Microplate reader (for Subheading 3.2.2).

2.3 Alanine Quantification

1. TCA solution: 10% trichloroacetic acid in DI water.
2. KCN solution: 40 mM potassium cyanide in DI water. Prepare a fresh solution once every week and store it in the refrigerator.
3. NDA solution: 8 mM naphthalene-2,3-dicarboxaldehyde in DI water. Prepare a fresh solution once every week and store it in the refrigerator.
4. Alanine standard solution: 1 mM L-alanine in DI water.
5. Borate buffer: 300 mM sodium borate in DI water, pH 9.0. Adjust the pH with 10 M NaOH.

6. NDA mix: 0.1 mM NDA, 1 mM KCN in 100 mM sodium borate buffer, pH 9. The solution is prepared from stock solutions of each ingredient and mixed immediately before use.
7. HPLC grade water (for Subheading 3.3.1).
8. HPLC grade methanol (for Subheading 3.3.1).
9. HPLC coupled with a fluorescence detector (for Subheading 3.3.1).
10. HPLC vials (for Subheading 3.3.1).
11. C-18 column, 150 mm CL × 4.6 mm ID, 5 μM particle size (for Subheading 3.3.1).
12. Black 96-well plate with a clear bottom (for Subheading 3.3.2).
13. Microplate reader fluorimeter (for Subheading 3.3.2).

3 Methods

The sulfurtransferase reaction of cysteine desulfurases is initiated upon binding the substrate cysteine to the PLP cofactor. This event leads to the formation of an external Cys-PLP Schiff base (Fig. 2). The electronic arrangement of the Cys-PLP adduct allows for the nucleophilic attack of the active site cysteine residue onto the cysteine substrate, resulting in the cleavage of the carbon-sulfur bond of the substrate and the final formation of the enzyme persulfide intermediate (CD-SSH) and alanine [22, 46]. The activated sulfur is then transferred to a sulfur acceptor via a persulfide sulfur transfer reaction or directly to a nascent Fe-S cluster, completing the reaction cycle [18, 43]. The methods described below have been developed to characterize the kinetic profile of reactions involving cysteine desulfurases by quantifying alanine and sulfide produced in these reactions.

3.1 Cysteine Desulfurase Assay

3.1.1 Cysteine Desulfurase Assays Under Nonreducing Conditions

Assays performed under nonreducing conditions in the presence or absence of sulfur acceptor molecules lead to the formation of Ala and accumulation of protein persulfides and polysulfides [8]. These latter products can be quantified through the sulfide assay, when reaction time points are reduced prior to detection [38](Fig. 3). The kinetic profile of reactions performed in the presence of sulfur acceptors typically displays a biphasic behavior (Fig. 4). The first phase is marked by a fast rate of alanine formation, attributed to rapid catalytic turnover in the presence of available sulfur acceptor substrate. The length and amplitude of this phase are stoichiometrically proportional to the amount of sulfur acceptor substrate present in the reaction [18, 38, 44]. Under nonreducing conditions and in the absence of subsequent downstream pathway components, persulfurated sulfur acceptors do not engage in subsequent catalytic cycles or, to a lesser extent, do not participate

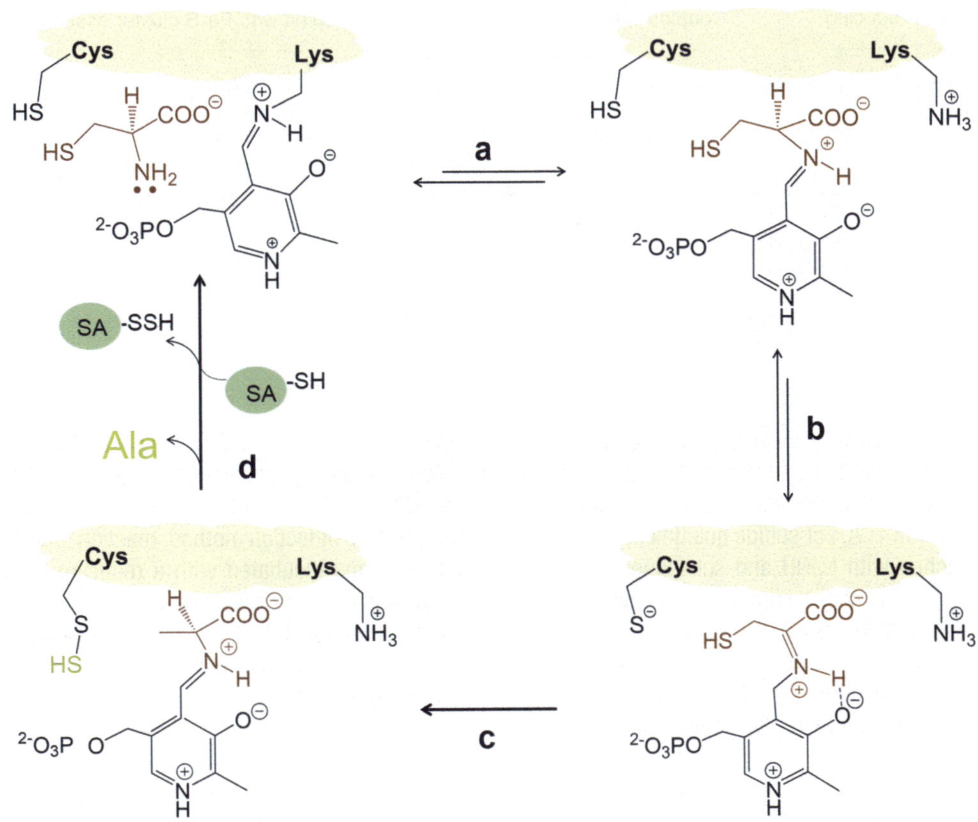

Fig. 2 Mechanism of cysteine desulfurase reaction. (**a**) The reaction proceeds with the binding of cysteine (substrate) to the PLP cofactor, forming the cys-PLP intermediate. (**b**) The formation of the enzyme persulfide intermediate is jumpstarted via deprotonation of the enzyme cys-thiol. (**c**) Deprotonated enzyme cys-thiol abstracts sulfur from cysteine via a nucleophilic attack. (**d**) Alanine is released, and the enzyme is regenerated for another catalytic cycle

with the same efficiency. At this point, the reaction transitions to a slow catalytic profile, as observed in the second phase of the reaction. The slow steady-state turnover presents similar rates of reactions performed without sulfur acceptors. The typical nonreducing reaction is performed as follows:

1. Mix 0.2 μM of cysteine desulfurase and 2 μM of the sulfur acceptor in MOPS buffer, and start the reaction by adding 0.5 mM of cysteine (*see* **Note 3**).
2. In determining biphasic profiles, take a range of time points to calculate the rates of both phases. Consider taking time points at 0.5, 1, 1.5, 2, 3, 6, 9, and 12 min. For an accurate rate determination, utilize the slope of at least three-time points in each phase.
3. For sulfide quantification, at each time point, take 800 μL of reaction aliquot and mix with 25 μL of NaOH solution to stop

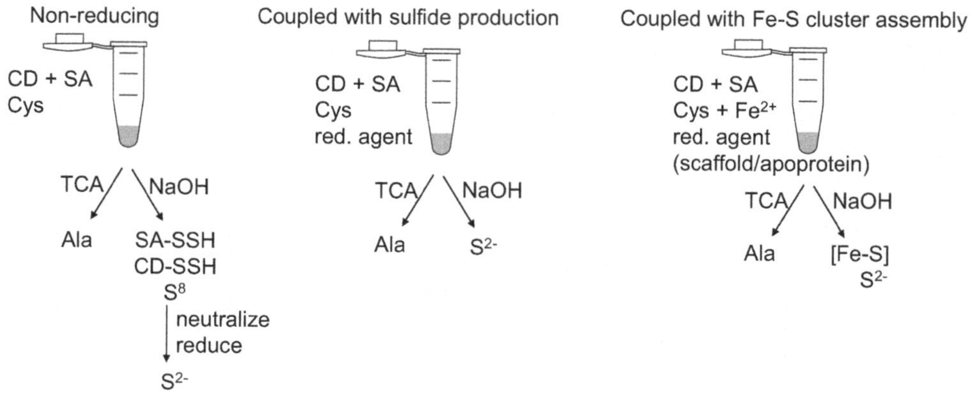

Fig. 3 Cysteine desulfurase assays. Nonreducing reactions involving cysteine desulfurase (CD), sulfur acceptor (SA), and the substrate cysteine (Cys) lead to the production of Ala, protein persulfides (SA-SSH and CD-SSH), and polysulfide (S_8). For alanine quantification using the Ala-NDA detection method, reaction time points are quenched with TCA. For sulfide quantification using the methylene blue detection method, reaction time points are quenched with NaOH and subsequently neutralized with Tris and incubated with a reducing agent to liberate S^{2-}. Reactions coupled to sulfide production are performed in the presence of a reducing agent generating sulfide during the course of the reaction. Reactions coupled to Fe-S cluster assembly are performed in the presence of Fe^{2+} and, when relevant, assay reactions include the presence of a scaffold and the final apo-protein. Fe-S clusters, free and/or associated, and S^{2-} generated in these reactions can be directly quantified through the methylene blue method

the reaction. Then readjust the pH by adding 200 μL of 1 M Tris buffer pH 8.0. Reduce persulfides and polysulfides by adding 3 μL of 1 M DTT solution prior to sulfide quantification (*see* Subheading 3.2 for detection method).

4. For Ala quantification, at each time point, take a 100 μL aliquot of the reaction and mix with 20 μL TCA solution to stop the reaction and proceed with Ala quantification (*see* Subheading 3.3 for detection method).
5. When characterizing biphasic reaction profiles, repeat **steps 1–4** using different equivalents of sulfur acceptors (e.g., 20 and 40 equivalents of sulfur acceptor in relation to cysteine desulfurase concentration).

3.1.2 Cysteine Desulfurase Assays Coupled with Sulfide Production

Assays performed in the presence of reducing agents lead to the direct formation of sulfides in solution (Fig. 3). Reducing agents are able to react with persulfides formed within the cysteine desulfurase and sulfur acceptors. This reaction setup is distinct from the setup of assays performed under nonreducing conditions, as it leads to regeneration of the sulfur acceptor and, in some cases, the cysteine desulfurase [21]. Under reducing conditions, persulfides formed within the cysteine desulfurase and sulfur acceptor are released into the solution in the form of sulfide [41, 42, 44]. It is important to consider that the involvement of some sulfur acceptor molecules may be obscured in reactions performed in the presence

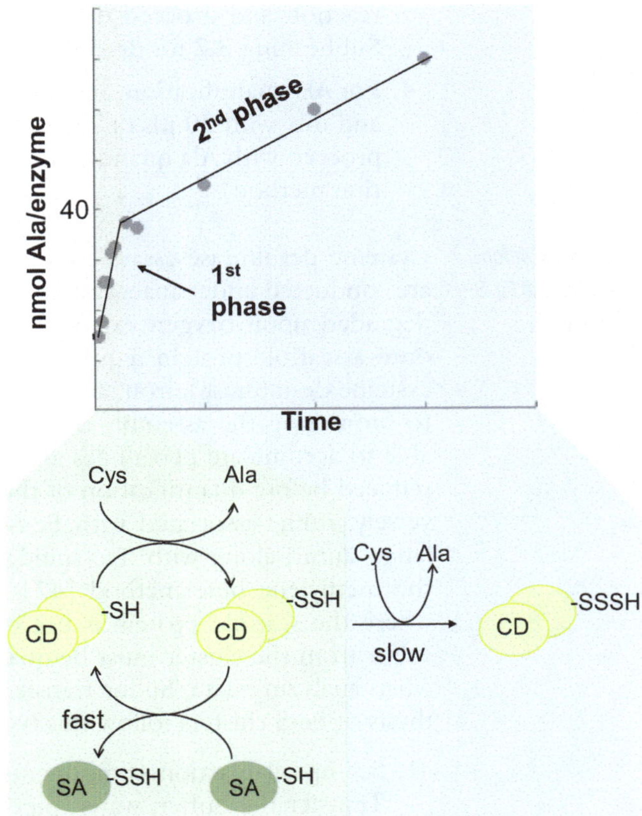

Fig. 4 Biphasic cysteine desulfurase reaction profile in the presence of a sulfur acceptor under nonreducing conditions. A rapid formation of alanine is observed in the presence of a sulfur acceptor (first phase; shaded in green). This burst is caused by a swift catalytic turnover of the cysteine desulfurase and is proportional to the equivalents of sulfur acceptor used in the reaction. After all the sulfur acceptor is persulfurated, the cysteine desulfurase slowly forms polysulfides and, ultimately, colloidal sulfur (second phase; shaded in blue). This data was obtained from the kinetic analysis of E. coli CsdA:CsdE:40CsdL sulfur transfer reaction

of artificial reductants. These agents may directly interact with the enzyme, effectively bypassing the sulfur transfer event of the second half of the reaction. Reducing reactions are performed following the standard protocol:

1. Mix 0.2 μM of cysteine desulfurase, 2 μM of the sulfur acceptor in MOPS buffer in the presence of 2 mM of DTT, and start the reaction by adding 0.5 mM of cysteine (*see* **Notes 3** and **4**).
2. Take a range of time points for an accurate rate determination similar to Subheading 3.1.1, **step 2**.
3. For sulfide quantification, at each time point, take 800 μL aliquot and mix with 25 μL of NaOH solution to stop the

reaction and proceed directly with sulfide quantification (*see* Subheading 3.2 for detection method).

4. For Ala quantification, at each time point, take a 100 μL aliquot and mix with 20 μL of TCA solution to stop the reaction and proceed with Ala quantification (*see* Subheading 3.3 for detection method).

3.1.3 Sulfur Mobilization Assays Coupled to the Fe-S Cluster Assembly

Cysteine desulfurase assays coupled to cluster assembly reactions are conducted under anaerobic conditions as Fe-S clusters are easily degraded upon oxygen exposure. The synthesis of Fe-S clusters onto a scaffold protein requires a source of sulfur (provided by a cysteine desulfurase), iron, and electrons [26, 42, 43]. In addition to promoting the assembly of Fe-S species, scaffold proteins are able to accumulate persulfides and polysulfides, which need to be reduced before quantification of the total sulfide mobilized. Conversely, sulfur associated with Fe-S clusters, also known as acid-labile sulfur, along with free sulfides, can be directly detected via the methylene blue method [47]. In cluster assembly reactions where the scaffold protein is not completely apo, the acid-labile sulfur from the cluster must be quantified and taken into account when analyzing data. Sulfur transfer reactions coupled to the synthesis of Fe-S clusters follow this typical protocol:

1. Set up all reaction components in the anaerobic glove box. Transfer the buffer, water, pipette tips, vials, and other glassware the day before the experiments. Prepare the cysteine and iron solutions freshly inside the glove box on the day of the experiment. Transfer the proteins from the freezer directly into the glove box (*see* **Note 2**).
2. Set up an Fe-S cluster assembly reaction by mixing 0.1 μM of cysteine desulfurase, 2 μM of sulfur acceptor, 20 μM of Fe solution in MOPS buffer in the presence of 2 mM of DTT, and start the reaction by adding 0.5 mM of cysteine (*see* **Note 5**).
3. Take a range of time points for an accurate rate determination while samples remain in the glove box (*see* **Note 6**).
4. For sulfide quantification, at each time point, take 800 μL aliquot and mix with 25 μL of NaOH solution to stop the reaction and proceed directly with sulfide quantification (*see* Subheading 3.2 for detection method).
5. For Ala quantification, at each time point, take a 100 μL aliquot and mix with 20 μL of TCA solution to stop the reaction and proceed with Ala quantification (*see* Subheading 3.3 for detection method).

3.2 Sulfide Detection Method

Sulfide is produced upon reducing persulfides or acid treatment of Fe-S clusters. It can be quantified in colorimetric assays that couple

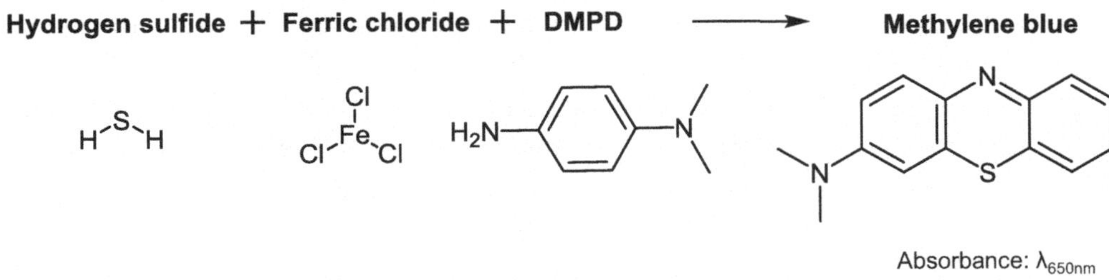

Fig. 5 Methylene blue reaction. Sulfide liberated during the acid treatment of Fe-S clusters or reduction of persulfide bound to cysteine desulfurase/sulfur acceptor can be detected via methylene blue formation

sulfide release to the formation of methylene blue [48] (Fig. 5). In reactions performed in the presence of reducing agents, persulfides formed within cysteine desulfurases and sulfur acceptors are directly reduced, releasing free sulfide. In reactions performed in the absence of reducing agents, mobilized sulfur in the form of persulfide and polysulfide can be quantified by stopping reaction time points and reducing these samples prior to quantification of total mobilized sulfur (Fig. 3). The rates of sulfide production are calculated by the slopes of at least three reaction time points. Methylene blue formed in sulfur transfer reactions can be detected in a standard spectrophotometer (Subheading 3.2.1) or plate reader (Subheading 3.2.2).

3.2.1 Cuvette Assay

1. Select each reaction time point containing 800 μL aliquot and 25 μL of NaOH, and add 50 μL of ZnOAc. Close the vials with a stopper (*see* **Note 7**).
2. Once all the time points have been collected, develop the reaction. Using a syringe, add 100 μL of DMPD and 100 μL of $FeCl_3$ to each vial and invert a few times.
3. Allow the reaction to proceed for 20 min to complete formation of methylene blue. Centrifuge individual reaction times points if the reaction is cloudy and read the absorbance at 650 nm with a spectrophotometer (*see* **Note 8**).
4. Prepare a standard curve of sodium sulfide under the same conditions to quantify the amount of sulfide produced in the reaction at each time point. A linear standard curve ranges from 5 to 100 nmol of sulfide (*see* **Note 9**).

3.2.2 Plate Assay

The volume of reagents, especially proteins, required for the cuvette assay is quite large and not always feasible. This detection method can be adapted to smaller reaction volumes in a clear 96-well plate.

1. Using a 96-well plate, stop each 200 μL reaction time point with 6.25 μL of the NaOH solution and 12.5 μL of the ZnOAc solution.

2. Once all time points are collected, seal the wells containing reaction time points with parafilm (*see* **Note 7**).
3. Using a syringe, add 25 μL of DMPD and 25 μL of $FeCl_3$, and reseal with parafilm after adding each reagent.
4. Carefully mix the reactions and allow methylene blue to form for 20 min.
5. If the sample is cloudy, centrifuge each time point for 5 min and pipette 200 μL of developed reaction into another clean, adjacent well. Determine the absorbance at 650 nm using a microplate reader (*see* **Note 8**).
6. Prepare a standard curve of sodium sulfide under the same conditions and use it as a reference. A linear standard curve ranges from 2 to 50 nmol of sulfide (*see* **Note 9**).

3.3 Alanine-NDA Detection Method

The reaction profile of cysteine desulfurases can be probed by quantifying the rate of alanine production determined via an NDA-derivatization reaction that forms a fluorescent product [23, 30, 38, 49, 50](Fig. 6). This detection method is not affected by the presence of reducing agents. Unlike the sulfide assay, this assay can be performed under nonreducing conditions or used to develop reactions performed with any reducing agent, including TCEP. Reaction time points are quenched with TCA and subsequently reacted with cyanide and NDA under basic conditions. Under these conditions, the primary amine of alanine reacts with NDA to form a fluorescent 1-cyano-2-alkyl-benz[f]isoindole adduct (*see* **Note 10**). This detection method is typically employed with HPLC separation of NDA-adducts coupled to fluorescence detection (Subheading 3.3.1). This method is very sensitive; as low as 0.1 pmol of alanine can be detected in a sample read.

3.3.1 NDA Assay with HPLC-Fluorescence Detection

1. Prepare reaction time points containing 100 μL of reaction aliquot and stop with 20 μL of 10% TCA, as described in Subheading 3.1 (*see* **Note 9**).

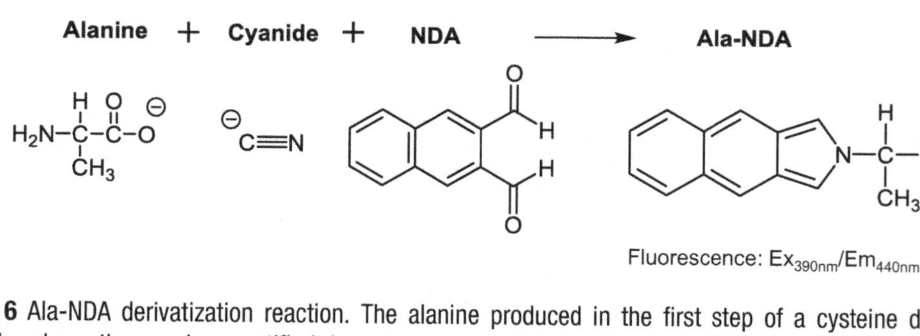

Fig. 6 Ala-NDA derivatization reaction. The alanine produced in the first step of a cysteine desulfurase-catalyzed reaction can be quantified through fluorescence detection of Ala-NDA adduct

2. Once all reaction time points are quenched, spin tubes for 5 min and transfer 100 μL of supernatant into HPLC vials (*see* **Note 1**).
3. Add 1 mL of freshly prepared NDA mix (0.1 mM NDA and 1 mM KCN in 100 mM borate buffer pH 9) and incubate vials in the dark for 1 h (*see* **Notes 11–13**).
4. Inject 10 μL of each sample onto a C-18 column connected to an HPLC coupled to a fluorescence detector (Ex_{390nm}/Em_{440nm}) at a flow rate of 0.5 mL/min. Run a gradient of 0–100% B over 12 min using 5% water/MeOH (Solvent B) and 5% MeOH/water (Solvent A). The Ala-NDA adduct elutes at 8 min.
5. Prepare a calibration curve of alanine under the same conditions to quantify the integrated fluorescence peak area of reaction time points. A linear standard curve ranges from 1 to 50 pmol of Ala-NDA adduct injected in each HPLC run.

3.3.2 NDA Assay with Direct Fluorescence Detection

The alanine vial assay protocol can be adapted for direct quantification using a black 96-well plate and plate reader fluorimeter. In this method layout, the background signal is significantly higher; as a consequence, this method is about ten times less sensitive than the HPLC method (Subheading 3.2.1). For direct fluorescence detection, the standard curve and the sample must be derivatized and read simultaneously as the fluorescence levels change with time of incubation.

1. Prepare reaction time points containing 50 μL of reaction aliquot stopped with 5 μL of 10% TCA and mix gently by pipetting up and down.
2. Once all reaction time points are completed, set up a standard curve of alanine under the same conditions (*see* **Note 14**).
3. Add 200 μL of freshly prepared NDA mix to each sample and standard to allow the Ala-NDA to form.
4. Allow the reaction to develop in the dark for an hour before measuring the fluorescence at Ex_{390nm}/Em_{440nm} in a microplate fluorescence reader.

4 Notes

1. For the detection of alanine, it is important for enzymes to be isolated in a buffer that does not contain a primary amine since these functional groups also react with NDA thereby interfering with the derivatization reaction. For the detection of sulfide, other buffers can be used, including Tris-HCl, phosphate, bicine, and HEPES. An important variant is the pH of reaction

assays. The reaction profile is dependent on pH with a defined pK_a and varies with the enzyme of study. The maximum activity of the *B. subtilis* SufS, YrvO, and NifZ is achieved at pH 8.

2. Experiments involving sulfur transfer reactions coupled to the assembly of Fe-S clusters require conditions devoid of oxygen. In these experiments, proteins are purified anaerobically. The frozen aliquots may be thawed under argon or nitrogen bubbling gas or directly transferred to the anaerobic chamber for Fe-S synthesis assays. The buffers should be degassed for at least 30 min while stirring with alternate sparging of nitrogen or argon gas cycles. Iron and cysteine solutions should be prepared inside the glove box on the day of the experiment.
3. For the method described, the activity of the *B. subtilis* SufS containing 10 equivalents of SufU in the presence of 2 mM DTT and 0.5 mM cysteine at pH 8 is 13 min^{-1} (280 nmol/min/mg). However, different cysteine desulfurases display distinct catalytic rates varying from 0.01 to 100 min^{-1}. Their turnover rate is dependent on the presence and concentration of sulfur acceptors, type of reducing agent, pH, and concentration of cysteine. Thus, the amount of enzyme used in the assay needs to take into consideration the activity of the enzyme and the sensitivity of the detection method used.
4. Reagents such as buffer, DTT, iron, and glutathione interfere with methylene blue quantification. Tris (2-carboxyethyl) phosphine (TCEP) reduces methylene blue and cannot be used as a reducing agent in sulfide assays. Physiological reducing agents, such as the thioredoxin/thioredoxin reductase system, provide a controlled physiological reductant system for effective reduction of persulfides.
5. The quantity of each reagent can be adjusted based on the reactivity of the cysteine desulfurase. To limit the formation of free Fe-S species, limit the concentration of Fe to no more than 10 molar equivalents of the scaffold present in the reaction.
6. For anaerobic assays, reactions can be developed outside the anaerobic chamber.
7. The $FeCl_3$ and DPMD solutions are acidic and lead to the formation of hydrogen sulfide gas. Thus, the formation of methylene blue in reaction time points must be performed in closed reaction containers or wells sealed with parafilm to prevent loss of this analyte.
8. Quenching the reaction with NaOH solution along with developing reactants leads to protein precipitation. For assays performed with higher protein concentrations, the sample may become cloudy, interfering with absorbance readings. In

those cases, centrifuge the samples for 5 min and use the supernatant for the absorbance measurements.

9. Prepare the standard curve with the same buffer, reducing agent, and the same cysteine concentration used in the assay. These components affect the absorbance of methylene blue.
10. NDA generally reacts with primary amines. Thus, buffers that have primary amines, such as Tris, should be avoided. Though cysteine has a primary amine, reaction with NDA results in an unstable adduct with lower fluorescence intensity.
11. Centrifuge reaction time points and transfer the supernatant into the vial before developing to prevent clogging the HPLC column.
12. NDA and KCN solutions can be prepared once every week, but the NDA mix should be prepared right before the derivatization reaction.
13. The fluorescence of the Ala-NDA adduct decays after 4 h, and hence, reaction time points must be developed accordingly. While several assays can be carried simultaneously, reaction with the NDA mix can be spaced based on the HPLC running time.
14. When performing assays using different concentrations of Cys, complete Ala standard curve using solutions containing the same concentration of Cys used in the assays.

Acknowledgments

The research and methodology developed in the PDS laboratory was supported by the National Science Foundation (1716535). A.M.E. was a recipient of a T32 NIH predoctoral fellowship (GM 95440-9).

References

1. Johnson DC, Dean DR, Smith AD et al (2005) Structure, function, and formation of biological iron-sulfur clusters. Annu Rev Biochem 74:247–281
2. Beinert H, Holm RH, Munck E (1997) Iron-sulfur clusters: nature's modular, multipurpose structures. Science 277:653–659
3. Py B, Barras F (2010) Building Fe-S proteins: bacterial strategies. Nat. Rev Microbiol 8:436–446
4. Peters JW, Broderick JB (2012) Emerging paradigms for complex iron-sulfur cofactor assembly and insertion. Annu Rev Biochem 81:429–450
5. Lee SC, Lo W, Holm RH (2014) Developments in the biomimetic chemistry of cubane-type and higher nuclearity iron-sulfur clusters. Chem Rev 114:3579–3600
6. Malkin R, Rabinowitz JC (1966) The reconstitution of clostridial ferredoxin. Biochem Biophys Res Commun 23:822–827
7. Dos Santos PC, Dean DR (2014) A retrospective on the discovery of [Fe-S] cluster biosynthetic machineries in Azotobacter vinelandii. In: Rouault T (ed) Iron-sulfur clusters in chemistry and biology. Verlag Walter de Gruyter, Berlin

8. Zheng L, White RH, Cash VL et al (1993) Cysteine desulfurase activity indicates a role for NIFS in metallocluster biosynthesis. Proc Natl Acad Sci U S A 90:2754–2758
9. Zheng L, Cash VL, Flint DH et al (1998) Assembly of iron-sulfur clusters. Identification of an iscSUA-hscBA-fdx gene cluster from Azotobacter vinelandii. J Biol Chem 273:13264–13272
10. Takahashi Y, Nakamura M (1999) Functional assignment of the ORF2-iscS-iscU-iscA-hscB-hscA-fdx-ORF3 gene cluster involved in the assembly of Fe-S clusters in Escherichia coli. J Biochem (Tokyo) 126:917–926
11. Lill R, Kispal G (2000) Maturation of cellular Fe-S proteins: an essential function of mitochondria. Trends Biochem Sci 25:352–356
12. Ellis KE, Clough B, Saldanha JW et al (2001) Nifs and Sufs in malaria. Mol Microbiol 41:973–981
13. Lagier MJ, Tachezy J, Stejskal F et al (2003) Mitochondrial-type iron-sulfur cluster biosynthesis genes (IscS and IscU) in the apicomplexan Cryptosporidium parvum. Microbiology 149:3519–3530
14. Outten FW (2015) Recent advances in the Suf Fe-S cluster biogenesis pathway: beyond the Proteobacteria. Biochem Biophys Act 1853:1464–1469
15. Braymer JJ, Freibert SA, Rakwalska-Bange M et al (2020) Mechanistic concepts of iron-sulfur protein biogenesis in biology. Biochim Biophys Acta, Mol Cell Res 1868:118863
16. Maio N, Rouault TA (2020) Outlining the complex pathway of mammalian Fe-S cluster biogenesis. Trends Biochem Sci 45:411–426
17. Lukes J, Basu S (2015) Fe/S protein biogenesis in trypanosomes—a review. Biochem Biophys Act 1853:1481–1492
18. Dos Santos PC (2017) *B. subtilis* as a model for studying the assembly of Fe-S clusters in gram-positive bacteria. Methods Enzymol 595:185–212
19. Mueller EG (2006) Trafficking in persulfides: delivering sulfur in biosynthetic pathways. Nat Chem Biol 2:185–194
20. Black KA, Dos Santos PC (2015) Shared-intermediates in the biosynthesis of thio-cofactors: mechanism and functions of cysteine desulfurases and sulfur acceptors. BBA-Mol Cell Res 1853:1470–1480
21. Zheng L, White RH, Cash VL et al (1994) Mechanism for the desulfurization of L-cysteine catalyzed by the *nifS* gene product. Biochemistry 33:4714–4720
22. Blahut M, Wise CE, Bruno MR et al (2019) Direct observation of intermediates in the SufS cysteine desulfurase reaction reveals functional roles of conserved active-site residues. J Biol Chem 294:12444–12458
23. Selbach B, Earles E, Dos Santos PC (2010) Kinetic analysis of the bisubstrate cysteine desulfurase SufS from *Bacillus subtilis*. Biochemistry 49:8794–8802
24. Kessler D (2006) Enzymatic activation of sulfur for incorporation into biomolecules in prokaryotes. FEMS Microbiol Rev 30:825–840
25. Hidese R, Mihara H, Esaki N (2011) Bacterial cysteine desulfurases: versatile key players in biosynthetic pathways of sulfur-containing biofactors. Appl Microbiol Biotechnol 91:47–61
26. Zheng C, Dos Santos PC (2018) Metallocluster transactions: dynamic protein interactions guide the biosynthesis of Fe-S clusters in bacteria. Biochem Soc Trans 46:1593–1603
27. Mihara H, Esaki N (2002) Bacterial cysteine desulfurases: their function and mechanisms. Appl Microbiol Biotechnol 60:12–23
28. Black KA, Dos Santos PC (2015) Abbreviated pathway for biosynthesis of 2-Thiouridine in *Bacillus subtilis*. J Bacteriol 197:1952–1962
29. Nakamura R, Hikita M, Ogawa S et al (2020) Snapshots of PLP-substrate and PLP-product external aldimines as intermediates in two types of cysteine desulfurase enzymes. FEBS J 287:1138–1154
30. Selbach BP, Chung AH, Scott AD et al (2014) Fe-S cluster biogenesis in gram-positive bacteria: SufU is a zinc-dependent sulfur transfer protein. Biochemistry 53:152–160
31. Raulfs EC, O'carroll IP, Dos Santos PC et al (2008) In vivo iron-sulfur cluster formation. Proc Natl Acad Sci U S A 105:8591–8596
32. Dai Y, Outten FW (2012) The *E. coli* SufS-SufE sulfur transfer system is more resistant to oxidative stress than IscS-IscU. FEBS Lett 586:4016–4022
33. Outten FW, Wood MJ, Munoz FM et al (2003) The SufE protein and the SufBCD complex enhance SufS cysteine desulfurase activity as part of a sulfur transfer pathway for Fe-S cluster assembly in Escherichia coli. J Biol Chem 278:45713–45719
34. Loiseau L, Ollagnier-De Choudens S, Lascoux D et al (2005) Analysis of the heteromeric CsdA-CsdE cysteine desulfurase, assisting Fe-S cluster biogenesis in Escherichia coli. J Biol Chem 280:26760–26769
35. Kambampati R, Lauhon CT (1999) IscS is a sulfurtransferase for the in vitro biosynthesis of 4-thiouridine in Escherichia coli tRNA. Biochemistry 38:16561–16568
36. Ikeuchi Y, Shigi N, Kato J et al (2006) Mechanistic insights into sulfur relay by multiple

sulfur mediators involved in thiouridine biosynthesis at tRNA wobble positions. Mol Cell 21:97–108

37. Lehrke M, Rump S, Heidenreich T et al (2012) Identification of persulfide-binding and disulfide-forming cysteine residues in the NifS-like domain of the molybdenum cofactor sulfurase ABA3 by cysteine-scanning mutagenesis. Biochem J 441:823–832
38. Selbach BP, Pradhan PK, Dos Santos PC (2013) Protected sulfur transfer reactions by the Escherichia coli Suf system. Biochemistry 52:4089–4096
39. Martin HL, Black KA, Dos Santos PC (2015) Functional investigation of *Bacillus subtilis* YrkF's involvement in sulfur transfer reactions. Peptidomics 34:54
40. Cartini F, Remelli W, Dos Santos PC et al (2010) Mobilization of sulfane sulfur from cysteine desulfurases to the Azotobacter vinelandii sulfurtransferase RhdA. Amino Acids 41 (1):141–150
41. Bridwell-Rabb J, Fox NG, Tsai CL et al (2014) Human frataxin activates Fe-S cluster biosynthesis by facilitating sulfur transfer chemistry. Biochemistry 53:4904–4913
42. Webert H, Freibert SA, Gallo A et al (2014) Functional reconstitution of mitochondrial Fe/S cluster synthesis on Isu1 reveals the involvement of ferredoxin. Nat Commun 5:5013
43. Gervason S, Larkem D, Mansour AB et al (2019) Physiologically relevant reconstitution of iron-sulfur cluster biosynthesis uncovers persulfide-processing functions of ferredoxin-2 and frataxin. Nat Commun 10:3566
44. Rajakovich LJ, Tomlinson J, Dos Santos PC (2012) Functional analysis of *Bacillus subtilis* genes involved in the biosynthesis of 4-Thiouridine in tRNA. J Bacteriol 194:4933–4940
45. Zheng C, Guo S, Tennant WG et al (2019) The Thioredoxin system reduces protein Persulfide intermediates formed during the synthesis of Thio-cofactors in *Bacillus subtilis.* Biochemistry 58:1892–1904
46. Tirupati B, Vey JL, Drennan CL et al (2004) Kinetic and structural characterization of Slr0077/SufS, the essential cysteine desulfurase from *Synechocystis sp.* PCC 6803. Biochemistry 43:12210–12219
47. Beinert H (1983) Semi-micro methods for analysis of labile sulfide and of labile sulfide plus sulfane sulfur in unusually stable iron-sulfur proteins. Anal Biochem 131:373–378
48. Siegel LM (1965) A direct microdetermination for sulfide. Anal Biochem 11:126–132
49. Buhning M, Friemel M, Leimkuhler S (2017) Functional complementation studies reveal different interaction partners of Escherichia coli IscS and human NFS1. Biochemistry 56:4592–4605
50. Kim D, Singh H, Dai Y et al (2018) Changes in protein dynamics in Escherichia coli SufS reveal a possible conserved regulatory mechanism in type II cysteine Desulfurase systems. Biochemistry 57:5210–5217

Chapter 11

A Fast and Ratiometric Method for Quantification of Cysteine-Bound Persulfides Based on Alkylation and Gel-Shift Assays

Sylvain Gervason, Batoul Srour, and Benoit D'Autréaux

Abstract

Cysteine-bound persulfides (Cys-SSH) in proteins are sulfur carrier intermediates in the synthesis of essential cofactors such as iron-sulfur clusters, molybdenum cofactor, vitamin (thiamine), and thionucleosides (thiolated tRNA). Protein-bound persulfides are also used for signaling purposes as a carrier of the "H_2S" signal. Several methods have been developed to detect and quantify cysteine-bound persulfides in protein and monitor their exchange. The main challenge in developing these techniques is to discriminate persulfidated cysteine from cysteine and other cysteine modifications. It is also critical to develop ratiometric methods to quantify the level of persulfidation in the protein of interest. We describe here a ratiometric method to label and quantify protein-bound persulfides relying on alkylation and gel-shift assays. This method is based on the derivation of cysteine and persulfides with "heavy" alkylating agents, followed by specific cleavage of the sulfur-sulfur bond of the alkylated persulfide by a reducing agent and separation of the alkylated species by electrophoresis. A persulfide is thus revealed by the appearance of a species lacking one alkylation unit under reducing conditions. We call this alkylation-reduction band-shift (ARBS) assay. Moreover, the quantification of the bands corresponding to the persulfidated and non-persulfidated species in the same lane provides a ratiometric quantification allowing determination of the level of persulfidation of individual cysteine. Other cysteine modifications such as disulfides, sulfenic, sulfinic, sulfonic acids, nitrosothiols, and sulfenamides preclude alkylation. Thus, they may appear as false positives, but they are ruled out by the analysis under nonreducing conditions since these species do not behave as persulfides under these conditions.

Key words Persulfide, Sulfane, Sulfur, Alkylation, Maleimide, Ratiometric, Iron-sulfur cluster, Thionucleosides, Molybdenum cofactor, Persulfidation, Sulfhydration, H2S, Signaling, Band-shift, Cysteines modifications, Disulfides, Sulfenic, Sulfinic, Sulfonic acids, Nitrosothiols, Sulfenamide

1 Introduction

A growing interest in persulfide biology has emerged with the discovery that they function as sulfur carrier intermediates in the synthesis of iron-sulfur clusters, thionucleosides, molybdenum cofactor, and thiamine [1, 2]. In the last decades, a growing interest in the study of persulfides has emerged with the discovery that

Patricia C. Dos Santos (ed.), *Fe-S Proteins: Methods and Protocols*, Methods in Molecular Biology, vol. 2353, https://doi.org/10.1007/978-1-0716-1605-5_11,

protein-bound persulfides are also used for signaling purposes as a carrier of the "H_2S" signal [3–5]. In proteins, persulfides are formed with the thiol moiety of cysteine residues (Cys-SSH). They belong to a particular type of disulfides in which one sulfur is not linked to a carbon but to a hydrogen as in thiols. The generic formula of persulfides is R(R′)(R″)CSSH, where R, R′, and R″ are any type of substituents, and the terminal sulfur (the sulfhydryl) is called sulfane sulfur. As for disulfides, the sulfur–sulfur bond of persulfides reacts with thiols, which allows the exchange of sulfur atoms between cysteine residues via trans-persulfidation reactions [3].

Several methods have been developed to detect and quantify persulfidated cysteines in proteins, both in cellulo and in vitro [3–5]. Mass spectrometry is one of the most powerful techniques to detect these species but is not suitable for routine and in vivo studies [6, 7]. Radioactive labeling by ^{35}[S] has been successfully used, but is not a ratiometric method allowing quantitative detection and presents hazard issues [8–11]. Since the modification induced by persulfidation is too small (+32 Da) to be detected by conventional electrophoresis techniques and its absorption coefficient (~400 M^{-1} cm^{-1} at 340 nm) is too low for sensitive detection by electronic absorption spectroscopy [12–14], most detection techniques rely on the derivation of persulfides. The major challenge in developing these techniques is to discriminate a persulfide from a cysteine thiol since both are nucleophiles. It is also challenging to discriminate this modification from other modified forms of cysteine, such as disulfide bonds, sulfenic, sulfinic, sulfonic acids, nitrosothiols, and sulfenamides. Another issue is to obtain ratiometric quantification that is essential to measure the level of persulfidation of a particular cysteine residue. A ratiometric method entails that both the persulfidated and non-persulfidated forms are quantified simultaneously to afford the most accurate results. The last challenge is to determine the number of persulfidated cysteines, which requires either a determination of the total number of persulfides per protein or the separation and the identification of each persulfidated form. Finally, a protocol with a limited number of steps warrants better reproducibility and facilitates routine studies. Our laboratory has developed the alkylation-reduction band-shift (ARBS) assay, which is a specific, ratiometric, and rapid technique to quantify protein persulfides and perform kinetic studies routinely [7].

1.1 Principle of the ARBS Assay

This assay is based on redox-sensitive labeling of persulfides using "heavy" alkylating agents to allow separation of each persulfidated species by gel electrophoresis. The terminology "heavy" is used here to describe an alkylating agent carrying a substituent sufficiently heavy to increase the molecular weight of the protein and enables separation of poly-alkylated species of different sizes. The

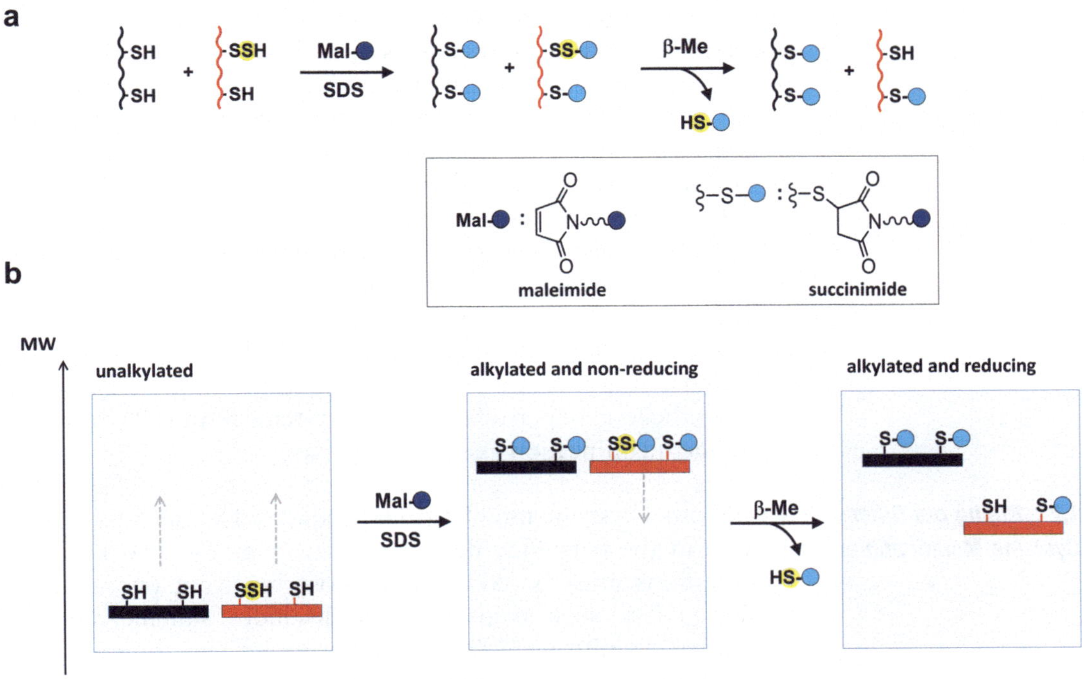

Fig. 1 Alkylation assay principle. (**a**) Scheme of the ARBS assay. The proteins are first reacted with a heavy alkylating agent, a maleimide bound to a heavy substituent (deep blue balls) under denaturing conditions (SDS). The alkylating agent reacts with all sulfhydryl (cysteine residues, persulfide, and free L-cysteine) leading to formation of a succinimide-sulfur link carrying the heavy substituent (light blue balls). In a second step, the reaction mixture is reduced by β-mercaptoethanol. (**b**) Expected migration profiles on SDS-PAGE at each step of the ARBS assay. An example is provided for a protein containing two cysteine residues, and that is partially persulfidated on one of them. Before alkylation (**unalkylated**), the persulfidated (red bar) and non-persulfidated (black bar) protein co-migrate at the expected size of the protein. Upon alkylation, both cysteines and persulfidated cysteines are alkylated by the heavy maleimide, which increases the size of the protein by two alkylation unit. Under these conditions (**alkylated and nonreducing**), the persulfidated and non-persulfidated alkylated protein still co-migrate. Upon reduction of the S–S bond of the alkylated persulfide by β-mercaptoethanol, a sulfur-succinimide-substituent moiety is lost, which decreases the size of the protein by one alkylation unit (**alkylated and reducing**). Then, the bands corresponding to the persulfidated and non-persulfidated alkylated proteins are quantified by light densitometry in the same well, which provides a ratiometric measurement of the persulfidation level

labeling part relies on two successive reactions (Fig. 1a). First, the protein mixture is incubated under denaturing conditions with a maleimide compound, which is a thiol-specific alkylating agent (*see* **Note 1**). This reaction leads to the alkylation of both cysteine and cysteine-bound persulfides in the form of sulfur-bound succinimide group linked to the sulfur of cysteines and the sulfane sulfur of persulfides. To differentiate cysteine and persulfides, the sulfur–sulfur bond of the persulfide is cleaved by a reducing agent such as β-mercaptoethanol (*see* **Note 2**), which leads to the release of the sulfur-succinimide moiety with its substituent. As reducing agents are not able to cleave the sulfur-carbon bond within the sulfur-succinimide attached to cysteine residues, the release of a sulfur-succinimide-substituent moiety is specific to the presence of a persulfide. Then, the reaction mixture is analyzed by electrophoresis on gel (SDS-PAGE) to separate the alkylated forms of the proteins (Fig. 1b). The substituent linked to the maleimide must be sufficiently heavy to allow visualization of its loss by SDS-PAGE.

Moreover, the mass loss is directly related to the number of persulfides in the protein by counting the number of mass-unit lost using a ladder of partially alkylated species (*see* Subheading 3.1 for its generation). Then, the bands corresponding to the fully alkylated protein (non-persulfidated) and partially alkylated species (persulfidated species) are quantified by light densitometry, which provides the level of persulfidation of a cysteine residue. We call this persulfide quantification assay (ARBS assay).

1.2 Ruling out Other Cysteine Modifications

Several other cysteine modifications produce effects akin to persulfidation in the persulfide detection assay (Fig. 2). Two kinds of modifications must be considered: those involving one cysteine residue (1-Cys) such as inter-protein disulfides, sulfenic, sulfinic, sulfonic acids, nitrosothiols, and sulfenamides and the modifications involving two cysteine residues on the same protein (2-Cys), the only representative being intramolecular disulfide bonds (*see* **Note 3**). As both 1-Cys and 2-Cys modifications preclude full alkylation at the first step, they will appear as partially alkylated species, which could constitute false positives for persulfides. However, the analysis of the reaction before reduction will discard false positives since all these modifications have very specific behaviors under nonreducing conditions.

1.2.1 Characterization of 1-Cys Modifications

Among 1-Cys modifications, sulfenic, sulfinic, sulfonic acids, nitrosothiols, and sulfenamides produce distinct features on gels compared to inter-protein disulfides (Fig. 2). Under nonreducing conditions, inter-protein disulfides appear as high molecular weight complexes between the protein of interest (Fig. 2, **pink species**) and another protein or the same protein in the case of a homodimer (Fig. 2, **orange species**), while protein carrying sulfenic, sulfinic, sulfonic acids, nitrosothiol, or sulfenamide remain monomeric under both reducing and nonreducing conditions (Fig. 2, **green species**). Upon reduction, the intermolecular disulfide bond will be cleaved to yield a species missing one alkylation mass-unit that could be confounded with a persulfide (Fig. 2, compare pink and red species). But the presence of a complex of high molecular weight under nonreducing conditions that disappears upon reduction will provide a proof that this is a false positive. Similarly, proteins carrying 1-Cys modifications will appear as species missing one or more alkylation units under reducing conditions, which again could be confounded with persulfides (Fig. 2, compare green and red species). However, under nonreducing conditions,

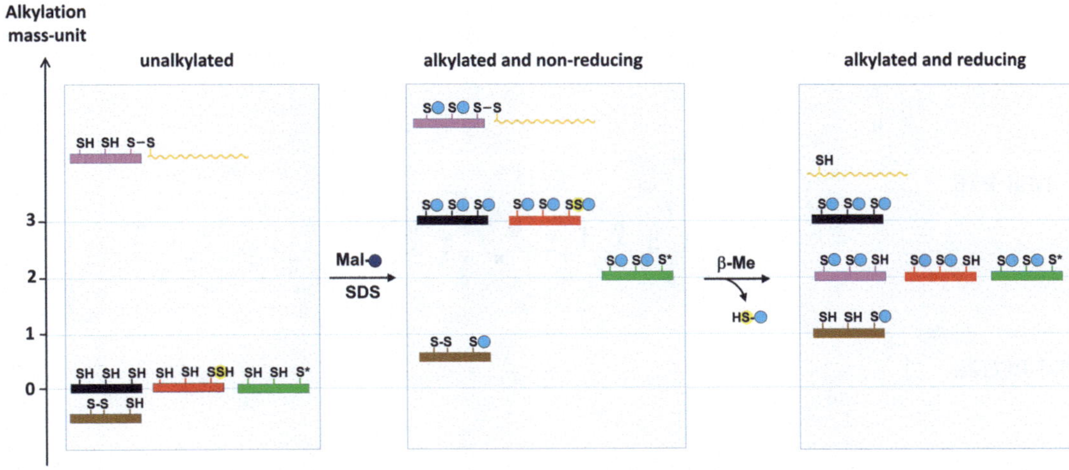

Fig. 2 Expected migration profiles for other cysteine modifications in the ARBS assay. The migration profiles in the ARBS assay for the most common cysteine modifications is presented with a protein containing 3 cysteine residues either not modified (black species), modified on one of cysteine residue by persulfidation (red species), or by the formation of a sulfenic, sulfinic, sulfonic acids, a nitrosothiol, or a sulfenamide (green species with an asterisk on S to indicate the modification) or by the formation of an intermolecular disulfide bond (pink species linked to orange species) and modified on two cysteines by the formation of an intramolecular disulfide bond (brown species). The migration profiles are analyzed before alkylation (**unalkylated**) and after alkylation by a maleimide derivative (**alkylated**), under nonreducing (**alkylated and nonreducing**) and reducing (**alkylated and reducing**) conditions. Under reducing conditions, 1-Cys modified proteins (pink and green species) co-migrate with persulfidated proteins on 1 cysteine (red species), and 2-Cys modified proteins (brown species) co-migrate with proteins persulfidated on 2 cysteines (not represented), which constitute false positives. Under nonreducing conditions, 1-Cys and 2-Cys modified proteins do not co-migrate with the persulfidated protein (red species), which identifies them as false positives

these species will already appear as partially alkylated while a persulfidated protein is fully alkylated. Thereby, again, the analysis of nonreducing conditions will provide a means to discard these false positives.

1.2.2 Characterization of 2-Cys Modifications

Proteins oxidized in the form of an intramolecular disulfide, which is the only representative of 2-Cys modifications, contains two unreactive cysteines toward alkylating agents. These species will be lacking two alkylation unit under reducing conditions and thus could be confounded with the presence of two persulfides (Fig. 2 **brown species**). However, as for 1-Cys modification, the analysis under nonreducing conditions will reveal that this species is not fully alkylated, while a persulfide will be. In addition, the presence of a disulfide can alter protein migration, especially if the disulfide involves distant cysteines. Upon reduction of the disulfide, its migration will be further affected, which could provide another way to pinpoint the presence of intramolecular disulfides.

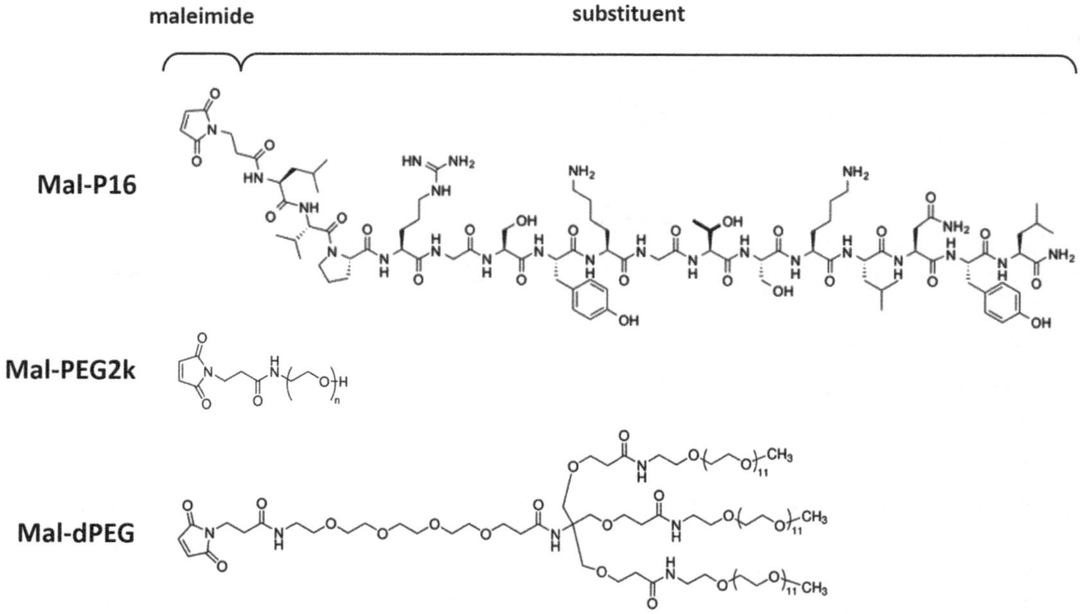

Fig. 3 Structure of selected "heavy" maleimide derivatives. Heavy maleimide derivatives are generated by connecting a heavy substituent to the nitrogen atom of the heterocycle ring of the maleimide moiety. Here, the structures of "heavy" maleimide derivatives are presented with three different substituents: a 16-mer peptide (Mal-P16, MW = 1.95 kDa), a PEG polymer of mean size of 2 kDa (Mal-PEG2k, MW ≈ 2 kDa), and a branched chained poly-PEG substituent (Mal-dPEG, MW = 2.36 kDa)

1.3 Choice of Alkylating Agents

Heavy maleimides are created by linking a substituent to the nitrogen atom of the heterocycle ring of the maleimide moiety (Fig. 3). Three maleimide derivatives with different substituents are depicted here (Fig. 3). A maleimide fused to a homemade 16-mer peptide (Leu-Val-Pro-Arg-Gly-Ser-Tyr-Lys-Gly-Thr-Ser-Lys-Leu-Asn-Tyr-Leu), resulting in a 1.95 kDa compound named Mal-P16, and two commercially available compounds, one with a polyethylene glycol (PEG) polymer tail of mean size of 2 kDa (Mal-PEG2k) and another one with a branched chained poly-PEG substituent (Mal-dPEG$_4$-(m-dPEG$_{12}$)$_3$), resulting in a 2.36 kDa derivative named Mal-dPEG. The main goal in using a peptide (Mal-P16) instead of PEG derivatives (Mal-PEG, Mal--dPEG) is to ensure that migration in SDS-PAGE is not aberrant and directly correlated to the size of the substituent. However, the dPEG substituent provides reliable results too (Fig. 4a). The choice of the most suitable alkylating agent relies on two main criteria: (1) its ability to produce the thinnest bands, which is linked to the nature of the substituent; and (2) the resolution of the alkylation pattern, that is, the separation between each alkylated species, which is linked to the size of the protein of study, the number of cysteine residues, and the size of the maleimide derivative. This must be evaluated on a case-by-case basis by generating an alkylation profile prior to any assay.

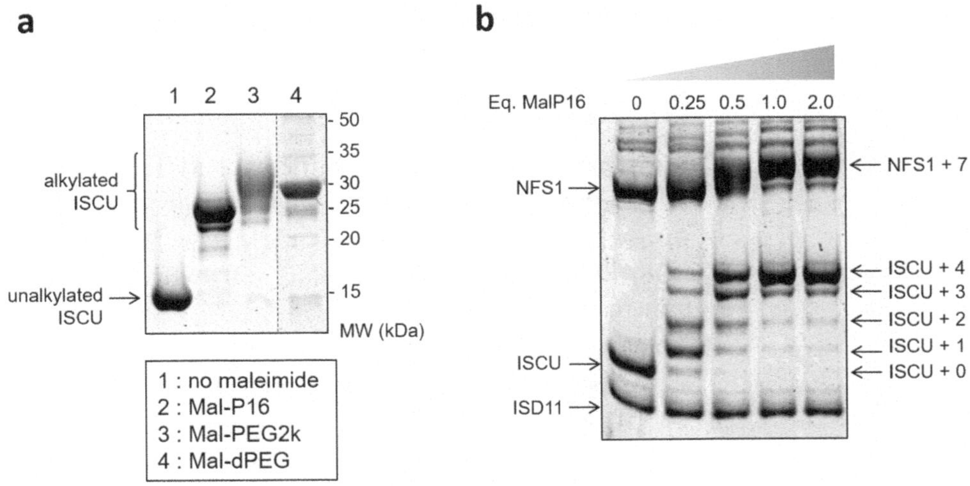

Fig. 4 Alkylation profiles. (**a**) Alkylation profiles of mouse ISCU with Mal-P16, Mal-PEG2k, and Mal-dPEG. The reactions were performed by adding 2 μL of the alkylating reagent solution at 4.0 mM to 50 μL of the ISCU protein at 20 μM, which corresponds to 2 molar equivalents relative to the total number of cysteines since ISCU contains four cysteine residues. The Mal-PEG2k shows very fussy bands due to its heterogeneity, while Mal-P16 and Mal-dPEG display much thinner bands (*see* **Note 4**). The estimated molecular weights of the main alkylated species with Mal-P16 and Mal-dPEG are 25 kDa and 28 kDa, respectively, thus very close to those expected for the alkylation of all four cysteine residues of ISCU at 23.5 kDa and 25 kDa, which indicates that the most intense bands correspond to fully alkylated ISCU. (**b**) Titration of the cysteines in the mouse NFS1-ISD11-ACP-ISCU (NIAU) complex by Mal-P16. NFS1 contains seven cysteines, ISD11 no cysteine and the cysteine residue of ACP is not available. Thus, a total of 11 reactive thiols are present in the NIAU complex. The reactions were performed by adding 0, 1, 2, 4, and 8 μL of the 2.75 mM maleimide reagent solution (Mal-P16/SDS) to 50 μL of the 20 μM NIAU complex, which corresponds to 0.25, 0.5, 1, and 2 molar equivalents of Mal-P16 versus total thiols. The alkylated species are denoted ISCU + n and NFS1 + n, where n is the number of alkylated cysteines. The reaction performed with sub-stoichiometric amounts of Mal-P16 (0.25 equivalent) shows five bands that were attributed to unalkylated ISCU (ISCU+0) and ISCU alkylated on one (ISCU+1), two (ISCU+2), three (ISCU+3), and four (ISCU+4) cysteine residues. The titrations show that 1 molar equivalent is sufficient to fully alkylate both ISCU and NFS1, which corresponds to the minimum required, thus indicating that the reaction is efficient

2 Materials

1. Phosphate buffer: 50 mM sodium phosphate, 150 mM sodium chloride, pH 8.
2. Protein sample: purified protein prepared in phosphate buffer (*see* **Note 5**).
3. Mal-P16 (synthesized at the laboratory [7]), Mal-PEG2k (Sigma), Mal-dPEG (Sigma) stock solutions: 5 mM solution prepared in deionized water and stored at −20 °C.
4. Maleimide reagent solution: alkylating agent at the desired concentration, 1% sodium dodecyl sulfate (SDS) freshly prepared in deionized water from stock solutions of the alkylating agent.

5. TCA: 20% trichloroacetic acid prepared in water and stored at 4 °C.
6. Resuspension buffer: 50 mM sodium phosphate, 8 M urea, 150 mM NaCl, pH 8.
7. 5× nonreducing gel-loading buffer: 0.1 M Tris–HCl pH 6.8, 10% SDS, 0.05% bromophenol blue, 50% glycerol.
8. 5× reducing gel loading buffer: 0.1 M Tris–HCl pH 6.8, 10% SDS, 0.05% bromophenol blue, 50% glycerol, 0.5 mM β-mercaptoethanol.
9. SDS-PAGE: 14% and 10% polyacrylamide gels, 0.1% SDS. Gels are prepared following standard procedures with acrylamide: bis-acrylamide ratio of 19:1.
10. Gel imager: Gel imager for band quantification such as Odyssey scanner (Li-Cor) (*see* **Note 6**).
11. Coomassie staining solution: 1 g/L Coomassie blue R 250, 45% ethanol, 45% water, 10% glacial acetic acid.
12. Coomassie destaining solution: 80% water, 10% ethanol, 10% glacial acetic acid.
13. L-Cysteine solution: 10 mM L-cysteine prepared in deionized water.

3 Methods

All the assays presented here were developed with the mouse iron-sulfur cluster assembly machinery, which encompasses NFS1-ISD11-ACP (NIA), a cysteine desulfurase complex that generates persulfide on NFS1 by desulfurization of L-cysteine and transfers its persulfide to ISCU within the NFS1-ISD11-ACP-ISCU (NIAU) complex, in a reaction stimulated by frataxin (FXN) in the NFS1-ISD11-ACP-ISCU-FXN (NIAUF) complex [7, 15]. Nevertheless, these methodologies can be adapted to the study of any other proteins involved in persulfide formation and transfer.

3.1 Generation of Alkylation Profiles

The protocols described below are used to probe the alkylation profile of the protein of interest (Fig. 4). These experiments help to choose the most suitable alkylating agent, assess its efficiency, and provide a ladder of the alkylated species that will be used in the persulfidation assays to count the number of persulfidated cysteines. The protocol is presented here for the ISCU protein and the NIAU complex, but can be applied to any other proteins.

3.1.1 Selection of Alkylating Agents

1. Reduce the protein of interest with 10 molar equivalents of DTT for 30 min at room temperature, then desalt the protein on a Sephadex G-25 column equilibrated with phosphate buffer.

2. Initiate the reaction by mixing 50 μL of the protein at a final concentration of 20 μM with a twofold excess of the maleimide reagent regarding the total number of cysteines (*see* **Note 7**) and incubate the reaction for 30 min at room temperature.
3. Add 5 μL of reducing loading buffer to 20 μL of the reaction mixture and incubate for 30 min at room temperature.
4. Load a volume containing 200 pmoles of the protein solution onto 14% acrylamide gel (*see* **Notes 5** and **8**). Run the gel for 2 h at 90 V, stain the gel with Coomassie blue solution, and destain. Scan the gel with an imager.
5. Analyze the results (Fig. 4a). Compare the alkylation profiles obtained with the maleimide derivatives and select the alkylating agent generating the thinnest bands (*see* **Note 4**). Check that the size of the alkylated species on the gel is consistent with the expected size according to the total number of cysteines; otherwise, it means that some of the cysteine residues of the protein are oxidized (*see* Subheading 1.2). In this case, repeat **step 1** with higher DTT concentrations to try to get rid of reversible oxidations.

3.1.2 Generation of a Ladder and Assessment of the Efficiency of the Alkylating Agent

1. Reduce the protein of interest with 10 molar equivalents of DTT for 30 min at room temperature, then desalt the protein on a Sephadex G-25 column equilibrated with phosphate buffer.
2. Initiate the reaction by mixing 50 μL of the protein of interest at 20 μM with 0.25, 0.5, 1, and 2-molar equivalents of the maleimide reagent regarding the total number of thiols (*see* **Note** 7). Incubate the reaction for 30 min at room temperature.
3. Add 5 μL of loading buffer to 20 μL of the reaction mixture and incubate for 30 min at room temperature.
4. Load 200 pmoles of the protein solution onto 14% acrylamide gel (*see* **Notes 5** and **8**). Run the gel for 2 h at 90 V and stain with the commasie blue and then destain using acetic acid +ethanol+ water solution. Scan the gel with an imager.
5. Analyze the results (Fig. 4b). First, count the number of bands corresponding to the alkylated species in the reactions performed with sub-stoichiometric amounts of the alkylating agent. The number of bands is expected to be equals to the total number of cysteines with an additional band corresponding to the unalkylated species. In any other cases, a cysteine is probably oxidized (*see* Subheading 1.2). In this case, repeat **step 1** with higher DTT concentrations to try to get rid of reversible oxidations. Note the position of each alkylated species, which will be used to determine the number of persulfidated cysteines (*see* Subheading 3.2). Check how many

equivalents of the alkylating agent are required to saturate all the thiols, which should be close to the stoichiometry regarding the total number of thiols in the solution (*see* **Note 9**). Assess the resolution of the alkylation procedure: the bands corresponding to partially alkylated species must not overlap. Otherwise, use an alkylating agent of higher molecular weight.

3.2 Detection and Quantification of Persulfides, Monitoring the Kinetic of Persulfide Transfer

Two protocols are presented here, one for persulfide detection and quantification and another for the kinetics of trans-persulfidation reactions. The reactions are analyzed under reducing and nonreducing conditions to discriminate persulfides and false positives (*see* Subheading 1.2). The mouse NIA complex in which a persulfide is formed by NFS1 and the mouse NIAU and NIAUF complexes in which the persulfide is transferred to ISCU are used to illustrate the assays, but the protocols apply to any other proteins [7, 15].

3.2.1 Persulfide Detection and Quantification

1. Reduce the proteins and the complexes with 10 molar equivalents of DTT prior to any reaction and desalt them on a Sephadex G-25 column equilibrated with phosphate buffer.
2. Prepare 100 μL of the cysteine desulfurase complexes (NIA and NIAUF) at 20 μM and split the reaction into two parts of 50 μL. Add 2 μL of L-cysteine at 1 mM to one part, which corresponds to 2 molar equivalents of L-cysteine, and incubate the reaction at 25 °C for 1 min (*see* **Note 10**). Leave the other part without L-cysteine.
3. Add twofold excess of the maleimide derivative solution regarding the total number of thiols (*see* **Note** 7) to 20 μL of each reaction mixture and incubate the reaction for 30 min at room temperature.
4. Split each reaction into two parts of 10 μL and add 3 μL of loading buffer with β-mercaptoethanol to one part and 3 μL of loading buffer without β-mercaptoethanol to the other part. Incubate 30 min at room temperature.
5. Load 50 pmoles of the NIA complex and 200 pmoles of the NIAUF complex onto 10% and 14% SDS-PAGE, respectively (*see* **Notes 5** and **8**).
6. Run gels at 90 V, 2 h for a 14% gel, and 5 h for a 10% gel. Reveal the proteins by Coomassie staining/destaining. Scan the gels with an imager.
7. Quantify individual bands using an imager (*see* **Note 6**).
8. Analyze the results (Fig. 5a, b). Mass loss must be observed under reducing conditions only in the reaction performed with L-cysteine. Compare the position of the bands with the ladder to determine how many cysteines are persulfidated. No mass loss must be observed under nonreducing conditions; otherwise, it means that it is a false positive (Subheading 1.2). No

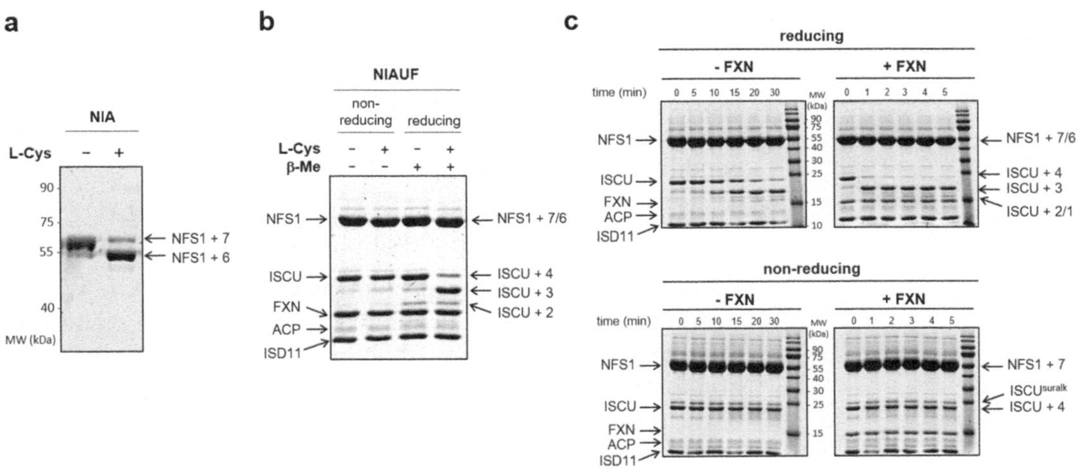

Fig. 5 ARBS assays for persulfide formation on NFS1 and transfer to ISCU. (**a** and **b**) ARBS assay on NFS1 in the NIA complex (**a**) and ISCU in the NIAUF complex (**b**) in the absence and presence of L-cysteine analyzed using Mal-dPEG (**a**) and Mal-P16 (**b**). (**a**) The reaction is analyzed under reducing conditions. In the absence of L-cysteine, NFS1 is alkylated on all its cysteines (NFS1 + 7 species) and on 6 cysteines when L-cysteine is added (NFS1 + 6 species), which indicates formation of a persulfide. Resolution (spacing between NFS1 + 6 and NFS1 + 7) is satisfactory with Mal-dPEG. (**b**) The reaction is analyzed under reducing and nonreducing conditions. ISCU is alkylated on all its cysteine residues in the absence of L-cysteine (ISCU + 4 species) and loses one alkylation unit in the presence of L-cysteine (ISCU + 3 species), which indicates that a persulfide is transferred to ISCU (lane 3 and 4). ISCU keeps all its cysteine alkylated under nonreducing conditions both in the presence and in the absence of L-cysteine (lanes 1 and 2), which confirms that the loss of one alkylation unit is due to the reduction of a persulfide by β-mercaptoethanol. An ISCU + 2 species is present under reducing conditions but not under nonreducing conditions. This could not be attributed to oxidised cysteine because IscU reaches full alkylation (ISCU+4) under nonreducing conditions. However, this band is also present in the absence of L-cysteine, thus ruling out the possibility of a second persulfide. It may thus correspond to an intramolecular disulfide (*see* Subheading 1.2). (**c**) ARBS assays for kinetics of persulfide transfer to ISCU in the NIAUF complex in the presence (+FXN) and absence (−FXN) using Mal-P16 and analyzed under reducing and nonreducing conditions. The gels show the progressive appearance of the ISCU + 3 species corresponding to persulfidated ISCU on a single cysteine (top panels). This transfer is accelerated in the presence of FXN. The analysis under nonreducing conditions (lower panels) shows mainly the ISCU + 4 species, which validates that the ISCU + 3 species corresponds to a persulfide. Under nonreducing conditions, an alkylated species with lower mobility than the ISCU + 4 appears. This species is attributed to over alkylated ISCU on a non-cysteine residue, probably nitrogen, since maleimides also react with nitrogen (*see* **Note 5**) and is denoted $ISCU^{suralk}$. The formation of this species is most likely due to the presence of Mal-P16 that is not quenched under nonreducing conditions

mass loss must neither be observed in the absence of L-cysteine since it is the substrate of the NIA complex to generate a persulfide; otherwise, it is a false positive (Subheading 1.2).

3.2.2 Monitoring the Kinetic of Persulfide Transfer by SDS Quenching

The same protocol as persulfide detection and quantification is used to assess the kinetics of persulfide formation or transfer, but in this case, a larger volume of the reaction mixture is prepared, and samples are withdrawn at desired times. The reaction is quenched by mixing with the alkylating agent in SDS. The alkylating agent blocks all available thiols while SDS rapidly denatures the proteins.

1. Reduce the proteins with 10 molar equivalents of DTT prior to any reaction and desalt them on a Sephadex G-25 column equilibrated with phosphate buffer.
2. Prepare 150 μL of the protein solutions (NIAU, NIAUF) at a final concentration of 20 μM in phosphate buffer.
3. Withdraw 20 μL of the protein mixture and mix it with a twofold excess of the maleimide derivative solution relative to the total number of thiols (*see* **Note 7**) and incubate the reaction for 30 min at room temperature. This will be used as the t_0 of the reaction.
4. Add 5.2 μL of L-cysteine aqueous solution at 1 mM to the 130 μL left of the protein solutions (NIAU, NIAUF) to initiate the reaction. Incubate at 25 °C.
5. Withdraw 20 μL from the reaction mixture at desired times and rapidly drop the solution in a tube containing twofold excess of the maleimide derivative solution regarding the total number of thiols (*see* **Note 7**). Immediately mix by three cycles of aspiration/ejection using the same pipette. Incubate the reaction for 30 min at room temperature.
6. Split each reaction into two parts of 10 μL each and add 3 μL of loading buffer with β-mercaptoethanol to one part and without β-mercaptoethanol to the other part. Incubate 30 min at room temperature.
7. Load 200 pmoles of each protein reaction on SDS-PAGE (*see* **Notes 5** and **8**).
8. Analyze the results (Fig. 5c). Check that mass loss is only observed under reducing conditions (Subheading 1.1). Otherwise, check for other cysteine modifications (Subheading 1.2). Quantify the bands corresponding to non-persulfidated and persulfidated proteins. Plot the ratio of the intensity of the bands corresponding to persulfidated and non-persulfidated species versus time. Simulate the curve with a relevant model to retrieve the rate constant.

3.2.3 Monitoring the Kinetic of Persulfide Transfer by TCA Quenching

An alternative method used for kinetic analysis is TCA precipitation. It is based on the denaturation by acidification to quench the reaction, which leads to precipitation of the proteins. The steps are summarized as following:

1. Reduce the proteins with 10 molar equivalents of DTT prior to any reaction and desalt them on a Sephadex G-25 column equilibrated with phosphate buffer.

2. Prepare 150 μL of the protein solutions (NIAU, NIAUF) at a final concentration of 20 μM in phosphate buffer.
3. Withdraw 20 μL of the protein mixture and mix it with 20 μL of cold TCA at 20%. Incubate on ice for 20 min.
4. In the meantime, add 5.2 μL of L-cysteine aqueous solution at 1 mM to the 130 μL of the protein (NIAU, NIAUF) solutions left to initiate the reaction. Incubate at 25 °C.
5. Withdraw 20 μL from the reaction mixture at desired times and mix it with 20 μL of cold TCA at 20%. Incubate on ice for 20 min.
6. After incubation of TCA-precipitated samples, centrifuge at 18,000 × *g* or 5 min at 4 °C.
7. Wash the pellets three times with cold acetone.
8. Leave the tubes standing in a fume hood for 30 min to evaporate residual acetone.
9. Resuspend the pellets in 20 μL alkylation buffer containing 2 molar equivalents of the maleimide derivative relative to the total number of thiols (*see* **Note** 7) and incubate the mixture for 30 min at room temperature.
10. Split each reaction into two parts of 10 μL each and add 3 μL of loading buffer with β-mercaptoethanol to one part and without β-mercaptoethanol to the other part. Incubate 30 min at room temperature.
11. Load 200 pmoles of each protein reaction on SDS-PAGE (*see* **Notes 5** and **8**).
12. Analyze the results. Similar kinetics must be observed when comparing the reaction quenched with SDS (Subheading 3.2.2) and TCA. Otherwise, it means that the quenching with SDS is not efficient.

4 Notes

1. Other types of alkylating agents can be used, such as iodoacetamide derivatives.
2. Other reducing agents of disulfides can be used, such as DTT and TCEP.
3. The main false positives expected are disulfides since persulfides react with reducing thiols to form disulfides. The other species (sulfenic, sulfinic, sulfonic acids, nitrosothiols, and sulfenamides) result from reaction with reactive oxygen and nitrogen species and are thus expected to occur under more specific conditions in the presence of oxidants.

4. The alkylation profiles show more fussy bands with Mal-PEG2k than Mal-P16 and Mal-dPEG (Fig. 4a). The Mal-PEG2k is not a pure compound but a mixture of polymers with a mean size of 2 kDa; thereby the size of the bands is linked to this mass heterogeneity. Mal-P16 and Mal-dPEG are defined compounds.
5. The maleimide group is prominently reactive with free sulfhydryls at pH 6.5–8.0, but also reacts with amines, albeit at a much slower rate. To prevent side reactions with amines from buffers and proteins, amine-based buffers are avoided, and the samples are not heated before loading on gels if the maleimide is not yet neutralized.
6. The quantification is based on densitometry measurements of the Coomassie-stained bands on gel using a routine instrument of the laboratory, which enables quantitative determination of persulfidation within the precision limit of the detection technique. Other methods based on fluorescent alkylating agents may enhance the sensitivity and accuracy of the assay.
7. The thiols include all cysteine residues from all proteins and small thiol compounds such as L-cysteine that is used as a substrate for cysteine desulfurase to generate a persulfide. The total concentration of thiols is calculated by summing the concentration of small thiols and the concentrations of each protein multiplied by the number of their cysteine residues.
8. The amount of protein to be loaded on gel varies in function of its size: about 200 pmoles for 10–30 kDa proteins and 50 pmoles for 40–80 kDa proteins. ISCU is analyzed on 14% acrylamide gels and NFS1 on 10% gels.
9. If the concentration of the alkylating agent needed to saturate all the cysteines residues is much higher than the stoichiometry regarding the total number of thiols, this may indicate either errors in the concentrations of the proteins or of the alkylating agent, presence of contaminating thiols, or a defective alkylating agent.
10. The formation of persulfide by NFS1 is completed in about 10 s and transferred to ISCU in less than 1 min in the presence of FXN [7], so that 1 min of incubation with L-cysteine is sufficient.

References

1. Black KA, Dos Santos PC (2015) Shared-intermediates in the biosynthesis of thio-cofactors: mechanism and functions of cysteine desulfurases and sulfur acceptors. Biochim Biophys Acta 1853:1470–1480
2. Leimkühler S, Bühning M, Beilschmidt L (2017) Shared sulfur mobilization routes for tRNA Thiolation and molybdenum cofactor biosynthesis in prokaryotes and eukaryotes. Biomol Ther 7:5

3. Filipovic MR, Zivanovic J, Alvarez B, Banerjee R (2018) Chemical biology of H2S signaling through Persulfidation. Chem Rev 118:1253–1337
4. Cuevasanta E, Moller MN, Alvarez B (2017) Biological chemistry of hydrogen sulfide and persulfides. Arch Biochem Biophys 617:9–25
5. Benchoam D, Cuevasanta E, Moller MN, Alvarez B (2020) Persulfides, at the crossroads between hydrogen sulfide and thiols. Essays Biochem 64:155–168
6. Smith AD, Agar JN, Johnson KA, Frazzon J, Amster IJ, Dean DR, Johnson MK (2001) Sulfur transfer from IscS to IscU: the first step in iron-sulfur cluster biosynthesis. J Am Chem Soc 123:11103–11104
7. Parent A, Elduque X, Cornu D, Belot L, Le Caer JP, Grandas A, Toledano MB, D'Autreaux B (2015) Mammalian frataxin directly enhances sulfur transfer of NFS1 persulfide to both ISCU and free thiols. Nat Commun 6:5686
8. Kambampati R, Lauhon CT (1999) IscS is a sulfurtransferase for the in vitro biosynthesis of 4-thiouridine in *Escherichia coli* tRNA. Biochemistry 38:16561–16568
9. Kambampati R, Lauhon CT (2000) Evidence for the transfer of sulfane sulfur from IscS to ThiI during the in vitro biosynthesis of 4-thiouridine in *Escherichia coli* tRNA. J Biol Chem 275:10727–10730
10. Lauhon CT, Skovran E, Urbina HD, Downs DM, Vickery LE (2004) Substitutions in an active site loop of *Escherichia coli* IscS result in specific defects in Fe-S cluster and thionucleoside biosynthesis in vivo. J Biol Chem 279:19551–19558
11. Bridwell-Rabb J, Fox NG, Tsai CL, Winn AM, Barondeau DP (2014) Human frataxin activates Fe-S cluster biosynthesis by facilitating sulfur transfer chemistry. Biochemistry 53:4904–4913
12. Bailey TS, Pluth MD (2015) Reactions of isolated persulfides provide insights into the interplay between H2S and persulfide reactivity. Free Radic Biol Med 89:662–667
13. Cavallini D, Federici G, Barboni E (1970) Interaction of proteins with sulfide. Eur J Biochem 14:169–174
14. Liu DK, Chang SG (1987) Kinetic study of the reaction between cystine and sulfide in alkaline solutions. Can J Chem 65:770–774
15. Gervason S, Larkem D, Mansour AB, Botzanowski T, Muller CS, Pecqueur L, Le Pavec G, Delaunay-Moisan A, Brun O, Agramunt J, Grandas A, Fontecave M, Schunemann V, Cianferani S, Sizun C, Toledano MB, D'Autreaux B (2019) Physiologically relevant reconstitution of iron-sulfur cluster biosynthesis uncovers persulfide-processing functions of ferredoxin-2 and frataxin. Nat Commun 10:3566

Part III

Biophysical Methods to Study Fe-S Clusters and Reactions Promoted by Metalloenzymes

Chapter 12

Circular Dichroism to Probe the Synthesis, Transfer, and Stability of Fe-S Clusters

Stefania Iametti, Francesco Bonomi, and Alberto Barbiroli

Abstract

All Fe-S proteins are characterized by distinctive circular dichroism (CD) features in the visible region of the spectrum due to chiral interaction between the cluster itself and the protein backbone. Therefore, the presence of a CD signal in the visible region relates to the presence of the cluster, whereas the disappearance of the signal refers to cluster breakdown or redox changes. The position of the CD features in the spectrum and the intensity of individual components of the CD signal show great variations among different Fe-S proteins. This feature can provide information on transfer processes between proteins, as well as on possible changes in cluster nuclearity. This method can also be used to detect changes in the chemical nature or spatial organization of cluster ligands that may be concurrent with cluster transfer and associated events.

Key words Circular dichroism, Fe-S proteins, Cluster assembly, Cluster transfer, Cluster stability

1 Introduction

Circular dichroism (CD) is a spectroscopic technique that measures the different absorption of left-hand circular and right-hand circular polarized light (LCP and RCP, respectively) in a dedicated instrument called a spectropolarimeter. These instruments use a xenon arc lamp, a monochromator, and a polarizing filter to select a monochromatic plane-polarized light that is then modulated in its LCP and RCP components, sent alternatively to the sample. A very sensitive photomultiplier measures the absorption of the differently polarized components by the sample. The measured parameter is the very minuscule difference of absorption $\Delta A = (A_L - A_R)$, for which the Lambert-Beer relationship remain true, being: $\Delta A = (\varepsilon_L - \varepsilon_R) \times C \times l$, where ε_L and ε_R are the molar extinction coefficients for LCP and RCP light, C is the molar concentration, and l is the optical path length in centimeters. Although molar circular dichroism is defined by $\Delta\varepsilon = (\varepsilon_L - \varepsilon_R)$, CD results are often reported—for historical reasons—in degrees of ellipticity.

Patricia C. Dos Santos (ed.), *Fe-S Proteins: Methods and Protocols*, Methods in Molecular Biology, vol. 2353, https://doi.org/10.1007/978-1-0716-1605-5_12,

Molar circular dichroism and molar ellipticity $[\theta]$ are related by the simple equation $[\theta] = 3298.2 \times \Delta\varepsilon$.

A CD signal is observed when a chromophore is optically active by being intrinsically chiral, covalently linked to a chiral center, or placed in an asymmetric environment [1]. The wavelengths at which CD signal is observed are the same wavelengths at which the interested chromophore absorbs light. The far-UV (180–250 nm) CD spectra of proteins can be used to estimate the fraction of a molecule having a given secondary structure (α-helix, β-sheet, β-turn, and random coil). The near-UV CD spectra (250–350 nm) of proteins provide information about the dipole orientation and the chemical environment of amino acids such as phenylalanine, tyrosine, tryptophan, and cysteine (including S-S bridges). The near-UV CD spectra may also provide structural information on the nature and environment of the prosthetic groups (organic or inorganic) in protein complexes.

Most relevant to the topics discussed here is the visible-range CD spectroscopy (300–700 nm), which provides a powerful technique for studying metal–protein interactions. CD has a very significant advantage over simple absorbance measurements, as CD spectra in the visible light region are only produced when a metal ion is in a chiral environment. This means that free metal ions in solution or in complexes with nonchiral molecules are not detected by CD, no matter how intense their absorbance is. In other words, CD allows the observation of only protein-bound metals and to monitor the extent and time course of a number of events relevant to the biological assembly of functioning metalloproteins. For example, this technique provides the resolution to determine the sequence of events that accompanies iron uptake by the simplest Fe-S proteins (i.e., rubredoxins, containing a mononuclear Fe $(SCys)_4$ center). In this case, CD was used to monitor iron-dependent refolding of the protein and the formation of a proper and functional structure [2, 3]. CD has also been used to assess the role of the cluster in ensuring proper structuring of proteins harboring a [2Fe-2S] cluster [4].

CD has been extensively used in the Fe-S research field. The CD signals of individual proteins are sensitive to cluster nuclearity, to the nature of the protein ligands (i.e., when ligands other than cysteine thiols are used for cluster coordination [5]), and to the redox state of the cluster [6]. UV/Vis CD spectra of Fe-S proteins containing metal centers of different nuclearity are presented in Fig. 1. All the unique CD features of metalloproteins may be exploited to study rates, yields, and mechanism of chemical or enzymatic assembly of Fe-S structures [7–11], in the presence/absence of accessory and regulatory proteins [12]. This approach can also probe interconversion among protein-bound clusters of different nuclearity [13, 14] or changes in the cluster ligands [5, 15, 16] that may be of physiological relevance. Along similar

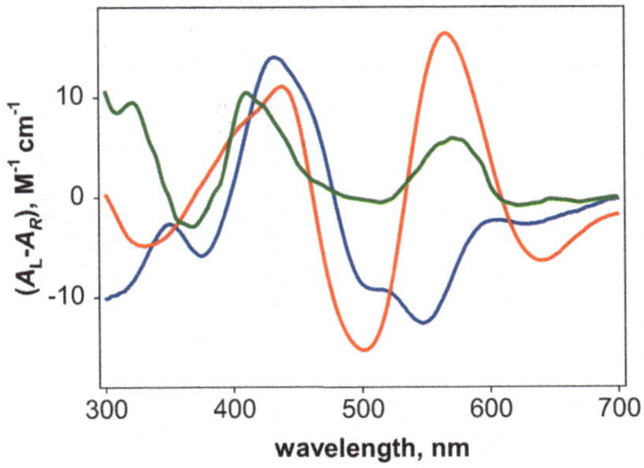

Fig. 1 UV/Vis CD spectra of Fe-S proteins. Representative CD spectra of various classes of Fe-S proteins: red, *Pyrococcus furiosus* rubredoxin, containing one [Fe (Cys)$_4$] center; blue, *Escherichia coli isc*-encoded ferredoxin, containing one [2Fe-2S(Cys)$_4$] cluster; and green, *Clostridium pasteurianum* ferredoxin, containing two [4Fe-4S(Cys)$_4$] clusters

lines, CD measurements allow the assessment of cluster stability toward biological, chemical, and physical agents that may perturb the protein structure or the protein/cluster interaction [17]. CD spectroscopy is suitable for detecting metal binding other than iron incorporated into Fe-S structures of various nuclearity [17–19]. Further information may be gathered by exploiting the fact that chemically identical Fe-S clusters may elicit signals of different intensity when harbored on different proteins. For instance, these latter features allow to monitor rates and yields of cluster transfer between different proteins, as these events occur with the many types of "scaffold proteins" involved in cluster assembly and transfer [20–24]. Selected examples of CD spectra for some of these "scaffold" proteins are presented in Fig. 2.

CD spectroscopy has also been used to assess the role of "structural distortion" of the aforementioned scaffold protein (s) by the molecular chaperones expressed by the many operons involved in the biological assembly of Fe-S structures [25, 26]. Specific chaperones and co-chaperones, along with other protein effectors, control the rate of the assembly process on scaffold proteins [27, 28]. Their use Mg^{2+}-dependent ATP hydrolysis to modulate the rate of transfer to acceptor apoproteins or other proteins acting as intermediates in the overall biosynthetic process, generating species that—once again—may differ in their CD properties [15, 26].

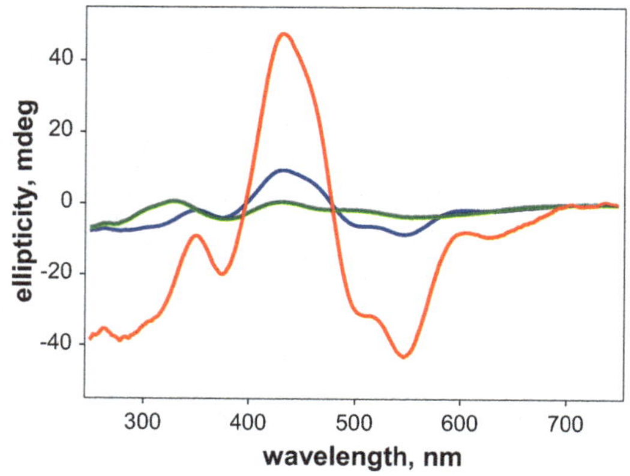

Fig. 2 UV/Vis CD spectra Fe-S *isc*-encoded proteins. Representative CD spectra of 45 μM solutions of the [2Fe-2S] forms of *Escherichia coli isc*-encoded scaffold proteins IscA (green), IscU (blue), and of the *isc*-encoded *Escherichia coli* [2Fe-2S] ferredoxin holoprotein (red)

2 Materials

Preparation of all solutions and sample handling must be carried out anaerobically, given the high sensitivity of Fe-S clusters to oxygen. The use of an all-glass high-vacuum manifold together with gas-tight syringes for sample handling is highly recommended, as is the use of Ar as the inert gas (*see* **Note 1**). In general, a vacuum manifold is much cheaper and less expensive to maintain than any anaerobic chamber and has the added advantage of great operational ease. There is no absolute need for using specialized and expensive Schlenk-type glassware. Still, the use of plastic lab ware for handling all solutions has to be avoided whenever possible due to the substantial gas permeability of most common plastics. Always work in a fume hood with appropriate personal protection. There are safety issues with some of the chemicals (e.g., sulfide) that should be handled with due care and disposed properly. Repeated cycles of vacuum/Ar are best suited for removal of oxygen from nonprotein solutions, but this procedure is difficult to apply in the case of protein solutions that are prone to frothing (the larger the protein, the worse the frothing), and thus to protein loss. In this case, keep the vacuum cycle as short as possible and quickly switch to a positive Ar pressure as soon as frothing begins.

2.1 Glassware, Other Hardware, and Liquid Transfer Equipment

1. Glass vials: screw-cap glass vials (without cap) or thick-walled 15 mL centrifuge tube to be used for small volumes (up to 5 mL).
2. Anaerobic flasks: Round-bottom flasks and/or Erlenmeyer bottles fit with a 14/20 female cone may be used for volumes higher than a few milliliters. They may be fitted with pierceable

rubber septa, which are cheap, convenient, and available in various sizes and materials (*see* **Note 2**).

3. Hypodermic needles: needles with a wide range of gauges (diameter) and lengths. Needles are conveniently connected to a 1 mL plastic syringe inserted in a thick-walled rubber tubing on a plain-style vacuum manifold outlet. Of course, the syringe plug is not used, and the syringe "ears" will have to be removed by using a suitable household cutter.
4. Luer-lock gas-tight syringes: Luer-lock gas-tight syringes suitable for transferring volumes from 0.005 to 5 mL. These syringes come in a variety of sizes and are identified by a Teflon plug on the piston tip (*see* **Notes 3** and **4**).
5. Stainless-steel cannulae: Long needles with both ends sharpened. They are used for the anaerobic transfer of larger volumes of liquid by pulling a vacuum in the receiving vessel and leaving a positive Ar pressure in the one containing the liquid to be delivered. Solutions are kept under Ar once degassed or transferred.
6. Anaerobic cuvettes: One milliliter semi-micro cuvettes (10 mm light path) with a red, ribbed Suba-Seal septum (size 9) replacing the original Teflon stopper. Unless otherwise specified, all spectroscopic measurements are carried out in anaerobic cuvettes.
7. CD spectrometer.

2.2 Buffer Systems (See Note 5)

1. Standard Tris buffer: 50 mM Tris–HCl, pH 8.0. To make 50 mL of buffer, weigh 0.3 g of Trizma base and dissolve in 40 mL of deionized water, and adjust pH with 1 M HCl before completing the volume to 50 mL.
2. Standard HEPES buffer: 50 mM HEPES-Na, pH 7.0. To make 50 mL of buffer, weigh 0.6 g of HEPES, dissolve in 40 mL of deionized water, and adjust pH with 1 M NaOH before completing the volume to 50 mL.
3. Standard Tris buffer with 2-mercaptoethanol: 50 mM Tris–HCl, pH 8.0 containing 5 mM of 2-mercaptoethanol. To prepare 50 mL, degas the standard Tris buffer by repeated vacuum/Ar cycles before adding 0.018 mL of 13.4 M 2-mercaptoethanol (*see* **Note 6**).
4. Standard HEPES buffer with 2-mercaptoethanol: 50 mM HEPES-Na, pH 7.0 containing 5 mM of 2-mercaptoethanol. To prepare 50 mL, degas the standard HEPES buffer by repeated vacuum/Ar cycles before adding 0.018 mL of 13.4 M 2-mercaptoethanol (*see* **Note 6**).
5. Mg/K buffer: Standard HEPES buffer with 150 mM KCl and 5 mM $MgCl_2$. For 50 mL of buffer, weigh 0.56 g of KCl and

24 mg of anhydrous $MgCl_2$. Place both salts in a flask, add a magnetic stirring bar. Measure 50 mL of the standard buffer and pour over the salts and close with a rubber septum. Start magnetic stirring and make the solution anaerobic by repeated vacuum/Ar cycling.

2.3 Solutions

Recipes given below are for preparing 20 mM stock solutions—unless otherwise indicated—of different common chemicals. In all cases, weigh an amount of material close to what you may need and adjust the buffer volume using a simple proportion. To prepare these solutions, weigh enough of the desired chemical to make 3–5 mL of a solution of the required concentration and transfer the solid into a properly labeled thick-walled glass centrifuge or glass vial tube fitted with an appropriate septum. Remove oxygen by a couple of quick vacuum/Ar cycles. Then, working under Ar, use a gas-tight syringe to add the appropriate volume of degassed buffer (or degassed water, in the case of iron salts, *see* **Note 7**) to the solid. Use a couple of quick vacuum/Ar cycles and gentle shacking to completely dissolve the chemical. Always leave the prepared solutions under positive Ar pressure.

1. Cysteine solution: 3.5 mg/mL cysteine hydrochloride monohydrate (MW 175.63 g/mol) prepared in standard Tris buffer.
2. Dithionite solution: 3.5 mg/mL sodium dithionite (MW 174.11 g/mol) prepared in standard Tris buffer.
3. Dithiothreitol solution: 3 mg/mL dithiothreitol (DTT, MW 154.25 g/mol) prepared in standard Tris buffer.
4. 20% TCA: 20% trichloroacetic acid solution prepared in water.
5. Sulfide solution: 0.9 mg/mL lithium sulfide (MW 45.95 g/mol) prepared in standard Tris buffer (*see* **Note 8**). Sulfide solution should be prepared daily. Please note that all soluble sulfides are poisonous and stinky: exert due caution when handling and disposal.
6. Iron solution: 5.25 mg/mL ferric ammonium citrate (nominal MW 261.98 g/mol) prepared in water, corresponding to an iron content of 20% (*see* **Note 7**). Aqueous solutions of this chemical have a pH around neutrality and are stable for several hours in the dark. The solution should be prepared daily.
7. ATP solution: 20 mM adenosine 5′-triphosphate disodium salt (MW 551 g/mol) solution prepared in water, corresponding to 11 mg/mL. This solution is used in the assembly/transfer experiments and has to be prepared daily (*see* **Note 9**).
8. Urea solution: 8 M urea solution prepared in standard Tris buffer. To make 10 mL solution, weigh 4.8 g of spectroscopy-grade urea (MW 60.06 g/mol) and transfer the solids into a 15 mL graduated centrifuge tube with a stirring

bar at the bottom. Attach a septum onto the tube, make the content of the tube anaerobic, and anaerobically add enough of the appropriate buffer—under continuous magnetic stirring—to reach the 10 mL mark. *See* **Note 10** for the stability of urea solutions.

9. Guanidine hydrochloride solution: 6 M guanidine hydrochloride prepared in standard Tris buffer. To make a 10 mL solution, weigh 5.7 g of guanidine hydrochloride (MW 95.54 g/mol). Transfer the solids in a 15 mL graduated centrifuge tube with a stirring bar at the bottom. Attach a septum onto the tube, make the content of the tube anaerobic, and anaerobically add enough of the appropriate buffer, under continuous magnetic stirring, to reach the 10 mL mark.
10. EDTA solution: 5.8 mg/mL ethylenediaminetetraacetic acid (MW 292.24 g/mL) prepared in standard Tris buffer.
11. $ZnCl_2$ solution: 2.7 mg/mL zinc chloride (MW 136.30 g/mL) prepared in water.
12. $CdCl_2$ solution: 3.6 mg/mL cadmium chloride (MW 183.32 g/mL) prepared in water.

2.4 Ferredoxin Holoproteins and Apoproteins

It is suggested to store any Fe-S holoprotein as small aliquots of concentrated protein at subzero temperature (liquid nitrogen or −80 °C freezers are best, but standard −24° freezers are more than adequate).

1. *C. pasteurianum* 8Fe-8S ferredoxin: 1–3 mM *C. pasteurianum* [8Fe-8S] ferredoxin. The protein is prepared as described in [8] and stored as 50–100 μL aliquots in liquid nitrogen or −80 °C freezer. Thaw an aliquot of the protein stock just before use.
2. [2Fe-2S] ferredoxin: 1–5 mM 2Fe-2S ferredoxin holoprotein. The proteins are prepared as described ([7] for plant-type ferredoxin, [19] adrenodoxin, and [16] *isc*-encoded *E. coli* ferredoxin) and stored as 50–100 μL aliquots in liquid nitrogen or −80 °C freezer. Thaw an aliquot of the protein stock just before use.

2.5 Enzymes and Accessory Proteins (See Note 11)

1. Cysteine desulfurase: 1–2 mM of IscS. The enzyme is prepared as described in [6, 12] and references therein, and stored as 50–100 μL aliquots in liquid nitrogen or −80 °C freezer. Thaw an aliquot of the protein stock before use.
2. IscA: 1–5 mM IscA. The protein is obtained as apoprotein as described in [20] and references therein and stored as 50–100 μL aliquots in liquid nitrogen or −80 °C freezer. Thaw an aliquot of the protein stock before use.

3. IscU: 1–5 mM IscU. The scaffold is obtained as apoprotein as described in [21, 22] and references therein and stored as 50–100 μL aliquots in liquid nitrogen or −80 °C freezer. Thaw an aliquot of the protein stock before use.
4. HscA/HscB: Molecular chaperones HscA and HscB are prepared as described in [25, 26] and references therein and stored as 50–100 μL aliquots of concentrated protein in liquid nitrogen or −80 °C freezer. Thaw an aliquot of the protein stock before use.

3 Methods

3.1 Procedures for Preparing Ferredoxin Apoproteins from Holoproteins

1. Thaw an aliquot of the ferredoxin holoprotein.
2. Dilute the ferredoxin holoprotein to about 1–2 mg/mL with buffer or water. No anerobiosis is required. The resulting solution corresponds to about 0.15–0.3 mM protein in the case of small bacterial ferredoxins (such as *Clostridium pasteurianum* 8Fe-8S ferredoxin) and to about 0.08–0.15 mM protein in the case of 2Fe-2S ferredoxins.
3. Take UV/Vis absorption spectrum of the diluted protein solution to assess holoprotein concentration. The molar extinction coefficients are 30,000 M^{-1} cm^{-1} for 2 × [4Fe4S] *C. pasteurianum* ferredoxin (at 390 nm) and around 12,000 M^{-1} cm^{-1} for most [2Fe-2S] ferredoxins (at 420 nm).
4. Take a reference CD spectrum of the ferredoxin holoprotein (*see* **Note 12**). Please note that **steps 2** and **3** may be carried out only once when preparing protein aliquots to be stored in frozen form.
5. Working under a fume hood, add a volume of 20% TCA equal to that of the diluted holoprotein solution. There is no need for anerobiosis in these steps. The solution will turn colorless and turbid and will release poisonous hydrogen sulfide. Some small-size ferredoxins are more resistant to TCA precipitation than others and may require doubling the volume of 20% TCA required for achieving quantitative precipitation of the apoprotein.
6. Spin down the protein precipitate for at least 15 min at 10,000 × *g*.
7. Wash the protein precipitate by resuspension in 1% TCA and centrifuge as above.
8. Wash the protein precipitate again in distilled water and centrifuge as described above.
9. Drain all the water carefully from the protein precipitate. Do not disturb the brittle precipitate by using pipettes or filter

paper for water removal, as this may result in substantial loss of apoprotein.

10. Close the centrifuge tube with a suitable rubber septum (*see* **Note 2**), connect to the Ar supply, and add a second needle to allow a flow of Ar that will wash away any residual O_2 from the sample headspace.
11. Dissolve the washed protein precipitate in a suitable volume of anaerobic buffer (50 mM Tris–HCl, 5 mM 2-mercaptoethanol, or 5 mM DTT, pH 8) to have a final protein concentration in the 0.1–0.2 mM range. Please make sure that none of the inlet and outlet needles contacts the liquid, and—by all means—do not apply vacuum! Please note that the apoprotein precipitate may take a while to dissolve completely. "Pump" the protein suspension back and forth in a gas-tight syringe to ensure thorough mixing. Check for any residual "flocks" in the solution. If complete dissolution is not achieved within 10–15 min at room temperature, add a single small crystal of Trizma Base to the suspension under an Ar flow. This little trick works almost invariably.
12. Take a UV absorbance spectrum of the dissolved apoprotein and calculate the apoprotein concentration from its absorbance at 280 nm. Use the appropriate entry in the ProtParam tools in the ExPASy portal (or any similar database) to retrieve the extinction coefficient required for calculation.

3.2 Chemical Cluster Assembly

1. Record a baseline CD spectrum (no need for anerobiosis) using the same cuvette to be used for further measurements, filled with 50 mM Tris–HCl, pH 8, containing 5 mM 2-mercaptoethanol. Maintain the same orientation of the cuvette with respect to the light beam in all subsequent CD measurements.
2. Discard the cuvette content and place rubber septum on cuvette (*see* **Note 2**). Remove oxygen from cuvette by repeated vacuum/Ar cycles. Leave the cuvette under Ar.
3. Anaerobically, add enough ferredoxin apoprotein (*see* Subheading 3.1 above for the holoprotein-apoprotein conversion) in the cuvette to reach a final concentration of 20–50 μM apoprotein in a total volume of about 0.8 mL. Add anaerobic buffer to the required final volume (0.65 mL/min, 1.2 mL max).
4. Take the "protein only" baseline spectrum (*see* **Note 12**).
5. Add iron from a stock 20 mM solution in water to reach a 3–10-fold molar excess over the protein, depending on the number and nuclearity of the cluster in the holoprotein. The solution should quickly acquire a rich reddish-brown color, typical of a $Fe^{(III)}(SR)_4$ complex. A green color indicates the

formation of iron(III) polysulfides, which impairs further experimentation (*see* **Note 13**).

6. Take "protein and iron" baseline spectrum.
7. Set the CD on "time course" measurement, using the wavelength where the CD signal is most intense (typically, 400–405 nm for [4Fe-4S] clusters and 435–440 nm for [2Fe-2S] clusters).
8. Start the chemical reconstitution reaction by injecting a volume of 20 mM lithium sulfide equal to that of the 20 mM solution of iron salts used in **step 5**.
9. Monitor ellipticity changes at an appropriate wavelength for 5–10 min (*see* **Note 14**), an accurate analysis of the time course of initial events is required.
10. Take spectra of the mixture at appropriate time intervals (*see* **Note 15**).
11. Take a final spectrum to assess yield once apparent equilibrium is reached. Compare the final CD spectrum with the spectrum of the original holoprotein.
12. Evaluate the reconstitution yield by comparing the CD spectrum of the reconstituted mixture with that of a "standard" holoprotein. If no "standard" holoprotein is available, use the highest yield reconstitution mixture as a reference (*see* **Note 16**).

3.3 Enzymatic Cluster Assembly

Enzymatic cluster assembly reactions are completed using sulfide-producing enzymes and, if required, accessory proteins that may affect the activity of sulfide producing enzymes. The pyridoxal-dependent cysteine desulfurase IscS or any of its many homologs in other species and cellular compartments are commonly used to generate sulfide in the presence of 2–5 mM cysteine as the substrate. These reactions are performed either in the presence or absence of additional thiols, such as 2-mercaptoethanol or DTT (*see* **Note 17**). Other protein components, such as proteins that may act as possible modulators of the activity of sulfide-producing enzymes [11, 26], may be added prior to the enzyme itself. Please note that testing the effects of other proteins requires a comparison of similar experiments run in the presence/absence of the "accessory" effector protein that is being investigated. To make everything consistent, run these experiments and the related controls as simultaneously as possible by using different cuvettes for reactions carried out in the absence/presence (or various concentrations) of the effector protein. It is suggested to offset the setup and start of each reconstitution reaction by about 20–30 min time intervals, which gives enough time for recording the initial phases of any reaction while getting ready to start a new reaction.

1. Follow the instructions provided in **steps 1–4** of Subheading 3.1 for preparing the acceptor apoprotein solution.
2. Add 2 mM cysteine from a 20 mM stock solution, prepared in standard Tris buffer as described under Subheading 2.2 (*see* **Note 17**).
3. Add iron as described in **step 5** of Subheading 3.2 and take baseline CD spectrum.
4. Add sulfide-producing enzyme at suitable catalytic concentrations and start time-course monitoring of ellipticity changes at a protein-specific wavelength to record the time course of the synthetic reaction.
5. Check the CD spectrum at appropriate intervals.
6. Take the final CD spectrum to assess yield in holoprotein once apparent equilibrium is reached and no further changes in the intensity of the CD signal are observed.

3.4 Cluster Transfer Experiments

3.4.1 "Stoichiometric" Cluster Transfer from Scaffold Holoproteins to Acceptor Apoproteins

1. Record a baseline CD spectrum (no need for anerobiosis) using the same cuvette to be used for further measurements, filled with buffer (typically, 50 mM Tris–HCl, 5 mM mercaptoethanol, or 5 mM DTT, pH around 8). Maintain the same cuvette orientation in all subsequent measurements.
2. Discard the cuvette content and place rubber septum on cuvette (*see* **Note 2**). Remove oxygen from cuvette by repeated vacuum/Ar cycles. Leave the cuvette under Ar.
3. In the same anaerobic cuvette, prepare scaffold holo-Fe-S-protein (e.g., [2Fe2S] IscU) starting from the appropriate apoprotein by using either a chemical or an enzymatic reconstitution procedure as described in Subheadings 3.2 and 3.3. By using only a slight molar excess of iron (5% with respect to stoichiometric ratio iron to scaffold) and a correspondingly slight excess of sulfide—in the case of chemical reconstitution—it is possible to avoid cumbersome and time-consuming isolation of the reconstituted scaffold protein prior to the addition of the acceptor apoprotein (see below).
4. Take CD spectrum to assess the reconstituted scaffold protein and determine when equilibrium is reached. The CD spectrum should also be used to define the "cluster equivalent" content of the reconstituted scaffold holoprotein (*see* Fig. 2 for reference spectra).
5. Prepare acceptor apoprotein solution as described under Subheading 3.1.
6. Start the cluster transfer reaction by adding an aliquot of the appropriate apoprotein acceptor. The added amount is typically stoichiometric (1:1 molar ratio) to the "cluster equivalent" present in the scaffold holoprotein based on their calculated

concentrations from their reported molar extinction coefficients [22].

7. Monitor the time course of ellipticity changes at an appropriate wavelength for 5–10 min, according to the spectral features of the acceptor protein.
8. Take spectra of the mixture at appropriate time intervals (*see* **Note 15**).
9. Once an apparent equilibrium is reached, take the final spectrum to assess yield and compare it with the CD spectrum of the original holoprotein, if available.

3.4.2 "Catalytic" Cluster Assembly by Scaffold-Type Proteins onto an Acceptor Apoprotein

1. Follow the instructions provided in **steps 1–4** of Subheading 3.1 as for preparing the acceptor apoprotein solution.
2. Add a proper amount of scaffold apoprotein to reach a final substoichiometric concentration with respect to the acceptor apoprotein (e.g., 2–0.2 μM), and add anaerobic 50 mM Tris–HCl buffer, pH 8.0, containing 5 mM 2-mercaptoethanol to the required final volume (1.2 mL maximum volume).
3. Take the "protein only" baseline spectrum (*see* **Note 12**).
4. Add iron and take iron-and-proteins baseline spectrum as detailed in **steps 5** and **6** under Subheading 3.2.
5. Set the CD on "time course" measurement using the wavelength where the CD signal is most intense, typically 400 nm for [4Fe-4S] clusters and 440 nm for [2Fe-2S] clusters.
6. Start the chemical reconstitution reaction by injecting a volume of 20 mM lithium sulfide equal to that of the 20 mM solution of iron salts used in the previous steps. Monitor time course and yield of cluster formation by following **steps 9–12** in Subheading 3.2.

3.5 Assessing the Role of Molecular Chaperones (and Nonenzymatic Accessory Proteins) in Cluster Assembly and Transfer

3.5.1 Effects on Cluster Assembly

1. Prepare the acceptor apoprotein solution as detailed in **steps 1–4** of Subheading 3.1, by using 50 mM HEPES/Na pH 7.0, containing 150 mM KCl, 5 mM $MgCl_2$, and 5 mM 2-mercaptoethanol as the buffer.
2. Anaerobically add a proper amount of any "accessory" proteins (e.g., chaperones and co-chaperones, such as bacterial HscA or HscB) in a 1:1 molar ratio among themselves and in a 1:1 molar ratio to the scaffold protein. Refer to Subheadings 3.2 or 3.3 above as for carrying out appropriate control runs in the absence of either the chaperone and/or the co-chaperone species. If using enzymes for sulfide production, also consider that some of the co-chaperones interfere with the activity of sulfide-producing enzymes [26].
3. Take the "protein only" baseline spectrum (*see* **Note 12**).

4. Follow **steps 5** and **6** of Subheading 3.2 as for adding the required chemicals and recording appropriate baselines.
5. Start reconstitution reaction by adding the sulfur source (*see* Subheadings 3.2 or 3.3) and–immediately after—add ATP to a final concentration of 2 mM (*see* **Note 9**). If using chemical reconstitution, start the reaction by adding lithium sulfide at final concentrations equal to that of the iron added previously. In the case of enzymatic reconstitution, start the reaction by adding the enzyme (considering that the sulfide-producing substrate, cysteine, should be present in the mixture) prior to ATP. Run appropriate "blanks" by assessing rates and yields of holoprotein formation in the absence of ATP.
6. Monitor ellipticity changes at an appropriate wavelength for at least 10 min.
7. Take spectra of the mixture at appropriate time intervals (*see* **Note 15**).
8. Once apparent equilibrium is reached, take a final spectrum to assess yield and compare it with CD spectrum of the original holoprotein.
9. Evaluate the role of chaperones proteins in reconstitution yield by comparing the CD data recorded with that of a mixture where all components are present but ATP.

3.5.2 Cluster Transfer

1. Prepare scaffold holo-Fe-S protein (e.g., [2Fe2S] IscU) starting from apoprotein and using one of the reconstitution procedures described above under Subheading 3.2 (chemical) or Subheading 3.3 (enzymatic). Please note that the buffer of choice for chaperone experiments is 50 mM HEPES/Na pH 7.0, containing 150 mM KCl, 5 mM $MgCl_2$, and 5 mM 2-mercaptoethanol as buffer.
2. Take CD spectrum to assess that scaffold holo-Fe-S protein is fully reconstituted and to define the cluster equivalent content in the reconstituted holoprotein.
3. Add chaperone proteins to a final molar concentration of each of them equal to that of the scaffold holoprotein.
4. Prepare acceptor apoprotein solution (*see* Subheading 3.1 above) in 50 mM HEPES/Na pH 7.0, containing 150 mM KCl, 5 mM $MgCl_2$, and 5 mM 2-mercaptoethanol.
5. Start the cluster transfer by adding the acceptor apoprotein in a 1/1 molar stoichiometry to the cluster present in the scaffold holoprotein. Immediately after, add ATP to a final concentration of 2 mM. Typically, ATP is added from a stock 20 mM solution in water (*see* **Note 9**).
6. Monitor ellipticity changes at an appropriate wavelength for at least 10 min (*see* **Note 14**).

7. Take spectra of the mixture at appropriate time intervals (*see* **Note 15**).
8. Once an apparent equilibrium is reached, take the final spectrum to assess yield and compare it with the CD spectrum of the original holoprotein.
9. Evaluate the effect of chaperones on transfer yield by comparing the CD data recorded with that of a mixture where all the components are present, but ATP.

3.6 Using CD to Assess Cluster Stability of Fe-S Proteins

3.6.1 Stability of Cluster Ligation to Chemical Denaturants (Chaotropes)

1. Prepare stock solutions of the buffer of choice at the maximum concentration of a chaotrope (e.g., 8 M for urea or 6 M for guanidine hydrochloride). Stock solutions of chaotrope must be prepared anaerobically (*see* **Note 10**), and typically do not contain 2-mercaptoethanol.
2. Record baseline CD spectra (no need for anaerobiosis) using the same cuvette to be used for further measurements, filled with the appropriate buffer at the concentrations of the chaotrope planned to be used in the titration. Take care of maintaining the same orientation of the cuvette in all subsequent measurements. If the spectra of the two stock solutions (without and with the chaotrope) do overlap, it is not necessary to acquire spectra at intermediate concentration.
3. Discard the cuvette content and place rubber septum on cuvette (*see* **Note 2**). Remove oxygen from cuvette by repeated vacuum/Ar cycles.
4. Anaerobically add enough protein in the cuvette to reach a final concentration of 20–50 μM in a total volume of 0.8 mL. Anaerobically add the appropriate volume of the two stock buffers in order to obtain the desired final concentration of urea (e.g., for 4 M urea, add 0.4 mL of stock 8 M urea solution and enough buffer to reach a final volume of 0.8 mL). The standard buffer must always be added first in order to avoid direct exposure of the protein to the high-chaotrope buffer. Incubate the cuvette for 15 min at room temperature to allow the thermodynamic equilibrium to be reached. It may be useful to run a CD time-course measurement at the appropriate wavelength (*see* **Note 15**) to verify that spectral changes are completed.
5. Take a spectrum of the protein/urea mixture.
6. Repeat **steps 2–4** above until spectra at all the desired concentration of chaotrope have been acquired (including no chaotrope) (*see* **Note 18**).
7. Identify the wavelength where the difference in CD signal is highest as a function of the increasing concentration of urea. Plot the signal intensity versus the concentration of urea. The

apparent C_m is usually defined as the concentration of urea at 50% of the total change and may be assessed, along with other thermodynamic parameters, by a number of routines in common data analysis software. In some instances, it may be useful to compare cluster stability data with data on the stability of the protein backbone itself, measured in the near- or far-UV region of the CD spectrum, as appropriate. For a detailed thermodynamic analysis, the information may be derived from "chemical denaturation curves," please *see* [29].

3.6.2 Stability Toward Iron-Substituting Metals or Metal Chelators (See **Note 19**)

1. Anaerobically prepare 20–50 μM holoprotein solution in the buffer of choice, in the absence of 2-mercaptoethanol (*see* Subheading 2.2 for buffers of different pH).
2. Take protein spectrum (*see* **Note 12**).
3. Add aliquots of the appropriate metal salt to reach a final substoichiometry concentration. Cd or Zn salts are commonly used from a stock solution in water (*see* Subheading 2.3). If testing the effects of metal chelators, use a similar approach adding the required chemical (e.g., EDTA from a stock solution in the buffer, *see* Subheading 2.3).
4. Wait 5 min or more to let the substitution experiments to be complete at room temperature. It may be useful to run a CD time-course measurement at the appropriate wavelength (*see* **Note 15**) to assess reaction rates and/or to verify that changes are completed.
5. Take the protein spectrum after every addition of metals or chelators (*see* **Note 12**).
6. Repeat **steps 5–7** after further adding the metal salts/chelating agent, until reaching a final concentration of 5 to 10-fold excess to the Fe-S cluster in starting holoprotein.
7. Compare the spectra of metal-substituted protein to the starting holoprotein spectra to define the metal cluster substitution yield (*see* **Note 16**).

3.6.3 Stability of Cluster Ligation to Temperature (See **Note 20**)

1. Anaerobically prepare 20–50 μM holoprotein solution in the buffer of choice, in the absence of 2-mercaptoethanol (*see* Subheading 2.2 for buffers of different pH).
2. Take protein spectrum (*see* **Note 12**).
3. Heat the protein solution by placing the cuvette in a cell holder connected to a Peltier-type element and by applying the desired temperature gradient (typically from 20 to 100 °C in 160 min, corresponding to a heating rate of 0.5 °C/min).
4. Monitor the cluster decay by following the decrease in the CD signal at the wavelength where the CD signal is most intense

(typically, 400 nm for [4Fe-4S] clusters, 440 nm for [2Fe-2S] clusters, and 492 nm for $Fe(SR)_4$ clusters).

5. After the final temperature is reached, take a spectrum of the "hot" protein solution. Evaluate the extent of cluster loss by comparison with the CD spectrum of the starting holoprotein.
6. Cool down the protein solution (and the cell holder) to room temperature. Take protein spectrum once cooled, and compare to starting holoprotein spectrum in order to define the reversibility of the observed changes.

4 Notes

1. Argon is much easier to separate from oxygen than nitrogen, making Ar of suitable quality almost as cheap as low-oxygen, high-quality nitrogen. In addition, Ar has a much higher density than nitrogen, and therefore stays on the bottom of any recipient, providing a "blanket" effect that minimizes the chances of occasional contact of the anaerobic solution with oxygen.
2. We recommend using ribbed red Suba-Seal size 9 septa for cuvettes (fits 8 mm necks) and ribbed red Suba-Seal size 13 septa for Eppendorf-style centrifuge tubes (fits 9.5 mm necks). Another option is small screw-cap glass vials (no caps) and natural rubber stoppers to fit 15 mm i.d. thick-walled centrifuge tubes, larger glass vials, or any container with a 14/20 standard female cone. We found natural (not colored) rubber to be more resistant to repeated piercing than other types of seals. Please exert care when inserting needles into the smallest-size septa, as the needle may end up in the thick part of the septum, making it impossible to deliver the liquid and—in the worst scenario—withdrawal of the needle. Place the cuvette in a stable holder and hold the needle (not the syringe) for proper insertion.
3. Do not use syringes with a fixed needle that often is both too short and too fragile. Hypodermic needles—available in a wide range of gauges and lengths—are an easy fit to Luer-lock gas-tight syringes. The same syringe may be used for measuring various solutions if properly cleaned between uses. When using any but the largest syringes, the dead volume of the needle and the Luer connection may be sensibly higher than the volume of the syringe itself. The recommended procedure for consistent and anaerobic delivery of a known volume of liquid involves: (1) filling the syringe with Ar (for instance, by using the headspace of a separate vial connected to the vacuum manifold); (2) remove the Ar-filled syringe and insert the needle tip into

the previously degassed liquid to be delivered, in a container connected to the Ar supply from the vacuum manifold; (3) briefly apply a vacuum; (4) reconnect to the Ar supply, so that the whole syringe will be filled with the liquid; and (5) take the required amount of liquid. Please note that the liquid in the syringe will remain anaerobic for at least 30 min.

4. Gas-tight syringes are best washed using a septum-fitted large container—such as a 500 mL round-bottom or a vacuum-filtration Erlenmeyer—connected to a vacuum line. Safely dispose of residual liquid in the syringe, insert the needle into the septum on the container connected to the vacuum line without removing it from the syringe, start the vacuum, and remove the plunger from the syringe body. Thoroughly wash the syringe and the needle (leave it on the syringe) with water, using a plastic wash bottle with a 0.2 mL (yellow) pipette tip pressure fit on the wash bottle spout. A subsequent quick rinse with methanol may help to speed up subsequent drying. Replace the plunger in the syringe, after washing the plunger separately with water and blotting it dry on paper.
5. As a general rule, phosphate buffers should not be used because of the possible formation of poorly soluble iron phosphates that, in turn, affect the actual iron concentration. The so-called Good's buffers (HEPES, Tris, and MOPS, according to the required pH range: 7, 8, or 9, respectively) should be preferred. Buffers may be prepared beforehand, stored as required and appropriate (plastic is fine for storage), and made anaerobic just before use. Do not add sodium azide or other antibacterials to any of the buffers.
6. Use a suitable lab pipette to add the required volume of 2-mercaptoethanol after quickly removing the septum on the buffer container under a positive Ar pressure.
7. Although sometimes of questionable quality, ferric ammonium citrate does not present all the instability and hydrolysis problems common to other iron salts, including the vastly used ferrous ammonium sulfate and ferric ammonium sulfate that are prone to precipitation, in particular when dissolved in buffer. Solutions of ferric ammonium citrate should be prepared daily. Use aluminum foil to protect ferric ammonium citrate solution from light until used.
8. Lithium sulfide is a very convenient replacement for the highly hygroscopic sodium sulfide: the yellow Li_2S powder is very stable if kept over silica gel in a desiccator at room temperature.
9. ATP is quite expensive, so take care of preparing just the amount of solution you will need. In our experience, freezing ATP solutions for subsequent use is not recommended. Avoid adding ATP to any iron salts prior to adding proteins.

10. Please note that highly reactive thiocyanate forms upon prolonged storage of urea solutions. Use them within 1 week and not refrigerate.
11. Enzymes and accessory proteins are described for the Isc system. However, equivalent and related proteins can be used in this method. There is no special recommendation here, if not for underscoring again the necessity to avoid the use of iron-chelating buffers, including phosphate and high-imidazole buffers, such as those used to isolate His-tagged proteins, whenever possible. However, this may not be an issue when adding minute volumes of a protein solution: up to 0.01 mL of "foreign" buffers will not cause problems if added to a solution with a final volume of about 1 mL.
12. Typical CD settings for recording a spectrum in the UV/Visible region on a Jasco 810 CD instrument are as follows: wavelength range 250–750 nm, scanning speed of 100 nm/min, data pitch 1 nm, bandwidth 1 nm, and response 0.25 s. Please make sure to follow the correct procedure when turning on any CD instrument, including ensuring that nitrogen is flowing all through the system. Although nitrogen is not required for the CD measurement itself in this spectral region, ensuring a proper flow of an inert gas prevents possible damage from the very intense light beam to the very delicate coatings of the instrument's optical mirrors.
13. Green iron polysulfides are polymeric ferric thiolates—often of poor solubility and marginal stability—that may be formed when: (1) there is not enough free thiols in the system (the Fe/SR^- ratio in these species is about 2, as compared to a Fe/SR^- ratio of 4 for a perfectly water-soluble species); and (2) the pH of the solution is too low (therefore, lowering the RS^- concentration even at nominal high thiol concentrations). Because of hydrolysis, this latter occurrence is common when using iron chlorides or sulfates as a source of iron.
14. Instrument settings for time-course measurements on a Jasco 810 CD: data pitch 0.5 s, bandwidth 2 nm, response 32 ms min, and 0.25 s max. Increasing the response lowers the noise level.
15. The time interval between subsequent CD spectra is usually limited to 5 min, given that scan time is about 4 min (from 300 to 700 nm at 100 nm/min) and considering that the instrument monochromator has to return to the original starting wavelength at the end of each scan. Most CD instruments can take care of serial measurements. This feature is distinct than averaging multiple spectra which is also offered by most CD spectrometers in an automatic fashion.
16. CD, differently from absorbance, only detects protein-bound clusters, and is insensitive to chromophores in a nonchiral

environment. Therefore, the isolation of the holoprotein is not required for its quantitation. However, —if desired—isolation of the reconstituted protein and/or its separation from excess reagents does not require strict anaerobiosis. Minimizing oxygen exposure during isolation lowers the risk of forming iron sulfide precipitates which are most likely formed in chemical reconstitution experiments. These blackish compounds stick to everything, may clog columns, filters, ultrafiltration devices, etc., and are extremely difficult to remove.

17. There has been some controversy about having both 5 mM 2-mercaptoethanol and 2 mM cysteine in buffers used for enzymatic cluster assembly studies, as 2-mercaptoethanol may act as a nonphysiological reductant in the conversion of cysteine sulfur to sulfide, thus making it difficult to assess the possible role of other proteins as for supplying the required electrons. The same thiol-related redox effects may also be relevant to cluster interconversion events since a reductive step is associated with dimerization of a [2Fe-2S] cluster into a [4Fe-4S] cluster [13].

18. CD spectra of the holoprotein are acquired on individual samples, prepared in the presence of increasing amounts of chaotrope. Most common chaotropes are urea and guanidine hydrochloride. Urea usually requires to acquire spectra at 1 M concentration incremental steps, whereas the more powerful guanidine hydrochloride is used in 0.5 M incremental steps. The smaller the increment between subsequent measurements, the more accurate the resulting denaturation curve.

19. Cluster stability toward metal substitution and iron chelators is strongly dependent on the redox state of the cluster. Since Fe(II) is much less thiophylic than Fe(III), cluster reduction strongly impairs the stability of iron ligation [16, 18]. On the other hand, all Fe-S proteins, but the mononuclear $Fe(RS)_4$ rubredoxins displaying an intense negative signal at 310 nm, have very weak CD signals in their reduced form. This characteristic makes this kind of measurement very difficult, also because reduction may be accompanied by changes in cluster nuclearity (for instance, conversion of [2Fe-2S] clusters into [4Fe-4S] ones on proteins with a suitably flexible structure, such as scaffold proteins [13]).

20. As reported here, these measurements imply the use of a Peltier-based heating/cooling element that ensures the highest accuracy and ease of operation, but this set up is not affordable. However, these measurements can be carried out by connecting the cell holder to a standard thermostatic bath fit with a circulation pump and by taking spectra of the protein every time you increase the temperature of the thermostatic bath by

5 or 10 °C. Just make sure to account for the—almost unavoidable—temperature difference between the settings on the thermostatic bath and the solution in the measuring cuvette. A small, portable thermocouple may come in handy for assessing the actual temperature in a cuvette containing a buffer or just water to be used in "test runs."

References

1. Kelly SM, Jess TJ, Price NC (2005) How to study proteins by circular dichroism. Biochim Biophys Acta 1751:119–139
2. Morleo A, Bonomi F, Iametti S, Huang VW, Kurtz DM (2010) Iron-nucleated folding of a metalloprotein in high urea: resolution of metal binding and protein folding events. Biochemistry 49:6627–6634
3. Bonomi F, Iametti S, Ferranti P, Kurtz DM, Morleo A, Ragg EM (2008) "iron priming" guides folding of denatured aporubredoxins. J Biol Inorg Chem 13:981–991
4. Pagani S, Vecchio G, Iametti S, Bianchi R, Bonomi F (1986) On the role of the 2Fe-2S cluster in the formation of the structure of spinach ferredoxin. Biochim Biophys Acta 870:538–544
5. Johnson DC, Dean DR, Smith AD, Johnson MK (2005) Structure, function, and formation of biological iron-sulfur clusters. Annu Rev Biochem 74:247–281
6. Freibert SA, Weiler BD, Bill E, Pierik AJ, Muhlenhoff U, Lill R (2018) Biochemical reconstitution and spectroscopic analysis of iron-sulfur proteins. In: David SS (ed) Fe-S cluster enzymes, part B. Methods in eEnzymology, vol 599. Elsevier, Cambridge, pp 197–226
7. Pagani S, Bonomi F, Cerletti P (1984) Enzymic-synthesis of the iron-sulfur cluster of spinach ferredoxin. Eur J Biochem 142:361–366
8. Bonomi F, Pagani S, Kurtz DM (1985) Enzymic synthesis of the 4Fe-4S clusters of *Clostridium pasteurianum* ferredoxin. Eur J Biochem 148:67–73
9. Fu WG, Jack RF, Morgan TV, Dean DR, Johnson MK (1994) NifU gene-product from *Azotobacter vinelandii* is a homodimer that contains 2 identical [2Fe-2S] clusters. Biochemistry 33:13455–13463
10. Yuvaniyama P, Agar JN, Cash VL, Johnson MK, Dean DR (2000) NifS-directed assembly of a transient [2Fe-2S] cluster within the NifU protein. Proc Natl Acad Sci U S A 97:599–604
11. Agar JN, Zheng LM, Cash VL, Dean DR, Johnson MK (2000) Role of the IscU protein in iron-sulfur cluster biosynthesis: IscS-mediated assembly of a [Fe2S2] cluster in IscU. J Am Chem Soc 122:2136–2137
12. Adinolfi S, Iannuzzi C, Prischi F, Pastore C, Iametti S, Martin SR, Bonomi F, Pastore A (2009) Bacterial frataxin CyaY is the gatekeeper of iron-sulfur cluster formation catalyzed by IscS. Nat Struct Mol Biol 16:390–396
13. Agar JN, Krebs C, Frazzon J, Huynh BH, Dean DR, Johnson MK (2000) IscU as a scaffold for iron-sulfur cluster biosynthesis: sequential assembly of [2Fe-2S] and [4Fe-4S] clusters in IscU. Biochemistry 39:7856–7862
14. Chandramouli K, Unciuleac MC, Naik S, Dean DR, Huynh BH, Johnson MK (2007) Formation and properties of [4Fe-4S] clusters on the IscU scaffold protein. Biochemistry 46:6804–6811
15. Li HR, Mapolelo DT, Dingra NN, Naik SG, Lees NS, Hoffman BM, Riggs-Gelasco PJ, Huynh BH, Johnson MK, Outten CE (2009) The yeast iron regulatory proteins Grx3/4 and Fra2 form heterodimeric complexes containing a [2Fe-2S] cluster with cysteinyl and histidyl ligation. Biochemistry 48:9569–9581
16. Bonomi F, Iametti S, Morleo A, Ta D, Vickery LE (2011) Facilitated transfer of IscU-[2Fe2S] clusters by chaperone-mediated ligand exchange. Biochemistry 50:9641–9650
17. Bonomi F, Fessas D, Iametti S, Kurtz DM, Mazzini S (2000) Thermal stability of *Clostridum pasteurianum* rubredoxin: deconvoluting the contributions of the metal site and the protein. Prot Sci 9:2413–24126
18. Conover RC, Park JB, Adams MWW, Johnson MK (1990) Formation and properties of a NiFe3S4 cluster in *Pyrococcus furiosus* ferredoxin. J Am Chem Soc 112:4562–4564
19. Iametti S, Uhlmann H, Sala N, Bernhardt R, Ragg E, Bonomi F (1996) Reversible, non-denaturing metal substitution in bovine adrenodoxin and spinach ferredoxin and the different reactivities of [2Fe-2S]-cluster-containing proteins. Eur J Biochem 239:818–826

20. Krebs C, Agar JN, Smith AD, Frazzon J, Dean DR, Huynh BH, Johnson MK (2001) IscA, an alternate scaffold for Fe-S cluster biosynthesis. Biochemistry 40:14069–14080
21. Wu SP, Cowan JA (2003) Iron-sulfur cluster biosynthesis. A comparative kinetic analysis of native and cys-substituted ISA-mediated $[2Fe\text{-}2S]^{2+}$ cluster transfer to an apoferredoxin target. Biochemistry 42:5784–5791
22. Bonomi F, Iametti S, Ta D, Vickery LE (2005) Multiple turnover transfer of [2Fe2S] clusters by the iron-sulfur cluster assembly scaffold proteins IscU and IscA. J Biol Chem 280:29513–29518
23. Mapolelo DT, Zhang B, Naik SG, Huynh BH, Johnson MK (2012) Spectroscopic and functional characterization of iron-sulfur cluster-bound forms of *Azotobacter vinelandii* (Nif)-IscA. Biochemistry 51:8071–8084
24. Shakamuri P, Zhang B, Johnson MK (2012) Monothiol glutaredoxins function in storing and transporting [Fe2S2] clusters assembled on IscU scaffold proteins. J Am Chem Soc 134:15213–15216
25. Chandramouli K, Johnson MK (2006) HscA and HscB stimulate [2Fe-2S] cluster transfer from IscU to apoferredoxin in an ATP-dependent reaction. Biochemistry 45:11087–11095
26. Bonomi F, Iametti S, Morleo A, Ta D, Vickery LE (2008) Studies on the mechanism of catalysis of iron-sulfur cluster transfer from IscU [2Fe2S] by HscA/HscB chaperones. Biochemistry 47:12795–12801
27. Wu SP, Mansy SS, Cowan JA (2005) Iron-sulfur cluster biosynthesis. Molecular chaperone DnaK promotes IscU-bound [2Fe-2S] cluster stability and inhibits cluster transfer activity. Biochemistry 11:4284–4293
28. Iametti S, Barbiroli A, Bonomi F (2015) Functional implications of the interaction between HscB and IscU in the biosynthesis of Fe-S clusters. J Biol Inorg Chem 20:1039–1048
29. Pace CN (1990) Measuring and increasing protein stability. Trends Biotechnol 8:93–98

Chapter 13

Native Mass Spectrometry of Iron-Sulfur Proteins

Jason C. Crack and Nick E. Le Brun

Abstract

Iron-sulfur clusters constitute a large and widely distributed group of protein cofactors that play key roles in a wide range of metabolic processes. The inherent reactivity of iron-sulfur clusters toward small molecules, for example, O_2, NO, or *free* Fe, makes them ideal for sensing changes in the cellular environment. Nondenaturing, or native, MS is unique in its ability to preserve the noncovalent interactions of many (if not all) species, including stable intermediates, while providing accurate mass measurements in both thermodynamic and kinetic experimental regimes. Here, we provide practical guidance for the study of iron-sulfur proteins by native MS, illustrated by examples where it has been used to unambiguously determine the type of cluster coordinated to the protein framework. We also describe the use of time-resolved native MS to follow the kinetics of cluster conversion, allowing the elucidation of the precise series of molecular events for all species involved. Finally, we provide advice on a unique approach to a typical thermodynamic titration, uncovering early, quasi-stable, intermediates in the reaction of a cluster with nitric oxide, resulting in cluster nitrosylation.

Key words Mass spectrometry, Nondenaturing mass spectrometry, Iron-sulfur clusters, Isotopic labeling, Cluster conversion, Native MS, Time-resolved MS, [2Fe-2S], [3Fe-4S], [4Fe-4S]

1 Introduction

Iron-sulfur (Fe-S) proteins are widespread in nature and exhibit a remarkable breadth of structural and functional diversity [1]. Many play key roles in electron transfer, chemical catalysis, and small molecule sensing [2–5]. They are characterized by the presence of an Fe-S cluster, an ancient protein cofactor composed solely of iron and inorganic sulfide [6]. Studies of Fe-S cluster proteins have been facilitated by a plethora of different structural, spectroscopic, and bioanalytical techniques, but the application of mass spectrometric methods have, until recently, been quite limited. Fe-S clusters are acid labile and were thus lost under the mildly acidic conditions employed during early electrospray ionization mass spectrometry (ESI-MS) investigations [7, 8]. In 1993, the ESI-MS spectrum of the *Chlorobium tepidum* high-potential ferredoxin (HiPIP) was found to exhibit three distinct charge states (m/z + 5 to +7).

Patricia C. Dos Santos (ed.), *Fe-S Proteins: Methods and Protocols*, Methods in Molecular Biology, vol. 2353,
https://doi.org/10.1007/978-1-0716-1605-5_13,

Deconvolution of the HiPIP spectrum revealed a molecular mass of the protein consistent with the presence of an intact [4Fe-4S] cluster [8]. Ferredoxins display a high degree of stability at physiological pH values and were the obvious choice to help define a set of conditions suitable for the detection of other less robust Fe-S cluster proteins. In most cases, the Fe-S cluster(s) survived ionization and remained associated with the protein, irrespective of whether they were [2Fe-2S] or [4Fe-4S] clusters [9–12].

The term *native* MS was introduced to unify the bewildering terminology (e.g., nondenaturing, noncovalent, native spray, macro- or supramolecular ESI-MS) that had arisen to describe the same fundamental approach to ESI-MS experiments [13, 14]. Almost two decades later, native MS is now used for accurate mass detection of intact proteins and protein complexes, and has been used extensively to study the dynamics of protein-protein interactions, and interactions of proteins with drugs, nucleic acids, sugars, lipids, and metals [15–39]. We have further developed the use of native MS to gain an insight into how Fe-S cluster containing transcriptional regulators sense their particular analyte [40–42].

Our approach is well illustrated by native MS studies of the [4Fe-4S] cluster regulator FNR and the effect of O_2 exposure on it. *Escherichia coli* is a facultative anaerobe that can, in the absence of O_2, preferentially use environmental nitrate as a terminal electron acceptor [3, 43–52]. Over many years, the [4Fe-4S] cluster of FNR was shown to undergo a [4Fe-4S] to [2Fe-2S] cluster conversion in response to O_2, driving a dimer to monomer transition with concomitant loss of DNA binding. Native MS was employed to investigate the cluster conversion process in more detail. Using a variant form (S24F) that is spectroscopically and mechanistically identical to the wild-type protein, but undergoes reaction more slowly [53], was key to being able to gather time-dependent data. Under anaerobic conditions, the monomer region was dominated by a [4Fe-4S] FNR peak at 29,898 Da (predicted mass 29,897 Da). In the dimer region, a peak at 59,796 Da corresponded to homodimeric [4Fe-4S] FNR. Exposure to dissolved atmospheric O_2 resulted in the formation of a variety of protein-bound clusters, including [3Fe-4S] (29,843 Da), [3Fe-3S] (29,811 Da), and [2Fe-2S] (29,720 Da) forms, as well as various persulfide forms (*see* **Note 1**). In the dimeric region, three low-intensity peaks were observed for dimers containing at least one [3Fe-4S] cluster, consistent with cluster conversion initiating the dimer to monomer transition (Fig. 1).

Analysis of native MS kinetic datasets led to the development of a model that describes the behavior of the main [4Fe-4S], [3Fe-4S], and [2Fe-2S] clusters species [42, 54]. Isotope substitution data, from a ^{34}S-substituted form of the [4Fe-4S] cluster, provided unambiguous confirmation of the assignments for the

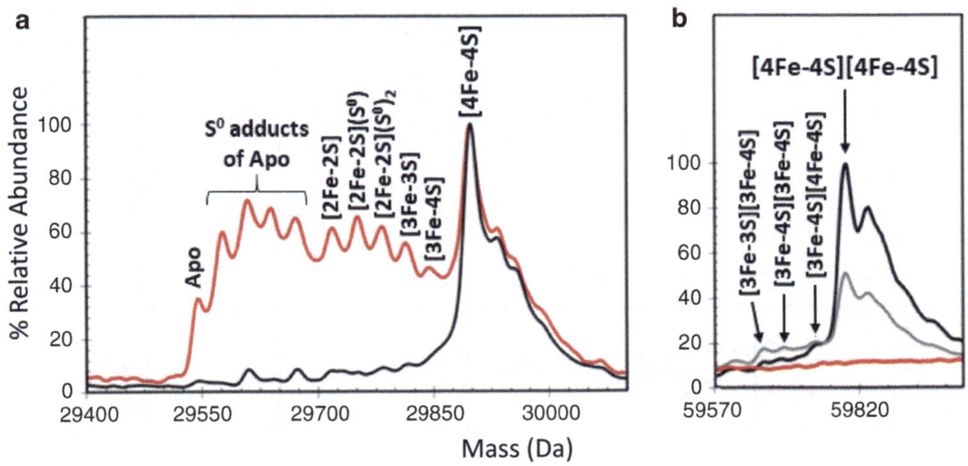

Fig. 1 Native MS of [4Fe-4S] FNR and the effect of O_2 exposure. Deconvoluted mass spectra of [4Fe-4S] S24F FNR in (**a**) the monomer region before (black line) and after (18 min exposure; red line) exposure to dissolved atmospheric O_2. This resulted in the formation of a variety of protein-bound clusters, including [3Fe-4S], [3Fe-3S], and [2Fe-2S] forms. Persulfide ligated [2Fe-2S] clusters, [2Fe-2S](S^0) and [2Fe-2S]$(S^0)_2$, were also observed. (**b**) The dimeric region before (black line), during (20 min, gray line), and after (80 min, red line) O_2 exposure [42]

cluster conversion intermediates and products [42, 55]. For further details, *see* [42]. Native MS is unique (in comparison to other techniques) in its ability to provide accurate mass data for mixtures containing multiple noncovalent complexes, including transiently stable intermediates, in both thermodynamic and kinetic experimental regimes [56]. Here, we provide practical guidance on how to perform such native MS experiments on Fe-S cluster proteins, starting from basic cluster assignments to more complex cluster reactivity measurements.

For readers new to the field or the application of native MS, Fig. 2 shows the basic anatomy of the mass spectrometer used in our laboratory (a Q-TOF), and the role of each stage is discussed in Subheading 3.5. Figure 2 also shows the route taken by the analyte, from ion generation in the source to ion separation/detection in the TOF stage. Here, we provide practical guidance for the preparation of native MS samples (Subheading 3.3), spectrometer setup (Subheading 3.5), and basic data analysis (Subheading 3.6). We then provide selected examples where we have used native MS to unambiguously determine the type of cluster coordinated to the protein framework (Subheading 3.7). We also demonstrate the use of time-resolved native MS to follow the kinetics of cluster reactions and provide guidance for analysis (Subheading 3.8).

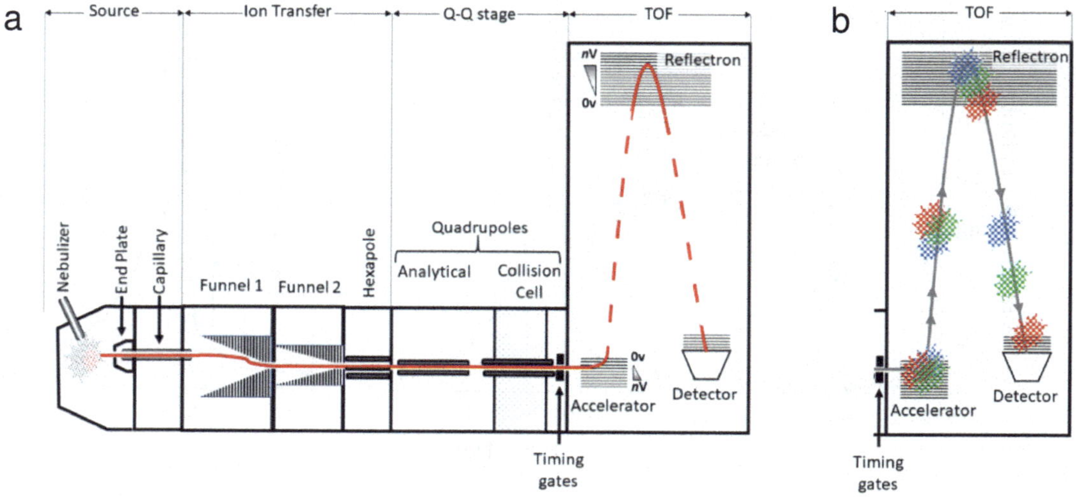

Fig. 2 Basic anatomy of a mass spectrometer. (**a**) A mass spectrometer can be divided into three fundamental components that function in ionization (the source and ion transfer), analysis (the Q-Q stage, TOF stage), and detection (located in the TOF stage). Each of the functions can be further subdivided, as discussed in the text (*see* Subheading 3.5). The red line represents the path taken by ions through the various stages of the mass spectrometer. The field-free drift region is indicated by a dashed red line. (**b**) TOF phase in more detail. Ion packets are accelerated through the TOF tube. The reflectron normalizes the kinetic energy of the ions and redirects them toward the detector. The time taken to reach the detector is then used to calculate the ions' mass to charge ratio (*see* Subheading 3.5, **step 9**). Ions with lower mass/higher charge (red) arrive sooner than ions with higher mass/lower charge (blue); intermediate ions are represented in green

2 Materials

2.1 Instrumentation

Several mass spectrometers from the major manufacturers meet the requirements for performing native MS. These include quadrupole time-of-flight (Q-TOF), quadrupole Orbitrap, and Fourier transform ion cyclotron resonance (FTICR) spectrometers. Newer hybrid instruments now incorporate ion mobility methodologies for enhanced separation of closely related species and access molecular shape/architecture information in addition to mass [39]. Electrospray ionization (ESI) is the method of choice for achieving *soft* ionization.

1. uHPLC coupled to a mass spectrometer: Mass spectrometer (*see* **Note 2**) associated with an ultra-high-performance liquid chromatography (uHPLC) system. Although not required for native MS, an associated uHPLC allows intact mass determination under denaturing conditions via liquid chromatography mass spectroscopy (LC-MS), *see* **Note 3**.
2. ESI nebulizer: ESI nebulizer (*see* **Note 4**) should be connected to a syringe port via PEEK tubing and associated fittings (*see* Subheading 2.3). Samples are introduced directly via air-tight syringe (*see* **Note 5**) and a basic syringe infusion pump. Wear gloves (*see* **Note 6**).

2.2 Sample Preparation

1. Protein solution(s): Protein solutions to be analyzed, at the highest purity possible [57, 58].
2. Stable isotopes: Iron-57 (≥95%) and sulfur-34 (≥98%) (Cambridge Isotope Laboratories, Inc).
3. L-Cysteine.
4. Ferrous ammonium sulfate.
5. DL-dithiothreitol (DTT).
6. Ammonium acetate (≥98%).
7. Ammonium hydroxide solution (≥99%).
8. Glacial acetic acid (≥99%).
9. Zeba™ Spin desalting column, 7 K MWCO (Thermo Scientific).
10. PD-10 desalting columns, with Sephadex G-25 resin, 5 K MWCO (Cytiva).
11. PD MiniTrap G-25, 5 K MWCO (Cytiva).
12. LC-MS grade water, ≤50 ppb common metal ions, 0.2 μm filtered.
13. LC-MS grade formic acid, ≤0.5 ppm common metal ions.
14. LC-MS grade acetonitrile (≥99%), ≤100 ppb common metal ions.
15. A suitable reversed phased column.
16. Microcentrifuge.

2.3 Native Mass Spectrometry

1. Mass spectrometer: Mass spectrometer, with associated services, equipment, and control/analysis software (*see* Subheading 2.1), installed as per the manufacturer's specifications.
2. ESI-L low-concentration tuning mixture (Agilent Technologies).
3. Cesium iodide, mass spectroscopy grade.
4. Perfluoroheptanoic acid (tridecafluoroheptanoate, $C_7F_{13}O_2$).
5. Gas-tight syringes, 22-gauge blunt needle.
6. Spare ESI nebulizer needles (*see* **Note 4**).
7. Polyether ether ketone (PEEK) tubing.
8. Valco syringe fitting, 1/16 in., for 22-gauge needles.
9. Stainless steel unions, 1/16 in.
10. Selection of finger-tight PEEK fittings, 1/16 in.
11. Basic syringe infusion pump.
12. Access to an anaerobic chamber capable of maintaining O_2 at <2 ppm.

13. A UV-visible absorbance spectrophotometer capable of measuring in the range 250–700 nm.

3 Methods

3.1 In Vitro Fe-S Assembly and Isotope Labeling

We have previously described, in this series, a detailed methodology applicable to the preparation of a wide variety Fe-S cluster containing transcriptional regulators [58]. While it is usually advantageous to combine protein expression with in vivo Fe-S cluster assembly, in vitro reconstitutions can be used where this is not possible. Below is a brief protocol for in vitro reconstitution and the preparation of isotopically labeled Fe-S clusters [57, 58], *see* **Note 7**.

1. Transfer previously prepared apoprotein (*see* [58]) to an anaerobic cabinet and equilibrate.
2. Prepare the following solutions: Solution 1, 50 mM (12.8 mg/mL) L-cysteine, 125 mM (41.7 mg/mL) dithiothreitol. Solution 2, 20 mM (78.5 mg/10 mL) ferrous ammonium sulfate.
3. Add an appropriate aliquot of solution 1 to the apoprotein to give a final concentration of 1 mM L-cysteine and 2.5 mM dithiothreitol. For ^{34}S-labeled clusters, use ^{34}S-L-cysteine or ^{34}S-sulfide (for detailed methodology, *see* [55]).
4. Add an appropriate volume of solution 2 to the apoprotein. For [4Fe-4S] clusters, add seven-fold excess of Fe^{2+} ions over the apoprotein. For ^{57}Fe-labeled clusters, use ^{57}Fe dissolved in HCl (for detailed methodology, *see* [55]).
5. To initiate Fe-S cluster reconstitution, add purified NifS [59] to a final concentration of 225 nM. Seal the reaction mixture and incubate between 25 and 37 °C for up to 5 h.
6. Pass the reconstitution mixture through a 0.2-μm syringe filter to remove any particulate matter.
7. Separate the reconstituted protein from low-molecular-weight reactants using one of two methods. The first, which is ideal for small-scale preparations (up to 2.5 mL), involves rapid gel filtration using a PD10 column (Cytiva). The second method, which is ideal for larger preparations (up to 25 mL), involves the use of an ion-exchange column to simultaneously clean and concentrate the reconstituted protein (*see* [55] for further details). The reconstituted protein is eluted in a buffer containing ~1 M NaCl.

3.2 Calibration

For improved accuracy, the mass spectrometer should be calibrated prior to starting any native MS work. Wear gloves (*see* **Note 6**). A good calibrant should cover the desired *m/z* range expected for your sample, have stable masses in both positive and negative modes, and not adhere to the surfaces of the of the nebulizer or ESI spray chamber.

1. Choose an appropriate calibrant. We routinely use a commercial ESI-L low-concentration tuning mixture that provides 10 calibration points spanning 118–2721 *m/z* in positive mode. We also use ~10 mM cesium tridecafluoroheptanoic acid (CsTFHA) solutions to provide calibration points up to 10,000 *m/z* [60] (*see* **Note 8**).
2. Directly infuse the ESI-L low-concentration tuning mixture (tune mix) into the spectrometer. Once ionization is stable, follow the manufacturer's guidelines for calibrating the spectrometer. In the case of CsTFHA, first calibrate the instrument with tune mix, flush the nebulizer with 50% (v/v) acetonitrile/water, then introduce the CsTFHA calibrant, and record data for 60 s (*see* **Note 8**). This provides a CsTFHA calibration file for performing an offline calibration using the manufacturer's data analysis software [61]. CsTFHA calibrations can be performed monthly or less frequently, as required.

3.3 Sample Preparation for Native Mass Spectrometry

Prior to performing native mass spectrometry, all nonvolatile salts (e.g., NaCl, KCl, phosphate, etc.) and buffer components (Tris, HEPES, MES, etc.) need to be removed from the protein sample and replaced with volatile buffer components. Aqueous solutions of ammonium acetate are widely used for native MS (for other reagents, *see* **Note 9**), with reported concentrations ranging from 5 mM to 2 M and a near neutral pH [15, 56]. Samples may be exchanged into the volatile buffer by dialysis, ultrafiltration, or gel filtration techniques (*see* **Notes 10** and **11**) [15, 56]. Try using 200 mM ammonium acetate pH 8.0 as a starting point for beginning native MS studies. The following describes sample preparation using a PD-10 column (gravity protocol).

1. Weigh out the desired amount of ammonium acetate (>98% purity) directly into glassware previously cleaned with glacial acetic acid and rinsed with LC-MS grade water to remove any residual small molecules (*see* **Note 6**). For a 200-mM solution, weigh out 1.54 g of ammonium acetate and dissolve in 100 mL of LC-MS grade water. As prepared, the solution will have a pH of ~6.9. The addition of 100 μL of 5 M ammonium hydroxide will result in a pH of ~8.0 (*see* **Note 12**).
2. Equilibrate the PD-10 column with 25 mL of 200 mM ammonium acetate pH 8.0. Discard the eluant.
3. Apply a maximum of 2.5 mL of sample to the column. Ideally, take 0.5 mL of a concentrated protein solution (~200–500 μM), dilute it to 1 mL with the same ammonium acetate buffer, and apply to the top of the column.

4. Let the protein sample completely enter the column. Fe-S proteins are deeply colored, and so their progress can be readily tracked by eye. Add an additional 1.5 mL of ammonium acetate buffer. Discard the eluant.
5. At this point, the sample should be located toward the bottom of the column. Spreading of the color across the entirety of the column or retention of the color close to the top of the column would suggest abnormal behavior (*see* **Note 13** for troubleshooting advice).
6. Place a microcentrifuge tube under the column and elute with 3.5 mL of buffer and collect the eluant. PD-10 columns have a dilution factor of ~1.4 and a desalting efficiency of at least 90%. Therefore, the first 2 mL of the eluant will contain the bulk of the desalted sample. The remaining portion of the eluant may contain enough nonvolatile salts/buffer components to make native mass spectrometry problematic (collect as a separate 1.5 mL aliquot). The same principle can be applied to PD mini Trap G-25 columns. For Zeba spin desalting columns, dilute the sample by half with ammonium acetate buffer prior to loading and follow the manufacturer's instructions.
7. Measure the absorbance spectrum of the eluted protein sample. The concentration of the resulting sample can be determined from the Fe-S charge transfer bands near 400 nm and a generic extinction coefficient of ~4 mM^{-1} cm^{-1} per $Fe(RS)_n$ atom present [62]. Where possible, the extinction coefficient for the protein of interest should be used: [4Fe-4S] FNR ε_{406nm} 16.22 mM^{-1} cm^{-1}, [4Fe-4S] NsrR ε_{406nm} 13.30 mM^{-1} cm^{-1}, [4Fe-4S] RirA ε_{383nm} 13.46 mM^{-1} cm^{-1}, [2Fe-2S] RsrR ε_{455nm} 3.45 mM^{-1} cm^{-1} [45, 63–65]. This completes the preparation of a stock solution of desalted protein, which should provide enough for multiple native MS experiments.

3.4 Introducing the Sample into the Mass Spectrometer

1. Flush the nebulizer assembly. After calibrating the mass spectrometer and before the introduction of a sample, flush the flow path with 50% (v/v) aqueous acetonitrile. Repeat with water, and then 5 mL of anaerobic ammonium acetate. The use of two gas-tight syringes to maintain a continuous flow limits O_2 build up between samples (*see* **Note 14**).
2. Dilute the protein stock. An aliquot of the desalted stock solution should be diluted with ammonium acetate buffer to between 2 and 30 μM of cluster. When performing native MS for the first time, it is advantageous to prepare 0.5–1 mL of this diluted sample. This provides enough material (and time) for parameter optimization (*see* Subheading 3.5). The amount of sample prepared can be reduced once optimal ionization conditions have been determined.

3. In an anaerobic chamber, load the prepared sample into a gas-tight syringe, remove it from the chamber, and directly introduce into the spectrometer. Try infusing at 0.3 mL/h with 0.8 bar nebulizer pressure, using positive ion mode (*see* **Note 15**). The aim is to create a stable spray capable of achieving ionization. In most cases, some degree of ionization is observed with the following parameters: capillary voltage 4500 V, offset 500 V, dry gas flow rate 4 L/min, dry gas temperature 180 °C. Further optimization of these parameters is usually required (*see* Subheading 3.5).

3.5 Parameter Optimization (Tuning)

Having obtained some degree of ionization, it is usually worthwhile investing some time in exploring different parameters. The aim of tuning is to obtain the maximum possible signal intensity and resolution while minimizing in-source or collisional damage. This is achieved by optimizing the source, funnel, hexapole, quadrupole, and collision cell settings, as well as timing adjustments. When tuning the instrument, it is advisable to optimize each setting in turn along the ion path, from the front to the back of the instrument (Fig. 1).

1. Adjust the gas flow. Samples are introduced to the mass spectrometer through the nebulizer. The nebulizer pressure has a direct effect on the dispersion of the infused solution into a fine mist of droplets. Heated dry gas (N_2), flowing in the opposite direction to the nebulizer spray, is used to aid solvent (water) evaporation and to prevent uncharged material from entering the spectrometer. Lower dry gas flow rates may favor the entry of larger, less charged ions into the spectrometer. Initially, try using 0.8 bar nebulizer pressure and a 4-L/min dry gas flow rate, with a dry gas temperature of 180 °C. Solvent evaporation shrinks the droplets further until the Rayleigh limit is reached and coulombic repulsion results in droplet fission. This process continues until the radii of the droplet shrink to a few nanometers, promoting the release of the analyte into the gas phase, with analyte ions drawn toward the endplate (reviewed in [66, 67]).
2. Select the source voltage. An electric field (>1 kV) between the endplate and nebulizer initiates the electrospray ionization (ESI) process, causing the mist of droplets to become charged. The source can be operated in positive or negative ion modes. In positive ion mode, an applied voltage difference (0.5 kV) between negatively charged endplate and capillary draws the ions toward the entrance of the capillary and into the ion transfer stage. Start in positive mode with a capillary voltage of 4.5 kV. Ionization at lower voltages (2.5–3.5 kV) should be compared to that at higher voltages (≥4 kV) to minimize in-source damage to the Fe-S clusters. In negative ion mode,

the capillary and endplate are positively charged. The capillary transfers a mixture of analyte ions, drying gas, and traces of solvent to the first vacuum stage. Subsequent parts of the transfer region further separate the analyte ions from the drying gas (Fig. 2).

3. Adjust the funnel voltages: Funnels 1 and 2 are a major part of the ion transfer stage. Their primary purpose is to guide incoming analyte ions toward the later stages, separating them from the residual solvent and drying gas. An applied RF voltage (voltage peak to peak, Vpp) to each of the funnels assists in the formation of an ion beam. The default values for Funnels 1 and 2 RF are 400 *V*pp. Higher funnel voltages will focus more ions into the nascent ion beam, improving ion transfer efficiency. In most cases, adjustment of the funnel voltages has a little noticeable effect; it becomes more noticeable for larger, lower charge ions ($\geq$3000 *m/z*) (hereafter referred to as larger ions, *see* **Note 16**). Some newer instruments now allow the pressure of Funnel 1 to be increased, resulting in enhanced desolvation, higher ion transfer efficiency, and improved detection of high-molecular-weight protein complexes.
4. Check the isCID voltage. This is functionally identical to the cone voltage (orifice voltage or declustering potential) and can be used to bring about in-*s*ource *c*ollision-*i*nduced *d*issociation (isCID). Increasing the isCID voltage applies a potential difference between Funnels 1 and 2. All incoming ions are accelerated toward Funnel 2, increasing the likelihood of a collision and potential fragmentation. A value of 0 V prevents unwanted fragmentation. Increasing this value ($\leq$100 V) can improve the passage of larger ions by promoting gentle declustering of ions without fragmentation. However, as with the capillary voltage, it is important to check the applied voltage does not damage the Fe-S cluster or disrupt protein-protein interactions.
5. Adjust the hexapole RF voltage. The hexapole represents the end of the transfer stage. An applied RF voltage focuses the ion beam onto the hexapole axis. In general, higher voltages favor larger ions. The default value of 400 *V*pp is applicable in most situations.
6. Adjust quadrupole settings. In native MS experiments, the analytical quadrupole is seldom used to target ions. The default value permits the passage of ions $\geq$500 *m/z*. In some cases, it can be used as a mass filter, effectively excluding troublesome ions (up to 3000 *m/z*) from the resulting spectrum. The quadrupole can also be used to impart additional energy to ions. The default value of 5 eV is usually applicable in most cases, but can be increased by a few eV for larger ions.

7. Setup the collision cell. The collision cell operates at a higher pressure than the preceding compartments, resulting in a loss of kinetic energy. In native MS experiments, the collision cell is seldom used for CID. Hence, the collision voltage is usually kept to a minimum to prevent fragmentation. The default voltage of 10 V is usually applicable in most cases but can be increased somewhat for larger ions. As with the capillary and isCID voltages, it is important to check for Fe-S cluster damage or disruption of protein-protein interactions. An applied RF voltage focuses the ions onto the axis of the collision cell quadrupole, which, coupled with the high gas pressure (relative to the high vacuum stages) inside the collision cell, results in a loss of ion energy (ion cooling). The default value of 650 *V*pp is usually enough during the initial optimization; higher and lower voltages tend to favor larger and smaller ions, respectively.
8. Adjust the timings. The collision cell ends with an ion gate and transfer lens that temporarily blocks the transmission of ions into the TOF stage. The "transfer time" defines the period where ions are blocked from entering the TOF stage. After this period, ions are permitted into the TOF stage as defined by the "prepulse storage" time. Both timings are referenced to the next TOF pulse and affect the transferred *m/z* range. Default values for transfer time and prepulse storage are 120 μs and 10 μs, respectively. A longer transfer time increases the upper limit of *m/z* transmission, while a longer prepulse storage time increases the lower limit of *m/z* transmission (*see* **Notes 17** and **18**).

3.6 Basic Data Analysis

Most manufacturers provide bespoke data analysis packages with their instruments, through which users can retrieve, analyze, and process their data. Many include dedicated algorithms for routine compound identification, structural elucidation, and protein/peptide analysis and deconvolution. The maximum entropy (MaxEnt) deconvolution routine is one of the most widespread deconvolution algorithms, iteratively converging the observed charge state distribution present in the *m/z* spectrum into a deconvoluted neutral mass spectrum [68].

1. Setup the *m/z* spectrum for analysis. Two important user-controlled deconvolution parameters are the *m/z* range selected for deconvolution and the output neutral mass range. It is important to be aware that deconvolution can generate artifacts. Harmonic artifacts can occur when analyte charge states occur at positions that coincide with those for the same analyte with double (or even half) the true observed mass. One manifestation of this phenomenon is the appearance of pairs of adducts rather than a consecutive sequence. Satellite

artifacts occur when charge states become crowded, such that there is no clear boundary between the end of one charge state and the beginning of another, giving rise to charge state assignment ambiguities. In both cases, false masses may be represented in the resulting neutral mass spectrum [69]; *see* **Note 19**.

2. Repeat deconvolution. To be certain that there are no artifacts in the deconvoluted spectrum, it is advisable to reanalyze the charge states (m/z spectrum) using modified deconvolution parameters (either output mass range and/or the input m/z range); *see* **Note 19**.
3. Assign the molecular mass of a protein from the m/z spectrum. All the observed charges present on the ion result from an excess of protons during positive mode ESI (or a deficit of protons in negative mode). For apoproteins (via LC-MS or native MS), these masses are derived from the m/z spectra with peaks corresponding to $[M + zH]/z$, where M is the molecular mass of the protein, H is the mass of the proton, and z is the charge of the ion. However, additional considerations are required for the analysis of proteins containing cofactors/prosthetic groups, for example, metalloproteins, *see* **Note 20**.
4. Determine which association states of the protein are observed. Hydrophobic interactions, such as those that dominate dimer interfaces, can be disrupted upon transfer to the gas phase resulting in the appearance of monomers of proteins that are known to be dimeric in solution [70, 71]. Observation of the monomers, however, can be key to the elucidation of a mechanism because the presence of two clusters in a dimer makes it much more difficult to unambiguously assign intermediate species on the dimer (*see* Subheading 3.7).

3.7 Cluster Assignment

Native MS has been shown to provide definitive evidence of the cluster type [10, 40, 42, 63, 64, 72–74]. The sample requirements for native MS are significantly less demanding than those for Mössbauer or resonance Raman spectroscopy (50–200 μM vs. 1–2 mM of cluster) and do not usually require isotopic enrichment for cluster assignment, although this is advantageous. The basic principle of cluster assignment, together with key examples, is given below.

1. Assign the mass difference. Cluster assignment is usually achieved by measuring the difference between the apoprotein peak (by LC-MS or native MS) and holoprotein peak (by native MS); *see* **Notes 21–23** for examples.
2. Ideally, repeat measurements and mass assignment for isotopically substituted protein samples (Fig. 3), which include the

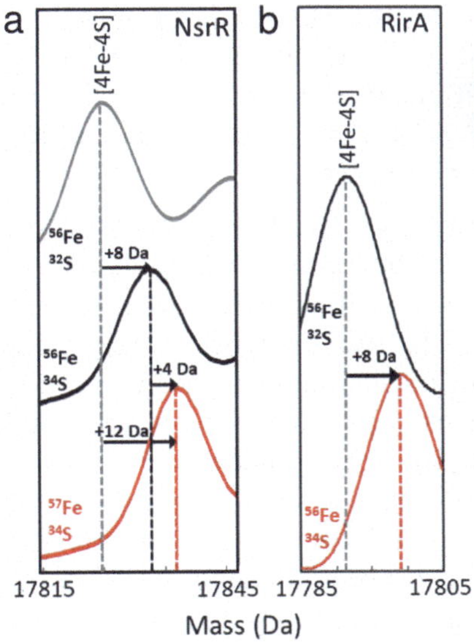

c

FeS Species	Predicted mass of natural abundance	^{57}Fe substituted	^{34}S substituted	^{57}Fe/^{34}S substituted
[4Fe-4S]	352	356 (+4)	360 (+8)	364 (+12)
[4Fe-3S]	320	324 (+4)	326 (+6)	330 (+10)
[3Fe-4S]	296	299 (+3)	304 (+8)	307 (+11)
[3Fe-3S]	264	267 (+3)	270 (+6)	273 (+9)
[3Fe-2S]	232	235 (+3)	236 (+4)	239 (+7)
[3Fe-S]	200	203 (+3)	202 (+2)	205 (+5)
[2Fe-4S]	240	242 (+3)	248 (+8)	250 (+10)
[2Fe-3S]	208	210 (+2)	214 (+6)	216 (+8)
[2Fe-2S]	176	178 (+2)	180 (+4)	182 (+6)
[2Fe-S]	144	146 (+2)	146 (+2)	148 (+4)

Fig. 3 Mass shifts observed for isotopically labeled (**a**) [4Fe-4S] NsrR and (**b**) [4Fe-4S] RirA. Isotopic labeling and mass shifts as indicated. (**c**) Predicted masses of natural abundance, ^{57}Fe, ^{34}S, and ^{57}Fe/^{34}S substituted Fe-S species. Expected mass shifts versus natural abundance are shown in parentheses. All masses are in Daltons

masses and isotopic shifts for ^{57}Fe, ^{34}S, and ^{57}Fe/^{34}S-enriched clusters, and a convenient reference list of expected masses for commonly encountered clusters and breakdown products.

3.8 Cluster Conversions

In many cases, Fe-S proteins undergo reaction as part of their biological function; in the case of Fe-S containing transcriptional regulators, this involves the reaction of the cluster with the relevant analyte(s) that results in a change in the cluster and, in turn, causes a protein conformational rearrangement that modulates DNA-binding properties. Being able to follow such reactions in real-time is potentially extremely powerful, and we have begun to use native MS to follow the cluster conversion reactions of a range of Fe-S containing transcriptional regulators. This has highlighted the remarkable power of this methodology to reveal precise mechanistic details.

1. For kinetic studies, establish whether the reaction you are interested in occurs at a rate that is accessible to native MS (effectively a few minutes). In principle, faster processes could be studied, but this presents problems of signal to noise. If the reaction of interest is too rapid, consider ways in which it can be slowed without altering the mechanism, for example, using a variant form of the protein [42].
2. Ensure, as far as is possible, that the kinetics of the process you are observing by MS match that in solution. To do this, a spectroscopic method that can follow the rate of change (but

does not necessarily reveal what the change is or what the species involved are) can be used. However, this must be done under exactly the conditions of the MS samples, that is, with the protein in whatever volatile buffer is being used for native MS studies [40, 42], *see* **Note 24**.

3. Acquire time-resolved MS intensity data. For global analysis, process the data to generate relative abundance plots of ion counts for the relevant species as a fraction of the total ion count for all relevant species. The usual practice of reporting MS intensities relative to the most abundant species is not sufficient for kinetic studies, particularly when the most abundant species may switch during the time course of the reaction. Using ion counts directly, and expressing the abundance of each species as a fraction of the total ion count for all relevant species, solves this issue and also permits changes in relative abundance to be followed without distortions due to variations in ionization efficiency that normally occur across a data collection run.
4. Fit the data using a suitable global kinetic analysis program. We have found Dynafit 4 (BioKin Ltd.) [75] to be extremely useful for such analyses; *see* **Notes 25** and **26** for examples.
5. In some multistep reactions where there may be stable intermediates on the reaction path, a thermodynamic titration approach can be useful. Decide on a native MS experimental setup that is appropriate for the system under investigation. In the case of reactions of Fe-S cluster proteins with nitric oxide (NO), a slow NO-releasing reagent such as dipropylenetriamine (DPTA) NONOate can be used as the NO source [73]. This releases NO at a sufficiently slow rate that a titration experiment can be conducted by adding an appropriate excess of DPTA NONOate to the protein. *See* **Notes 27** and **28** for examples.

4 Notes

1. *Escherichia coli* FNR was exposed to dissolved atmospheric O_2 (168 μM) for 18 min. In addition to the variety of protein-bound clusters observed, single (29,752 Da) and double (29,784 Da) persulfide-ligated forms of [2Fe-2S] cluster were detected in the monomer region, along with persulfide adducts of apoprotein.
2. Off the shelf, Q-TOF instruments are currently a good choice for performing native MS. With advancements in instrument hardware, even a modest Q-TOF instrument offers good resolution and a high mass range TOF analyzer at an attainable

price. Some Q-TOF instruments are also amenable to customization, permitting the efficient transmission of large protein complexes with *m/z* values ≥10,000 *m/z* [56, 76, 77].

3. The mass of the protein of interest should be checked by denaturing mass spectroscopy, that is, LC-MS, to verify the predicted mass. A uHPLC or HPLC system fitted with an appropriate reverse-phase column can be used. An aliquot of 50 μL of at least 30 μM protein should be diluted to 500 μL with 2% (v/v) acetonitrile, 0.1% (v/v) formic acid. In the absence of an HPLC system, use a spin desalting column to exchange the protein into MS-grade water. Then add 50% (v/v) acetonitrile, 0.1% (v/v) formic acid, and infuse directly into the ESI source. This will disrupt noncovalent interactions and unfold the protein of interest. Note that the ability to detect posttranslational modification may be hampered by the presence of residual buffer and salts [56].
4. In comparison to regular LC-MS, during native MS experiments, the ESI needle is more prone to fouling and will require periodic replacement more often.
5. This is sufficient to maintain anaerobic conditions during sample infusion.
6. Wear gloves during the preparation of native MS solvents, clean the source, calibrate, or introduce samples into the instrument. Polyethylene glycol (PEG) may be present in hand creams and hand soap preparations and may cause ion suppression. An aqueous mixture of 50% (v/v) acetonitrile and 1% ammonium hydroxide is useful for removing stubborn PEG contamination of glassware and HPLC columns [78, 79].
7. In most cases, up to 99% ^{34}S, ^{57}Fe, or $^{34}S/^{57}Fe$-substituted forms of Fe-S proteins can be achieved. This is often crucial for providing unambiguous confirmation of peak assignments for intermediates and products [40, 42, 55, 58, 73].
8. CsPFHA (~10 mM) can be prepared by combining 10 μL of 1 M cesium iodide with 2 μL of tridecafluoroheptanoic acid in 988 μL of aqueous 50% (v/v) propan-2-ol. It is worth mentioning that Cs (and to a lesser extent TFHA) temporarily foul the nebulizer and spray chamber. An excessive infusion may lead to fouling of the ion transfer stage. The following set of parameters are useful as a starting point for CsTFHA calibration: positive ion mode; capillary 4 kV; endplate offset 500 V; nebulizer pressure 0.8 bar; dry gas temperature 120 °C; dry gas flow 1.3 L/min; funnel 1400 *V*pp, funnel 2600 *V*pp; hexapole 800 *V*pp; ion energy 10 eV; collision cell voltage 5 V; collision cell RF 3800 *V*pp; transfer time 200 μs; and prepulse storage time 28 μs.

9. Alternatives volatile buffers to ammonium acetate are ammonium formate, ammonium bicarbonate, or triethylammonium bicarbonate. In comparison to ammonium acetate, ammonium bicarbonate offers improved pH stability. However, the likelihood of in-source unfolding may be increased with bicarbonate containing buffers [80, 81]. The ionic strength required to maintain stability in solution and achieve ionization will need to be optimized for each protein.
10. Dialysis and ultrafiltration membranes are commonly treated with polyethylene glycol. Some gel filtration (desalting) devices contain azide or other small molecules as a preservative. Extensive washing/equilibration is recommended prior to use.
11. We typically use gravity flow desalting columns designed to accommodate sample volumes in the range of 1.0–2.5 mL (PD-10, Cytiva Life Sciences) or 0.5–1 mL (PD midiTrap G25), and spin desalting columns (Zeba 0.5 mL, Thermo Scientific) designed to accommodate 30–130 μL. These columns are manufactured from chemically inert plastics, lack preservatives, such as azide, that might interfere with the analysis of metalloproteins, and are filled with resins that reproducibly exclude proteins of ≥5–7 kDa.
12. The tip of a pH probe may introduce chloride ions into the ammonium acetate solution. To check the pH, remove an aliquot (5 mL), measure the pH, then discard the aliquot. To adjust the pH, add an aliquot of 5 M ammonium hydroxide (or acetic acid) to the bulk of the ammonium acetate buffer. Recheck the pH, repeating until the desired pH is reached. Prior to entry into the anaerobic chamber, the ammonium acetate solution should be sparged with N_2. Freshly prepared ammonium acetate solutions give the best results for native MS and should not be used more than ~5 days after preparation.
13. Loss of color or smearing of the color across the column may indicate cluster loss or destruction, possibly by residual O_2. The stability of the Fe-S protein can be determined by monitoring the decay of the cluster absorption band(s) in anaerobic volatile buffer. If the sample was reconstituted, the smearing may result from residual low-molecular-weight Fe-S species. Retention of the color close to the top of the column may be indicative of protein binding to the gel matrix. In this case, try exchanging the protein into 1 M ammonium acetate. If it migrates normally, the resulting sample can be diluted to an appropriate ammonium acetate concentration (e.g., 200 mM) with dilute ammonium acetate (10 mM) or water. After loading, proteins should elute from the PD10 column within 3.5 mL; material eluting later will contain small molecules and salts. See product operating instructions for further advice

(Cytiva Code: 52130800). We typically write the date on which the PD10 column entered the glove box on the side of the column. This provides an idea of the amount of residual O_2 that may be encountered; the oldest columns ($\geq$4 days inside the glove box) will contain the least O_2. Ammonium acetate may ligate iron, triggering cluster disassembly. If this is suspected, try ammonium formate and/or ammonium bicarbonate.

14. Two syringes are used, one for infusing anaerobic ammonium acetate, one for infusing anaerobic protein samples, switching between them as required. The sample syringe can then be returned to the anaerobic box and cleaned, ready for the next sample. The oxygen permeability of PEEK tubing is 14 mL/ 250 cm^2 (1 atm/25 °C) over 24 h (data for tubing supplied by Upchurch Scientific).
15. In general, the positive-ion mode offers a better signal to noise than the negative ion mode. If ionization is poor, the negative mode should be investigated.
16. We define larger ions as protein ions with a low charge number and high m/z values, typically >3000 m/z.
17. The orthogonal accelerator receives the ion packet from the Q-Q stage and automatically applied voltages push the ion packet toward the TOF reflectron. A field-free region is created by maintaining zero potential difference between the exit of the accelerator and the reflectron entrance. The velocity of the ions in the field-free region is therefore solely dependent on the mass-to-charge ratio. The reflectron normalizes the spread in kinetic energy between the ions and redirects them through a second field-free region toward the detector. The detector converts an impacting ion into an electrical signal, which is amplified (via electron multiplication and electronic means) and transmitted to a digitizer for recording.
18. The time of flight for any given ion is proportional to the square root of the mass of the ion. Hence, lighter ions travel faster and reach the detector sooner than the heavier ions of the same charge (Fig. 1b). Precise measurement of the time elapsed between acceleration pulse and ion detection (t), in combination with the acceleration voltage (V) and length of the drift region (d), allows the m/z value of the detected ion (s) to be calculated; $^{m}/_{z} = 2eV\left(\frac{t^2}{d^2}\right)$, where m is the mass of the ion, z is the charge of the ion, and e is the elementary charge. This is calculated automatically by the spectrometer and associated data analysis software.
19. Newer charge state deconvolution routines and software packages have been developed that try to alleviate some of these issues; for further details, *see* [69, 82–86].

20. Metal containing cofactors usually carry a charge and may be present in more than one oxidation state. Therefore, the oxidation state of the metal-containing cofactor must be considered in order to properly assign a molecular mass from the m/z spectrum. For Fe-S proteins, the peaks correspond to $[M + [FeS]^x + (z - x)H]/z$, where M is the molecular mass of the protein, $[Fe\text{-}S]^x$ is the mass of the Fe-S cluster with x charge, H is the mass of the proton, and z is the charge of the ion. In this expression, the charge of the cluster, x, offsets the number of protons required to obtain an ion with z charge. Post deconvolution, the observed mass peak is usually offset from the predicted mass by the charge of the cluster, $[M + ([FeS]^x - x)]$. For example, a protein with a mass of M that binds a $[4Fe\text{-}4S]^{2+}$ cluster gives rise to a corresponding peak (M^{obs}) for the holoprotein according to $M^{obs} = (M + (352 - 2))$ [10, 36, 42]. This simply reflects charge compensation upon binding, that is, binding a charged cofactor normally results in dissociation of protons such that overall charge balance is maintained.
21. For example, the mass spectrum of anaerobically purified *Streptomyces coelicolor* NsrR contained a dominant peak at 17,474 Da, which corresponds to a monomeric form of the apoprotein (predicted mass 17,474 Da). An additional peak at 17,823 Da corresponded to the protein containing a [4Fe-4S] cluster. Broad peaks corresponding to dimeric NsrR were observed at ~34,942 Da, 35,294 Da, and 35,646 Da, corresponding to the apo-NsrR homodimer (predicted mass 34,942 Da) and dimeric NsrR containing one (35,294 Da) and two (35,646 Da) [4Fe-4S] clusters, respectively [65, 73].
22. For example, solutions of *Streptomyces venezuelae* RsrR displayed an unusual series of weak absorption bands across the 300–600 nm region, reminiscent of some [2Fe-2S] clusters [63]. The deconvoluted mass spectrum contained peaks in regions that corresponded to monomeric and dimeric forms of the protein. In the monomer region, a dominant peak was observed at 17,363 Da, which corresponds to the apoprotein (predicted mass, 17,364 Da). Additional peaks at 17,539 and 17,484 Da corresponded to the protein containing a [2Fe-2S] cluster and $Fe\text{-}S_2$ (a likely breakdown product of the [2Fe-2S] cluster), respectively. In the dimer region, a peak at 34,726 Da was observed, corresponding to the apo-RsrR homodimer (predicted mass, 34,728 Da). Additional peaks at 34,901 and 35,077 Da corresponded to dimers containing one and two [2Fe-2S] clusters, respectively [63, 87, 88].
23. For example, the iron sensor RirA, which shares homology with IscR, was hypothesized to be an Fe-S protein more than 10 years ago [89]. Recent anaerobic preparations of RirA from

Rhizobium leguminosarum revealed the presence of a [4Fe-4S] cluster with a labile fourth iron [64]. The deconvoluted mass spectrum contained peaks in regions that corresponded to monomeric and dimeric forms of the protein. In the monomer region, the mass spectrum contained a dominant peak at 17,792 Da, corresponding to [4Fe-4S] RirA (predicted mass, 17,792 Da). Minor, well-resolved peaks were observed on the low mass side, corresponding to traces of [4Fe-3S], [3Fe-4S], [3Fe-3S], [3Fe-2S], and [2Fe-2S] clusters. In the dimer region, a dominant peak at 35,585 Da corresponding to [4Fe-4S] RirA homodimer (predicted mass, 35,584 Da) was observed. A secondary peak on the low mass side corresponded to a dimer containing both [3Fe-4S] and [4Fe-4S] clusters (Fig. 4). The observation of a [3Fe-4S] cluster following sample preparations for native MS, which involves desalting (gel filtration), was consistent with solution observations [64, 90].

24. If the (at least initial) rates do not match, consider whether the MS measurement itself is causing the breakdown of Fe-S species (refer to Subheading 3.5).
25. DynaFit 4 (BioKin Ltd.) [75] employs nonlinear least-squares regression of kinetic data to generate fits based on multistep mechanisms. For example, RirA, a global iron sensor found in Rhizobia and related bacteria, was found to undergo a [4Fe-4S] to [2Fe-2S] cluster conversion reaction in response to low-iron conditions (induced by EDTA), suggesting it might be possible to follow the RirA cluster conversion via native MS [64]. The peak at 17,791 Da, corresponding to [4Fe-4S] RirA, gradually decayed away during the time course of the reaction. Additional peaks corresponding to protein-

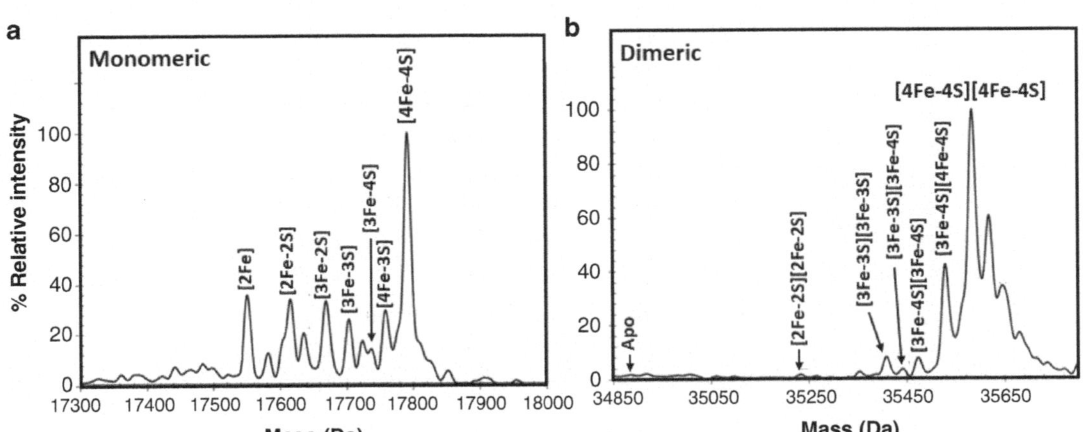

Fig. 4 Characterization of the RirA Fe-S cluster by native MS. Deconvoluted positive ion mode native MS of [4Fe-4S] RirA (~30 μM cluster in 250 mM ammonium acetate, pH 7.3) in the (**a**) monomer and (**b**) dimer mass regions [64]

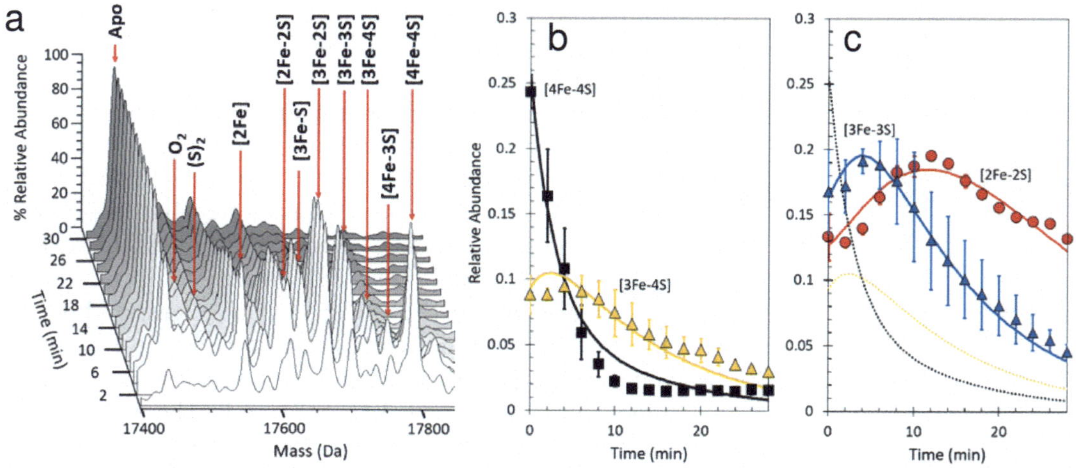

Fig. 5 Native MS of [4Fe-4S] RirA and the effect of low iron. Deconvoluted spectra of [4Fe-4S] RirA in the presence of 250 μM EDTA at 37 °C. (**a**) 3D plot of time-dependent changes in the native MS. (**b**) and (**c**) Plots of relative abundances of [4Fe-4S] cluster (black squares), [3Fe-4S] (yellow triangles), [3Fe-3S] (blue triangles), and [2Fe-2S] (red circles) species as a function of time postexposure to EDTA. Error bars show standard error for multiple ($n = 4$) datasets. Fits to the experimental data based on a multistep mechanism, modeled in Dynafit, are shown as solid lines. Broken lines in (**c**) correspond to the kinetic profile of the cluster species shown in (**b**) and are included to permit easy comparison between intermediates. For full details, *see* [40]

bound cluster fragments, including [4Fe-3S], [3Fe-4S], [3Fe-3S], [3Fe-2S], [3Fe-S], [2Fe-2S], and [2Fe-S] (17,586–17,762 Da), initially increased in intensity before decaying away to apo-RirA at later time points (Fig. 5). The same set of cluster intermediates were also observed using aerobic EDTA containing buffer [40]. The data showed that the [4Fe-3S] and [3Fe-4S] species were the first intermediates to reach their maximum abundance, consistent with them being early intermediates in the conversion process. The [3Fe-3S] species was the next intermediate to maximize, followed by the [3Fe-2S] species. The last intermediate to maximize in abundance was [2Fe-2S] RirA (Fig. 5b and c). The presence of cluster breakdown products in the starting spectrum (due to sample preparation/cluster fragility) was accounted for by allowing the starting abundance of each species to be offset from zero during DynaFit analysis [40, 91]. For further details, *see* Ref. 40.

26. Global analysis of data from time-dependent experiments on FNR and RirA led to a general multistep mechanism for the conversion of [4Fe-4S] clusters [40, 42] (Fig. 6).

27. For example, the *Streptomyces coelicolor* nitric oxide (NO) sensor [4Fe-4S] NsrR (*see* **Note 21**). is able to bind DNA and represses the cell's response to NO stress [65, 92–94]. Upon exposure to NO, [4Fe-4S] NsrR undergoes a rapid

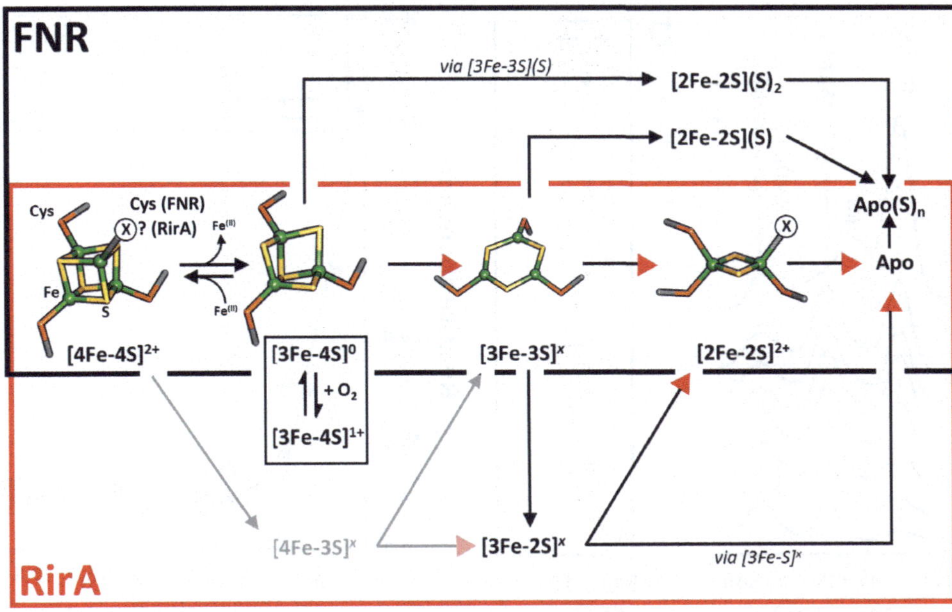

Fig. 6 Mechanistic scheme for the conversion of [4Fe-4S] FNR and RirA based on native MS. Starting from the [4Fe-4S] form, both regulators follow a central conversion process (overlapping black and red boxes). Under anaerobic conditions, [4Fe-4S] RirA cluster exists in a dynamic equilibrium with a reduced $[3Fe\text{-}4S]^0$ and a *free* Fe^{2+} ion. In FNR, this equilibrium favors a stable [4Fe-4S] cluster. Under aerobic conditions, $[3Fe\text{-}4S]^0$ RirA (scheme in red box) is rapidly oxidized to a $[3Fe\text{-}4S]^{1+}$ cluster, preventing reassociation of the *free* Fe^{2+} ion. The formation of [4Fe-3S] RirA is consistent with ionization damage and is shown grayed out. Other RirA cluster conversions downstream of the [3Fe-4S] cluster are accelerated by the presence of O_2 (red arrow heads). For FNR, the presence of O_2 converts the [4Fe-4S] into a $[3Fe\text{-}4S]^{1+}$ cluster (scheme in the black box). The [3Fe-4S] cluster passes through the planer [3Fe-3S] cluster on route to a $[2Fe\text{-}2S]^{2+}$ cluster, and eventually onto apoprotein. Distinctly different branch points exist for RirA and FNR, as indicated. For FNR, these give rise to the persulfide coordinated clusters, $[2Fe\text{-}2S](S)_n$, previously observed by resonance Raman. Here, for clarity, the Cys sulfur is colored orange, cluster sulfide is colored yellow, and iron is colored green. The fourth ligand (X) is unknown in RirA and a cysteine thiolate side chain in FNR

and complex nitrosylation reaction that results in the loss of DNA binding [95]. The end products of this reaction have been previously characterized, but the identification of protein-bound intermediates has remained a major challenge due to the rapid rate of reaction and spectroscopic similarity of iron nitrosyls to optical and vibrational techniques [95, 96]. Thus, native MS was employed to study the response of NsrR to NO. A slow NO-releasing reagent, dipropylenetriamine (DPTA) NONOate, was chosen as the NO source and limited reaction rate. To reduce sample complexity and assist data interpretation, we chose to study monomeric NsrR generated artificially in source upon transfer to the gas phase [73]. In the presence of NO, the [4Fe-4S] NsrR peak decayed away with the concomitant formation of new peak at 17,854 Da (+30 Da), consistent with the addition of a single NO molecule

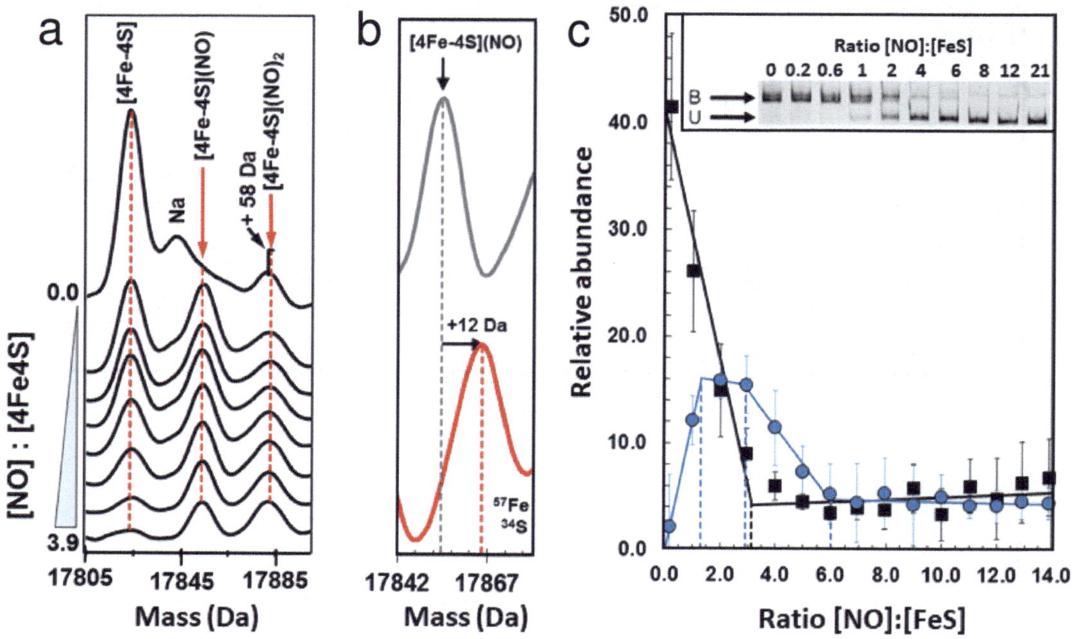

Fig. 7 Detection of the first intermediates of NsrR nitrosylation. DPTA NONOate-treated NsrR was incubated at 20 °C, releasing 0.21 [NO]:[Fe-S] per 5 min. (**a**) Exposure to NO resulted in the transient formation of single and double NO adducts (+30 and +60 Da) of the [4Fe-4S] cluster. (**b**) Isotopic shift analysis of mononitrosylated [4Fe-4S] NsrR. Natural abundance (gray line), $^{57}Fe/^{34}S$ enriched (red line). (**c**) Average ($n = 5$) relative abundance of [4Fe-4S] (black squares) and single NO adduct (blue circles) plotted as a function of the [NO]:[Fe-S] ratio. Error bars represent standard deviation. Solid lines indicate phases of the reaction, with breakpoint indicated by vertical broken lines; ***inset***, the effect of NO on DNA binding to NsrR regulated *hmpA1* promoter at increasing [NO]:[Fe-S] ratios, as indicated [73, 95]

to the [4Fe-4S] cluster. The [4Fe-4S](NO) peak initially increased in intensity before decaying away at [NO]:[4Fe-4S] ratios $\geq$3. The second species of NsrR, +60 Da heavier and very likely corresponding to [4Fe-4S]$(NO)_2$, was also observed to increase in intensity as NO was added, such that by [NO]:[4Fe-4S] $\approx$ 4, it was at least as intense as the +30 Da adduct (Fig. 7). A range of other iron nitrosyl species was also detected at higher ($\geq$4) levels of NO. These occurred in the mass regions that correspond to dinitrosyl iron complexes (DNICs, 17,588 Da, $[Fe(NO)_2(Cys)_2]$) and Roussin's red ester (RRE, 17,736 Da, $[Fe_2(NO)_4(S)]$) like protein adducts. Multiple Roussin's black salt (RBS)-like species were also detected with the general formula $[Fe_4(S)_3(NO)_x]$, where x = 3–6; RBS itself, $[Fe_4(S)_3(NO)_7]$, was not observed, though a putative intermediate in the conversion of RRE to RBS, $[Fe_4(S)_2(NO)_7]$, was [97]; for full details, *see* [73].

28. For example, the reaction of members of the [4Fe-4S] WhiB-like (Wbl) transcriptional regulator family [98] with NO has been an area of significant recent interest [73, 74, 98]. Native

MS was used to study the nitrosylation reactions of *Mycobacterium tuberculosis* WhiB1 and *S. coelicolor* WhiD. As observed for NsrR (*see* **Note 27**), the addition of NO led to the loss of the [4Fe-4S] WhiD peak. However, decay behavior was very different from that observed for NsrR, with the [4Fe-4S] WhiD peak decaying gradually throughout the titration, being lost entirely only at [NO]:[4Fe-4S] $\geq$10. This is consistent with earlier spectroscopic studies that indicated a concerted or near-concerted reaction. Cluster nitrosylation peaks were observed at +30 Da and + 60 Da in WhiD spectra and persisted over a broad range of NO ratios until the [4Fe-4S] WhiD peak decayed away. Isotope shift analysis (natural abundance vs. $^{57}Fe/^{34}S$ enriched cluster) revealed a mass difference of +12 Da, consistent with their assignment as [4Fe-4S](NO) and [4Fe-4S]$(NO)_2$, respectively. A further peak at +120 Da, corresponding to a tetranitrosylated cluster, [4Fe-4S]$(NO)_4$, was also observed [73, 99]. Native MS has also been used to investigate the interaction of Wbl proteins with the principal sigma factor (part of the cellular transcriptional machinery) and the role of NO in modulating this interaction. For further details, *see* [74, 100].

Acknowledgments

We thank the University of East Anglia for funding the purchase of the electrospray ionization mass spectrometer instrument. This work has been supported by a series of grants from the BBSRC over several years, most recently BB/P006140/1 and BB/P006140/1.

References

1. Beinert H, Meyer J, Lill R (2004) Iron–sulfur proteins. In: Lennarz WJ, Lane MD (eds) Encyclopedia of biological chemistry. Elsevier, New York, pp 482–489
2. Bian S, Cowan JA (1999) Protein-bound iron–sulfur centers. Form, function, and assembly. Coord Chem Rev 190–192:1049–1066
3. Crack JC, Le Brun NE (2018) Redox-sensing iron-sulfur cluster regulators. Antiox Red Signal 29:1809–1829
4. Balk J, Pilon M (2011) Ancient and essential: the assembly of iron-sulfur clusters in plants. Trends Plant Sci 16:218–226
5. Jeoung JH, Martins BM, Dobbek H (2020) Double-cubane [8Fe-9S] clusters: a novel nitrogenase-related cofactor in biology. Chembiochem 21(12):1710–1716. https://doi.org/10.1002/cbic.202000016
6. Johnson DC, Dean DR, Smith AD, Johnson MK (2005) Structure, function, and formation of biological iron-sulfur clusters. Annu Rev Biochem 74:247–281
7. Beinert H (1983) Semi-micro methods for analysis of labile sulfide and of labile sulfide plus sulfane sulfur in unusually stable iron-sulfur proteins. Anal Biochem 131:373–378
8. Moulis JM, Scherrer N, Gagnon J, Forest E, Petillot Y, Garcia D (1993) Primary structure of Chromatium tepidum high-potential iron-sulfur protein in relation to thermal denaturation. Arch Biochem Biophys 305:186–192
9. Petillot Y, Forest E, Meyer J, Moulis JM (1995) Observation of holoprotein molecular

ions of several ferredoxins by electrospray-ionization-mass spectrometry. Anal Biochem 228:56–63

10. Johnson KA, Verhagen MF, Brereton PS, Adams MW, Amster IJ (2000) Probing the stoichiometry and oxidation states of metal centers in iron-sulfur proteins using electrospray FTICR mass spectrometry. Anal Chem 72:1410–1418
11. Johnson KA, Amster IJ (2001) First observation by mass spectrometry of a 3+ oxidation state for a [4Fe-4S] metalloprotein: an ESI-FTICR mass spectrometry study of the high potential iron-sulfur protein from *Chromatium vinosum*. J Am Soc Mass Spectrom 12:819–825
12. Hernandez H, Hewitson KS, Roach P, Shaw NM, Baldwin JE, Robinson CV (2001) Observation of the iron-sulfur cluster in *Escherichia coli* biotin synthase by nanoflow electrospray mass spectrometry. Anal Chem 73:4154–4161
13. Heck AJ (2008) Native mass spectrometry: a bridge between interactomics and structural biology. Nat Methods 5:927–933
14. Leney AC, Heck AJ (2017) Native mass spectrometry: what is in the name? J Am Soc Mass Spectrom 28:5–13
15. Laganowsky A, Reading E, Hopper JT, Robinson CV (2013) Mass spectrometry of intact membrane protein complexes. Nat Protoc 8:639–651
16. Morgner N, Robinson CV (2012) Linking structural change with functional regulation-insights from mass spectrometry. Curr Opin Struct Biol 22:44–51
17. Belov ME, Damoc E, Denisov E, Compton PD, Horning S, Makarov AA, Kelleher NL (2013) From protein complexes to subunit backbone fragments: a multi-stage approach to native mass spectrometry. Anal Chem 85:11163–11173
18. Schmidt C, Robinson CV (2014) Dynamic protein ligand interactions—insights from MS. FEBS J 281:1950–1964
19. Mehmood S, Marcoux J, Gault J, Quigley A, Michaelis S, Young SG, Carpenter EP, Robinson CV (2016) Mass spectrometry captures off-target drug binding and provides mechanistic insights into the human metalloprotease ZMPSTE24. Nat Chem 8:1152–1158
20. Sahin C, Reid DJ, Marty MT, Landreh M (2020) Scratching the surface: native mass spectrometry of peripheral membrane protein complexes. Biochem Soc Trans 48 (2):547–558. https://doi.org/10.1042/BST20190787
21. Ventouri IK, Malheiro DBA, Voeten RLC, Kok S, Honing M, Somsen GW, Haselberg R (2020) Probing protein denaturation during size-exclusion chromatography using native mass spectrometry. Anal Chem 92 (6):4292–4300. https://doi.org/10.1021/acs.analchem.9b04961
22. Wittig S, Songailiene I, Schmidt C (2020) Formation and stoichiometry of CRISPR-cascade complexes with varying spacer lengths revealed by native mass spectrometry. J Am Soc Mass Spectrom 31:538–546
23. Wolff P, Ennifar E (2020) Native electrospray ionization mass spectrometry of RNA-ligand complexes. Methods Mol Biol 2113:111–118
24. Saikusa K, Kato D, Nagadoi A, Kurumizaka H, Akashi S (2020) Native mass spectrometry of protein and DNA complexes prepared in nonvolatile buffers. J Am Soc Mass Spectrom 31:711–718
25. Osterlund N, Moons R, Ilag LL, Sobott F, Graslund A (2019) Native ion mobility-mass spectrometry reveals the formation of beta-barrel shaped amyloid-beta hexamers in a membrane-mimicking environment. J Am Chem Soc 141:10440–10450
26. Adinolfi S, Puglisi R, Crack JC, Iannuzzi C, Dal Piaz F, Konarev PV, Svergun DI, Martin S, Le Brun NE, Pastore A (2017) The molecular bases of the dual regulation of bacterial iron sulfur cluster biogenesis by CyaY and IscX. Front Mol Biosci 4:97
27. Ross J, Lambert T, Piergentili C, He D, Waldron KJ, Mackay CL, Marles-Wright J, Clarke DJ (2020) Mass spectrometry reveals the assembly pathway of encapsulated ferritins and highlights a dynamic ferroxidase interface. Chem Commun 56:3417–3342. https://doi.org/10.1039/c9cc08130e
28. Scheller JS, Irvine GW, Stillman MJ (2018) Unravelling the mechanistic details of metal binding to mammalian metallothioneins from stoichiometric, kinetic, and binding affinity data. Dalton Trans 47:3613–3637. https://doi.org/10.1039/c7dt03319b
29. Ott DB, Hartwig A, Stillman MJ (2019) Competition between Al^{3+} and Fe^{3+} binding to human transferrin and toxicological implications: structural investigations using ultra-high resolution ESI MS and CD spectroscopy. Metallomics 11:968–981
30. Lermyte F, Everett J, Lam YPY, Wootton CA, Brooks J, Barrow MP, Telling ND, Sadler PJ, O'Connor PB, Collingwood JF (2019) Metal ion binding to the amyloid beta monomer studied by native top-down FTICR mass spectrometry. J Am Soc Mass Spectrom 30:2123–2134

31. Puglisi R, Boeri Erba E, Pastore A (2020) A guide to native mass spectrometry to determine complex interactomes of molecular machines. FEBS J 287:2428–2439. https://doi.org/10.1111/febs.15281
32. Lin CW, McCabe JW, Russell DH, Barondeau DP (2020) Molecular mechanism of ISC iron-sulfur cluster biogenesis revealed by high-resolution native mass spectrometry. J Am Chem Soc 142(13):6018–6029. https://doi.org/10.1021/jacs.9b11454
33. Jia M, Sen S, Wachnowsky C, Fidai I, Cowan JA, Wysocki VH (2020) Characterization of [2Fe-2S]-cluster-bridged protein complexes and reaction intermediates by use of native mass spectrometric methods. Angew Chem Int Ed Engl 59:6724–6728. https://doi.org/10.1002/anie.201915615
34. Gervason S, Larkem D, Mansour AB, Botzanowski T, Muller CS, Pecqueur L, Le Pavec G, Delaunay-Moisan A, Brun O, Agramunt J, Grandas A, Fontecave M, Schunemann V, Cianferani S, Sizun C, Toledano MB, D'Autreaux B (2019) Physiologically relevant reconstitution of iron-sulfur cluster biosynthesis uncovers persulfide-processing functions of ferredoxin-2 and frataxin. Nat Commun 10:3566
35. Woodall DW, Brown CJ, Raab SA, El-Baba TJ, Laganowsky A, Russell DH, Clemmer DE (2020) Melting of hemoglobin in native solutions as measured by IMS-MS. Anal Chem 92:3440–3446
36. Kay KL, Hamilton CJ, Le Brun NE (2019) Mass spectrometric studies of cu(I)-binding to the N-terminal domains of *B. subtilis* CopA and influence of bacillithiol. J Inorg Biochem 190:24–30
37. Zhou L, Kay KL, Hecht O, Moore GR, Le Brun NE (2018) The N-terminal domains of *Bacillus subtilis* CopA do not form a stable complex in the absence of their inter-domain linker. Biochim Biophys Acta 1866:275–282
38. Bennett SP, Soriano-Laguna MJ, Bradley JM, Svistunenko DA, Richardson DJ, Gates AJ, Le Brun NE (2019) NosL is a dedicated copper chaperone for assembly of the CuZ center of nitrous oxide reductase. Chem Sci 10:4985–4993
39. Chandler SA, Benesch JL (2018) Mass spectrometry beyond the native state. Curr Opin Chem Biol 42:130–137
40. Pellicer Martinez MT, Crack JC, Stewart MY, Bradley JM, Svistunenko DA, Johnston AW, Cheesman MR, Todd JD, Le Brun NE (2019) Mechanisms of iron- and O_2-sensing by the [4Fe-4S] cluster of the global iron regulator RirA. Elife 8:e47804. https://doi.org/10.7554/eLife.47804
41. Crack JC, Hamilton CJ, Le Brun NE (2018) Mass spectrometric detection of iron nitrosyls, sulfide oxidation and mycothiolation during nitrosylation of the NO sensor [4Fe-4S] NsrR. Chem Commun 54:5992–5995
42. Crack JC, Thomson AJ, Le Brun NE (2017) Mass spectrometric identification of intermediates in the O_2-driven [4Fe-4S] to [2Fe-2S] cluster conversion in FNR. Proc Natl Acad Sci U S A 114:E3215–E3223
43. Sutton VR, Mettert EL, Beinert H, Kiley PJ (2004) Kinetic analysis of the oxidative conversion of the [4Fe-4S]2+ cluster of FNR to a $[2Fe\text{-}2S]^{2+}$ cluster. J Bacteriol 186:8018–8025
44. Crack JC, Gaskell AA, Green J, Cheesman MR, Le Brun NE, Thomson AJ (2008) Influence of the environment on the $[4Fe\text{-}4S]^{2+}$ to $[2Fe\text{-}2S]^{2+}$ cluster switch in the transcriptional regulator FNR. J Am Chem Soc 130:1749–1758
45. Crack JC, Green J, Cheesman MR, Le Brun NE, Thomson AJ (2007) Superoxide-mediated amplification of the oxygen-induced switch from [4Fe-4S] to [2Fe-2S] clusters in the transcriptional regulator FNR. Proc Natl Acad Sci U S A 104:2092–2097
46. Crack Jason C, Jervis Adrian J, Gaskell Alisa A, White Gaye F, Green J, Thomson Andrew J, Le Brun NE (2008) Signal perception by FNR: the role of the iron–sulfur cluster1. Biochem Soc Trans 36:1144
47. Moore LJ, Mettert EL, Kiley PJ (2006) Regulation of FNR dimerization by subunit charge repulsion. J Biol Chem 281:33268–33275
48. Lazazzera BA, Beinert H, Khoroshilova N, Kennedy MC, Kiley PJ (1996) DNA binding and dimerization of the Fe-S-containing FNR protein from *Escherichia coli* are regulated by oxygen. J Biol Chem 271:2762–2768
49. Popescu CV, Bates DM, Beinert H, Munck E, Kiley PJ (1998) Mössbauer spectroscopy as a tool for the study of activation/inactivation of the transcription regulator FNR in whole cells of *Escherichia coli*. Proc Natl Acad Sci U S A 95:13431–13435
50. Khoroshilova N, Beinert H, Kiley PJ (1995) Association of a polynuclear iron-sulfur center with a mutant FNR protein enhances DNA binding. Proc Natl Acad Sci U S A 92:2499–2503
51. Khoroshilova N, Popescu C, Munck E, Beinert H, Kiley PJ (1997) Iron-sulfur cluster

disassembly in the FNR protein of *Escherichia coli* by O_2: [4Fe-4S] to [2Fe-2S] conversion with loss of biological activity. Proc Natl Acad Sci U S A 94:6087–6092

52. Kiley PJ, Beinert H (1999) Oxygen sensing by the global regulator, FNR: the role of the iron-sulfur cluster. FEMS Microbiol Rev 22:341–352
53. Jervis AJ, Crack JC, White G, Artymiuk PJ, Cheesman MR, Thomson AJ, Le Brun NE, Green J (2009) The O_2 sensitivity of the transcription factor FNR is controlled by Ser24 modulating the kinetics of [4Fe-4S] to [2Fe-2S] conversion. Proc Natl Acad Sci U S A 106:4659–4664
54. Zhang B, Crack JC, Subramanian S, Green J, Thomson AJ, Le Brun NE, Johnson MK (2012) Reversible cycling between cysteine persulfide-ligated [2Fe-2S] and cysteine-ligated [4Fe-4S] clusters in the FNR regulatory protein. Proc Natl Acad Sci U S A 109:15734–15739
55. Crack JC, Stewart MYY, Le Brun NE (2019) Generation of ^{34}S-substituted protein-bound [4Fe-4S] clusters using ^{34}S-L-cysteine. Biol Methods Protoc 4:byp015
56. Kondrat FD, Struwe WB, Benesch JL (2015) Native mass spectrometry: towards high-throughput structural proteomics. Methods Mol Biol 1261:349–371
57. Freibert SA, Weiler BD, Bill E, Pierik AJ, Muhlenhoff U, Lill R (2018) Biochemical reconstitution and spectroscopic analysis of iron-sulfur proteins. Methods Enzymol 599:197–226
58. Crack JC, Green J, Thomson AJ, Le Brun NE (2014) Techniques for the production, isolation, and analysis of iron-sulfur proteins. Methods Mol Biol 1122:33–48
59. Zheng L, White RH, Cash VL, Jack RF, Dean DR (1993) Cysteine desulfurase activity indicates a role for NIFS in metallocluster biosynthesis. Proc Natl Acad Sci U S A 90:2754–2758
60. König S, Fales HM (1999) Calibration of mass ranges up to m/z 10,000 in electrospray mass spectrometers. J Am Soc Mass Spect 10:273–276
61. Jackson VA, Mehmood S, Chavent M, Roversi P, Carrasquero M, Del Toro D, Seyit-Bremer G, Ranaivoson FM, Comoletti D, Sansom MS, Robinson CV, Klein R, Seiradake E (2016) Super-complexes of adhesion GPCRs and neural guidance receptors. Nat Commun 7:11184
62. Sweeney WV, Rabinowitz JC (1980) Proteins containing 4Fe-4S clusters: an overview. Annu Rev Biochem 49:139–161
63. Munnoch JT, Martinez MT, Svistunenko DA, Crack JC, Le Brun NE, Hutchings MI (2016) Characterization of a putative NsrR homologue in *Streptomyces venezuelae* reveals a new member of the Rrf2 superfamily. Sci Rep 6:31597
64. Pellicer Martinez MT, Martinez AB, Crack JC, Holmes JD, Svistunenko DA, Johnston AWB, Cheesman MR, Todd JD, Le Brun NE (2017) Sensing iron availability via the fragile [4Fe-4S] cluster of the bacterial transcriptional repressor RirA. Chem Sci 8:8451–8463
65. Crack JC, Munnoch J, Dodd EL, Knowles F, Al Bassam MM, Kamali S, Holland AA, Cramer SP, Hamilton CJ, Johnson MK, Thomson AJ, Hutchings MI, Le Brun NE (2015) NsrR from *Streptomyces coelicolor* is a nitric oxide-sensing [4Fe-4S] cluster protein with a specialized regulatory function. J Biol Chem 290:12689–12704
66. Susa AC, Xia Z, Tang HYH, Tainer JA, Williams ER (2017) Charging of proteins in native mass spectrometry. J Am Soc Mass Spectrom 28:332–340
67. Konermann L, Metwally H, Duez Q, Peters I (2019) Charging and supercharging of proteins for mass spectrometry: recent insights into the mechanisms of electrospray ionization. Analyst 144:6157–6171
68. Ferrige AG, Seddon MJ, Jarvis S, Skilling J, Aplin R (1991) Maximum entropy deconvolution in electrospray mass spectrometry. Rapid Commun Mass Spectrom 5:374–377
69. Bern M, Caval T, Kil YJ, Tang W, Becker C, Carlson E, Kletter D, Sen KI, Galy N, Hagemans D, Franc V, Heck AJR (2018) Parsimonious charge deconvolution for native mass spectrometry. J Proteome Res 17:1216–1226
70. Bich C, Baer S, Jecklin MC, Zenobi R (2010) Probing the hydrophobic effect of noncovalent complexes by mass spectrometry. J Am Soc Mass Spectrom 21:286–289
71. Volbeda A, Dodd EL, Darnault C, Crack JC, Renoux O, Hutchings MI, Le Brun NE, Fontecilla-Camps JC (2017) Crystal structures of the NO sensor NsrR reveal how its iron-sulfur cluster modulates DNA binding. Nat Commun 8:15052
72. Child SA, Bradley JM, Pukala TL, Svistunenko DA, Le Brun NE, Bell SG (2018) Electron transfer ferredoxins with unusual cluster

binding motifs support secondary metabolism in many bacteria. Chem Sci 9:7948–7957

73. Crack JC, Le Brun N (2019) Mass spectrometric identification of [4Fe-4S](NO)x intermediates of nitric oxide sensing by regulatory iron-sulfur cluster proteins. Chemistry 25:3675–3684
74. Kudhair BK, Hounslow AM, Rolfe MD, Crack JC, Hunt DM, Buxton RS, Smith LJ, Le Brun NE, Williamson MP, Green J (2017) Structure of a Wbl protein and implications for NO sensing by *M. tuberculosis*. Nat Commun 8:2280
75. Kuzmic P (1996) Program DYNAFIT for the analysis of enzyme kinetic data: application to HIV proteinase. Anal Biochem 237:260–273
76. Rose RJ, Damoc E, Denisov E, Makarov A, Heck AJ (2012) High-sensitivity orbitrap mass analysis of intact macromolecular assemblies. Nat Methods 9:1084–1086
77. Sobott F, Hernandez H, McCammon MG, Tito MA, Robinson CV (2002) A tandem mass spectrometer for improved transmission and analysis of large macromolecular assemblies. Anal Chem 74:1402–1407
78. Mendes MA, Chies JM, de Oliveira Dias AC, Filho SA, Palma MS (2003) The shielding effect of glycerol against protein ionization in electrospray mass spectrometry. Rapid Commun Mass Spectrom 17:672–677
79. Weaver R, Riley RJ (2006) Identification and reduction of ion suppression effects on pharmacokinetic parameters by polyethylene glycol 400. Rapid Commun Mass Spectrom 20:2559–2564
80. Hedges JB, Vahidi S, Yue X, Konermann L (2013) Effects of ammonium bicarbonate on the electrospray mass spectra of proteins: evidence for bubble-induced unfolding. Anal Chem 85:6469–6476
81. Honarvar E, Venter AR (2018) Comparing the effects of additives on protein analysis between desorption electrospray (DESI) and electrospray ionization (ESI). J Am Soc Mass Spectrom 29:2443–2455
82. Marty MT (2020) A universal score for deconvolution of intact protein and native electrospray mass spectra. Anal Chem 92:4395–4401
83. Reid DJ, Diesing JM, Miller MA, Perry SM, Wales JA, Montfort WR, Marty MT (2019) MetaUniDec: high-throughput deconvolution of native mass spectra. J Am Soc Mass Spectrom 30:118–127
84. Marty MT (2019) Eliminating artifacts in electrospray deconvolution with a SoftMax function. J Am Soc Mass Spectrom 30:2174–2177
85. Cleary SP, Li H, Bagal D, Loo JA, Campuzano IDG, Prell JS (2018) Extracting charge and mass information from highly congested mass spectra using Fourier-domain harmonics. J Am Soc Mass Spectrom 29:2067–2080
86. Lu J, Trnka MJ, Roh SH, Robinson PJ, Shiau C, Fujimori DG, Chiu W, Burlingame AL, Guan S (2015) Improved peak detection and deconvolution of native electrospray mass spectra from large protein complexes. J Am Soc Mass Spectrom 26:2141–2151
87. Volbeda A, Pellicer Martinez MT, Crack JC, Amara P, Gigarel O, Munnoch JT, Hutchings MI, Darnault C, Le Brun NE, Fontecilla-Camps JC (2019) The crystal structure of the transcription regulator RsrR reveals a [2Fe-2S] cluster coordinated by Cys, Glu and His Residues. J Am Chem Soc 141 (6):2367–2375. https://doi.org/10.1021/jacs.8b10823
88. Crack JC, Amara P, Volbeda A, Mouesca JM, Rohac R, Pellicer Martinez MT, Huang CY, Gigarel O, Rinaldi C, Le Brun NE, Fontecilla-Camps JC (2020) Electron and proton transfers modulate NA binding by the transcription regulator RsrR. J Am Chem Soc 142:5104–5116
89. Johnston AW, Todd JD, Curson AR, Lei S, Nikolaidou-Katsaridou N, Gelfand MS, Rodionov DA (2007) Living without fur: the subtlety and complexity of iron-responsive gene regulation in the symbiotic bacterium *Rhizobium* and other alpha-proteobacteria. Biometals 20:501–511
90. Behringer M, Plotzky L, Baabe D, Zaretzke MK, Schweyen P, Broring M, Jahn D, Hartig E (2020) RirA of *Dinoroseobacter shibae* senses iron via a $[3Fe\text{-}4S]^{1+}$ cluster co-ordinated by three cysteine residues. Biochem J 477:191–212
91. Kuzmic P (2009) DynaFit—a software package for enzymology. Methods Enzymol 467:247–280
92. Partridge JD, Bodenmiller DM, Humphrys MS, Spiro S (2009) NsrR targets in the *Escherichia coli* genome: new insights into DNA sequence requirements for binding and a role for NsrR in the regulation of motility. Mol Microbiol 73:680–694
93. Kommineni S, Yukl E, Hayashi T, Delepine J, Geng H, Moenne-Loccoz P, Nakano MM (2010) Nitric oxide-sensitive and -insensitive interaction of *Bacillus subtilis* NsrR with a ResDE-controlled promoter. Mol Microbiol 78:1280–1293

94. Volbeda A, Crack JC, Le Brun NE, Fontecilla-Camps JC (2020) Structure-function relationships of the NsrR and RsrR transcription regulators. In: Messerschmidt A (ed) Encyclopedia of inorganic and bioinorganic chemistry. Wiley, Hoboken, New Jersey
95. Crack JC, Svistunenko DA, Munnoch J, Thomson AJ, Hutchings MI, Le Brun NE (2016) Differentiated, promoter-specific response of [4Fe-4S] NsrR DNA binding to reaction with nitric oxide. J Biol Chem 291:8663–8672
96. Serrano PN, Wang H, Crack JC, Prior C, Hutchings MI, Thomson AJ, Kamali S, Yoda Y, Zhao J, Hu MY, Alp EE, Oganesyan VS, Le Brun NE, Cramer SP (2016) Nitrosylation of nitric oxide-sensing regulatory proteins containing [4Fe-4S] clusters gives rise to multiple iron-nitrosyl complexes. Angew Chem Int Ed Engl 55:14575–14579
97. Yeh SW, Tsou CC, Liaw WF (2014) The dinitrosyliron complex $[Fe_4(\mu_3\text{-}S_2)(\mu_2\text{-}NO)_2(NO)_6]^{2-}$ containing bridging nitroxyls: ^{15}N (NO) NMR analysis of the bridging and terminal NO-coordinate ligands. Dalton Trans 43:9022–9025
98. Bush MJ (2018) The actinobacterial WhiB-like (Wbl) family of transcription factors. Mol Microbiol 110:663–676
99. Victor E, Lippard SJ (2014) A tetranitrosyl [4Fe-4S] cluster forms en route to Roussin's black anion: nitric oxide reactivity of $[Fe_4S_4(LS_3)L']^{2-}$. Inorg Chem 53:5311–5320
100. Stewart MYY, Bush MJ, Crack JC, Buttner MJ, Le Brun NE (2020) Interaction of the *Streptomyces* Wbl protein WhiD with the principal sigma factor σ^{HrdB} depends on the WhiD [4Fe-4S] cluster. J Biol Chem 295:9752–9765

Chapter 14

Characterization of Paramagnetic Iron-Sulfur Clusters Using Electron Paramagnetic Resonance Spectroscopy

Troy A. Stich

Abstract

Continuous-wave (CW) electron paramagnetic resonance (EPR) spectroscopy is a powerful ally in characterizing the multitude of redox-active iron-sulfur cluster-containing ([Fe-S]) species present in biological samples. The technique detects only those clusters that are paramagnetic—having a nonzero total electron spin ($S > 0$)—thus, it can discriminate between clusters in different oxidation states. The low-temperature CW-EPR spectrum of an [Fe-S] yields the three magnetic g-values that serve as a fingerprint of its electronic structure. This chapter briefly describes the underlying theory that defines this electronic structure and provides a recipe for the acquisition and analysis of EPR spectra of [Fe-S] proteins.

Key words Iron-sulfur, [Fe-S], Electron paramagnetic resonance, EPR, ESR, continuous-wave EPR, CW-EPR, Electron transfer, Ferredoxin, Nitrogenase, Rieske, Radical SAM

1 Introduction

Iron-sulfur clusters [1, 2] play a number of roles in biology ranging from stabilizing protein tertiary structure to exercise control over gene expression (e.g., cytosolic aconitase) [3], to transferring electrons (e.g., ferredoxins) [4], to catalyzing chemical reactions (e.g., nitrogenase) [5, 6]. Owing to the redox-active nature of the constituent iron ions, many if not all [Fe-S] proteins can be prepared in a paramagnetic state through treatment with an appropriate oxidant or reductant. This paramagnetism allows such clusters to be characterized using electron paramagnetic resonance (EPR, sometimes known as ESR for electron spin resonance) spectroscopy [7, 8].

EPR spectroscopy measures the absorption of microwave light that induces an electron-spin flip, much as NMR spectroscopy measures radio wave absorption that leads to nuclear spin flips. The frequency of light that induces these spin flips is diagnostic of

Patricia C. Dos Santos (ed.), *Fe-S Proteins: Methods and Protocols*, Methods in Molecular Biology, vol. 2353,
https://doi.org/10.1007/978-1-0716-1605-5_14,

the magnetic structure of the molecule. The magnetic structure is a sensitive reporter of the [Fe-S] geometric structure. Thus, events like conformation changes, substrate/inhibitor binding, and modulation of the first-coordination sphere via site-directed mutagenesis can be followed by observing changes in the EPR spectrum.

1.1 Basics of the CW EPR Spectrometer

The most common EPR experiments on [Fe-S]-containing proteins are conducted at low temperature (below 20 K) using a continuous-wave (CW) spectrometer with incident radiation in the 9–10 GHz frequency range (so-called *X*-band). The basic spectrometer is depicted in Fig. 1 and minimally consists of an electromagnet, microwave source, resonant cavity, field modulation coils, lock-in amplifier, and a detector. The sample is placed into the center of the cavity, which is designed to match the wavelength and impedance of the incident microwave radiation. This matching maximizes the intensity of the oscillating magnetic field of the microwaves B_1 that drives the electron spin-flip transitions. In most cavities, the direction of the microwave B_1 overlaps with the vertical axis of the sample, which is also oriented perpendicular to the direction of B_0, the magnetic field produced by the electromagnet (so-called perpendicular mode, *see* **Note 1**). A helium cryostat (this could be a liquid helium flow system or a closed-cycle helium refrigerator/cryocooler) is placed within the cavity to allow for low-temperature data acquisition down to ≈4 K.

Unpaired electrons possess an intrinsic angular momentum called spin (S). The "spinning" negatively charged electron produces its own magnetic dipole moment, but in the absence of an external magnetic field, the orientation of this magnetic dipole in

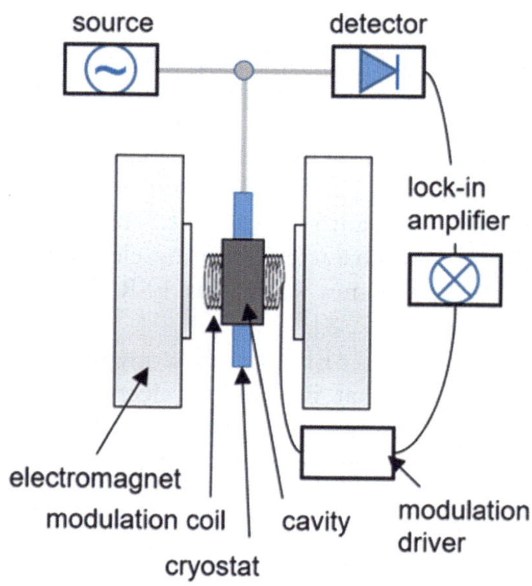

Fig. 1 Block diagram of CW-EPR spectrometer

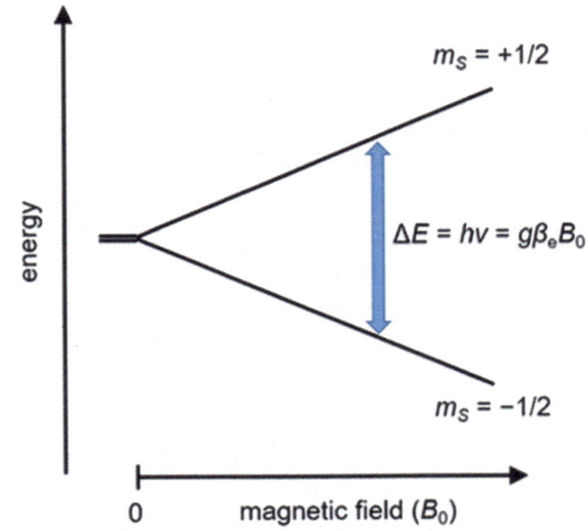

Fig. 2 The Zeeman effect for an $S = 1/2$ spin system. Monochromatic microwave radiation can only excite the electron spin-flip transition when the B_0-induced splitting of the spin-up and spin-down levels is equal to the energy of the incoming photon (blue arrow)

space has no influence on the total energy of the molecule in which it resides. More colloquially, we say that the "spin-up" ($m_S = +1/2$) and "spin-down" ($m_S = -1/2$) orientations are degenerate. However, by applying a magnetic field (B_0) produced by the electromagnet, the "spin-down" orientation drops in energy by the same amount that the "spin-up" orientation is destabilized. This is a manifestation of the Zeeman effect (Fig. 2). By tuning the strength of the applied B_0, we can match the energy difference (ΔE) between the $m_S = +1/2$ and $m_S = -1/2$ spin levels to the energy ($h\nu$) of the microwave radiation (approximately 0.3 cm^{-1} at X-band). Achieving this resonance condition allows for the radiation to be absorbed by the molecule, flipping its spin and resulting in a signal being sent to the detector.

The rate by which the spin levels change in energy is proportional to the spin quantum number (m_S), the strength of the magnetic field B_0, a collection of physical constants known as the electron Bohr magneton (β_e), and the electron g-value. The g-value for the free electron (one not bound to any atom) is one of the most precisely known physical constants ($g_e = 2.00231930436256$ [35]). For an X-band spectrometer employing 9.5 GHz microwave radiation, the value of B_0 at which resonance is achieved (339.0 mT) can be easily found using Eq. 1.

$$g = \frac{71.448\ \left(\frac{\text{mT}}{\text{GHz}}\right) \times \nu\ (\text{GHz})}{B_0\ (\text{mT})} \tag{1}$$

However, when the unpaired electron is found on an atom or molecule, the g-value changes as orbital angular momentum adds to the spin angular momentum [9, 10]. The new effective g-values reflect the internal magnetic environment of the molecule, much like the proton chemical shift, and can be used to infer structure. These g shifts can lead to resonance frequency changes that are tens of GHz different from the free-electron resonance frequency, but powerful microwave sources are only obtainable in narrow frequency ranges. So, to conduct an EPR experiment, the frequency of the radiant microwaves is fixed, and the applied magnetic field B_0 is swept. Thus, a change in g-value means that the resonant field position changes. Additionally, the magnetic fields produced within a molecule by its own electron cloud need not be identical in all three spatial directions (x, y, and z), a property referred to as g-anisotropy that is represented by an array of the three principal values, $g = [g_x, g_y, g_z]$ or $[g_1, g_2, g_3]$. Note that the x, y, and z directions of the g-matrix do not necessarily coincide with the geometric X, Y, and Z axes dictated by the molecular symmetry. A consequence of this anisotropy is that the EPR signal is spread over a larger field range (from g_{max} to g_{min}) meaning less molecules contribute to spectral intensity at a given field position leading to a reduction in sensitivity. For a powder or frozen-solution sample, all orientations of the molecule relative to the direction of B_0 are equally probable, but at a particular field position, only those few molecules contribute whose magnitude of their magnetic dipole moment parallel to B_0 is sufficient to achieve the resonance condition (Fig. 3). Certain NMR experiments attempt to recover intensity by averaging out the anisotropy in the chemical shift tensor; for example, liquid-phase rapidly spinning NMR and magic-angle spinning solid-state NMR experiments. Similarly, liquid-phase EPR can be performed to give only the isotropic g-shift from g_e, but this parameter is not very illuminating in terms of electronic structure, and often, EPR signals of [Fe-S] are not observable at elevated temperatures owing to fast relaxation (*see* Subheading 3).

Direct detection of microwave absorption is challenging as the population difference between the $m_S = +1/2$ and $m_S = -1/2$ spin levels is modest at $\approx$339 mT (recall that the Boltzmann constant is 0.695 cm^{-1} K^{-1}); though increased polarization can be achieved by decreasing the temperature (Curie Law) or by going to higher magnetic fields (EPR has been performed at $B_0 = 25$ T). However, by encoding the absorption signal on top of another signal that modulates at a fixed frequency, one can bring out the absorption well above any white noise. This is the job of the modulation coils that induce a second oscillating magnetic field of small magnitude (modulation amplitude MA is usually between 0.01 mT and 2.0 mT) along the same axis as the strong B_0 from the electromagnet. The frequency of this modulation is almost always set to 100 kHz. At any value for B_0, the spectrometer is actually measuring the difference in the absorption intensity at fields B_0 – MA and B_0 + MA. This difference essentially corresponds to the slope of the

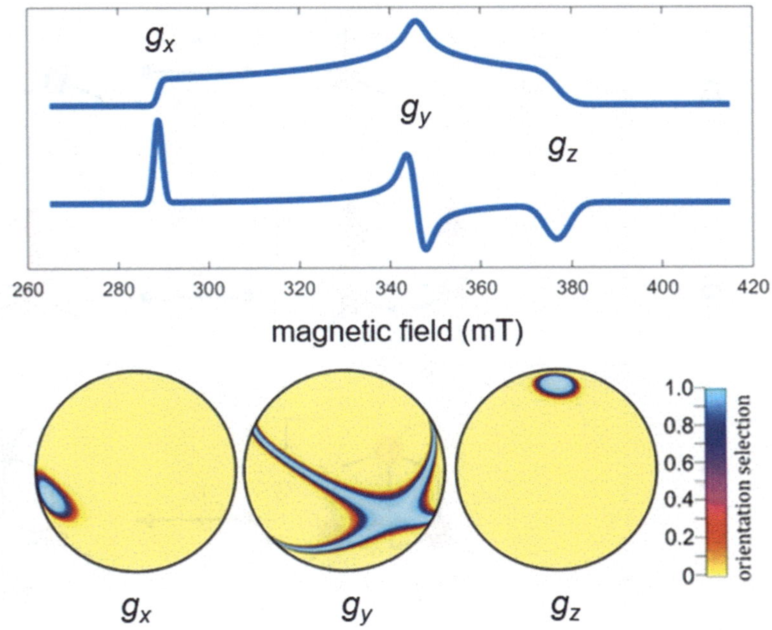

Fig. 3 The effect of g-anisotropy on EPR spectra. (Top) Absorption and corresponding first-derivative spectra of an $S = 1/2$ spin system with a rhombic set of g-values. Labels "g_x," "g_y," and "g_z" denote those regions of the spectrum where intensity arises from molecules whose g_x, g_y, and g_z axes, respectively, are parallel to the direction of the applied magnetic field (B_0). (Bottom) Unit spheres (yellow) represent all possible fixed orientations of molecules in a disordered sample (either powder or frozen solution). The regions shown in blue correspond to those select orientations of molecules that are contributing to the EPR spectrum at noted field positions. (Portions of this figure were adapted from [61])

absorption envelope and is why the output data is reported as the first-derivative of the absorption with respect to the field ($\delta\chi''/\delta B$, *see* **Note 2**). The derivative of the absorption signal is decoded from the 100 kHz carrier signal by the lock-in amplifier (also known as a phase sensitive detector) and sent to the data recorder. Guidelines for choosing experimental parameters will be given in Subheading 3.

1.2 Electronic Structure of [Fe-S] Clusters

In biological systems, iron ions can be found in the following oxidation states: 1+ (found in [FeFe] hydrogenases), 2+, 3+ (these are the most common), or 4+ (found in mononuclear and dinuclear oxidases/oxygenases). In the case of [Fe-S], only the 2+ and 3+ formal oxidation states are relevant. Nearly all Fe-sites in [Fe-S] are tetrahedrally coordinated, leading to generally high-spin configurations of the corresponding 6 and 5 valence electrons. Therefore, mononuclear [Fe-S] proteins like rubredoxin are $S = 5/2$ in the ferric state and $S = 2$ in the reduced, ferrous form

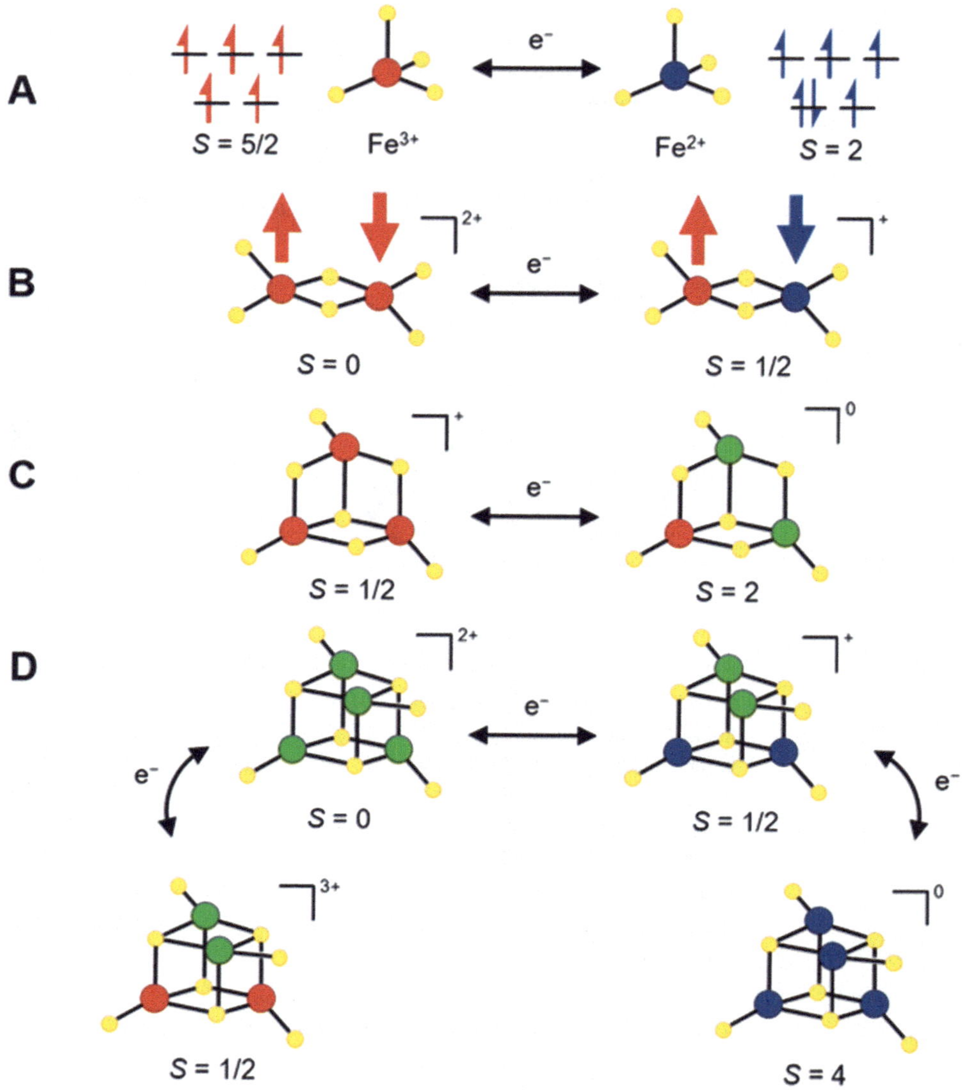

Fig. 4 Iron valence delocalization patterns for all-cysteinyl ligated [Fe-S]. Fe^{3+} is depicted in red, Fe^{2+} in blue, and delocalized pairs of $Fe^{2.5+}Fe^{2.5+}$ are shown in green. (Figure adapted from [62])

(Fig. 4a). Corresponding EPR spectra of these high-spin systems are very distinct from the $S = 1/2$ signals of oxidized [3Fe-4S] and reduced [2Fe-2S] and [4Fe-4S] that are the focus of this protocol, but the interested reader should explore the following reviews for relevant EPR methodology [11, 12]. Briefly, half-integer spin systems ($S = 1/2$, or $3/2$, or $5/2$, etc.) will have degeneracy of the pair of $\pm n/2$ spin-levels (where n is an odd integer) at $B_0 = 0$ T. Each of these so-called Kramers doublets will split as B_0 is increased (application of Eq. 1 multiplied by the m_S quantum number) and thus will inevitably meet the resonance condition at some field value. In the much more common perpendicular mode setup, the microwave B_1 will excite transitions between levels that differ by $\Delta m_S = \pm 1$. This is the general perpendicular-mode EPR selection rule. In the case of high-spin Fe^{3+}, one can see how this selection rule can lead to several new allowed transitions ($m_S = +1/2 \leftrightarrow m_S = -1/2$; $m_S = +1/2 \leftrightarrow m_S = +3/2$; etc.), some that appear at g-values very different from g_e. For integer-spin systems like high-spin Fe^{2+}, parallel-mode EPR can be employed (*see* **Note 1**). The observed transitions typically appear at low applied magnetic fields and thus correspond to large effective g-values.

One of the beauties of [Fe-S] clusters is their modularity and how small units can be assembled to form higher and higher nuclearity clusters. For example, dinuclear clusters consist of two sulfide-bridged tetrahedral iron sites reminiscent of the rubredoxin active site [13, 14]. Two dinuclear clusters ([2Fe-2S]) are combined to yield a [4Fe-4S], [15, 16], and two [4Fe-4S] can serve as a precursor for the [Fe_7MoS_9C] M-cluster of nitrogenase [17]. The paramagnetic versions of these clusters each possess a characteristic set of magnetic properties that can be used to trace [Fe-S] biosynthesis using EPR spectroscopy. First, it will help to have a cursory understanding of the origins of the observed magnetic behavior of higher nuclearity clusters.

1.3 EPR of [2Fe-2S] Clusters

For [2Fe-2S] coordinated by four cysteine thiolates (Fig. 4b), only two oxidation states can be accessed: oxidized $[2Fe\text{-}2S]^{2+}$ Fe(III, III) and reduced $[2Fe\text{-}2S]^{+}$ Fe(II, III). Charges for the clusters are determined by adding up the formal charges for all the iron ions and subtracting 2 for each bridging sulfide. The individual iron ions remain high spin, but their unpaired electrons communicate with one another via exchange coupling and, owing to the two strongly covalent sulfide bridges, pair up antiferromagnetically. So for $[2Fe\text{-}2S]^{+}$, there remain nine unpaired electrons, but the $S = 2$ ferrous site electron spin orients antiparallel to the ferric $S = 5/2$ electron spin yielding a net $S_{TOT} = 1/2$ ground state spin system. However, this is just the lowest energy level of a manifold of levels (sometimes called a spin ladder, Fig. 5) that span from $S_{TOT} = 1/2$ to $S_{TOT} = 9/2$, and the spacings between the levels are proportional to the strength of the exchange coupling (J) which is typically large ($|J| > 30$ cm^{-1}). The two ferric ions of oxidized [2Fe-2S] also pair up antiferromagnetically to give a diamagnetic $S_{TOT} = 0$ ground state. While the higher energy levels of the spin ladder are paramagnetic, they have not been found to contribute to the EPR spectra of [2Fe-2S]. The absence of the corresponding EPR features likely stems from the fact that the spin levels are sparingly populated even at elevated temperatures. And the high temperatures that would lead to significant population cause the microwave-excited spin-levels to relax back to equilibrium so quickly that corresponding EPR features are broadened beyond detectability.

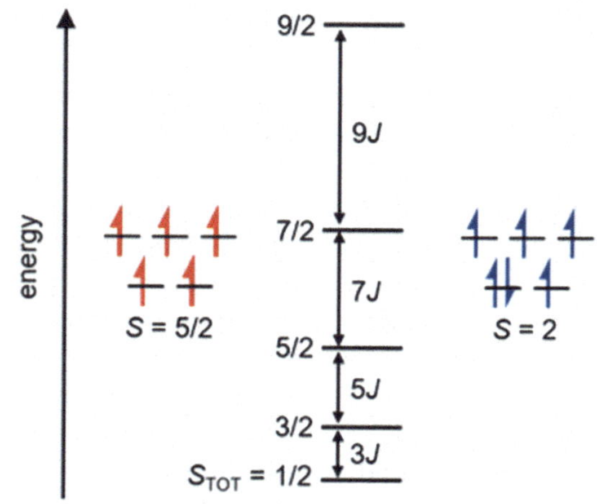

Fig. 5 Spin ladder for an antiferromagnetically coupled dimer of Fe^{3+} ($S_A = 5/2$) and Fe^{2+} ($S_B = 2$) ions. The energy spacing between levels is a function of the exchange coupling parameter (J). Unfortunately, there are different forms of the Heisenberg-Dirac-van Vleck Hamiltonian ($\hat{H}_{HDvV}$) used to predict these level energies that are employed in different sub-fields of chemistry and physics, so comparisons between reported J-values first need to verify which Hamiltonian was used. Here we used $\hat{H}_{HDvV} = -2JS_AS_B$. For [Fe-S] of nuclearity higher than two, these coupling schemes become much more complicated leading to a dramatic increase in density of states, and competing interactions that can lead to lowest energy levels have $S_{TOT} > 1/2$

Most $S = 1/2$, reduced [2Fe-2S] exhibit rhombic g-values (meaning $g_1 \neq g_2 \neq g_3$) that can vary quite a lot from one another. However, an interesting correlation was noted between the average g-value ($g_{av} = (g_1 + g_2 + g_3)/3$) and the type of coordinating ligands [18]. All-cysteinyl coordinated ferredoxins have a $g_{av} = 1.96$, whereas the $(His)_2(Cys)_2$-ligated Rieske clusters have $g_{av} = 1.88$–1.92. And the unusual $(His)(Cys)_3$-ligated cluster of MitoNEET has a $g_{av} = 1.947$ that lies in between these values. Other amino acid side chains have been shown to ligate [2Fe-2S], including Asp and Arg, but the effect on g_{av} is less consistent.

Cysteine substitutions can also affect the localization of formal charges and the communication of the unpaired electrons between the two iron sites. For example, Mössbauer and advanced EPR spectroscopic results have shown that the ferrous site of reduced Riseke proteins is the iron coordinated by the two histidine side chains, and the ferric site is coordinated by the two cysteines [19, 20]. These valences can swap when the cluster becomes deprotonated at high pH [21]. For a serine-substituted $(Cys)_4$ $[2Fe\text{-}2S]^+$, the iron ions cannot be distinguished from one another. Thus we adopt oxidation state descriptions of $Fe^{2.5+}$ for each [22, 23]. Further, the unpaired electrons exchange-couple ferromagnetically yielding the lowest energy $S_{TOT} = 9/2$ spin level on the spin ladder. However, the oxidized version of the $(Ser)(Cys)_3$ cluster has an

$S_{TOT} = 0$, providing a revealing example that charge delocalization and exchange-coupling are delicately balanced properties of [Fe-S] that are challenging to predict a priori.

1.4 EPR of [4Fe-4S] Clusters

Despite their nuclearity of four, [4Fe-4S] most often cycle between just two oxidation states: the oxidized $[4Fe\text{-}4S]^{2+}$, which nominally have an Fe(II, III, II, III) configuration and the reduced $[4Fe\text{-}4S]^{+}$ which have an Fe(II, II, II, III) configuration (Fig. 4d). Exceptions include the class of high-potential iron-sulfur cluster proteins (HiPIP) that can achieve the $[4Fe\text{-}4S]^{3+}$ oxidation state with a nominal Fe(II, III, III, III) arrangement [24], and the sole biological example of an all-ferrous $[4Fe\text{-}4S]^{0}$—the cluster in the iron protein NifH that powers the reduction of nitrogenase [25].

Results from Mössbauer spectroscopy revealed that $(Cys)_4$-ligated [4Fe-4S] tend to behave as two exchange-coupled [2Fe-2S] [26, 27]. Indeed, all four iron ions in oxidized $[4Fe\text{-}4S]^{2+}$ have essentially identical Mössbauer parameters despite having different formal valencies. This result is analogous to that observed for the $(Ser)(Cys)_3$-ligated [2Fe-2S] that showed equal sharing of the itinerant electron between the two iron ions leading to the adoption of a $Fe^{2.5+}$ oxidation state description for both centers. Additionally, each of the [2Fe-2S] dimers (sometimes called "rhombs" in this context) is considered to be in the $S = 9/2$ state, and then the two dimers coupled antiferromagnetically to give $S_{TOT} = 0$.

One-electron reduction to $[4Fe\text{-}4S]^{+}$ makes one of the rhombs diferrous ($S = 4$). This rhomb still antiferromagnetically couples to the mixed-valent rhomb ($S = 9/2$) to often give $S_{TOT} = 1/2$. The EPR spectra of $[4Fe\text{-}4S]^{+}$ typically exhibit two g-values slightly below g_e and one g-value somewhat above g_e (g_{max}). Interestingly, g_{max} seems to be particularly sensitive to changes in [4Fe-4S] ligation at one iron site. Some examples include Cys→His or Asp substitutions and nonproteinaceous ligand binding (water, CN^{-}, etc.) [28–30]. When g_{max} is larger than 2.1, this can be an indication that the lowest-lying $S = 3/2$ rung of the spin ladder is strongly mixing with the $S = 1/2$ level [31]. This $S_{TOT} = 3/2$ level can even become the ground state with effective g-values >5 [32–34].

Fully oxidized HiPIP clusters typically have one $Fe^{2.5+}Fe^{2.5+}$ dimer coupled to an $Fe^{3+}Fe^{3+}$ dimer giving an $S_{TOT} = 1/2$. The valence scheme shown in Fig. 4d is for only one isomer, but deconvolution of CW EPR spectra of seven different HiPIPs indicated that other arrangements of the electrons are possible [24]. The g-values for oxidized HiPIP clusters tend to be all greater than g_e and have axial symmetry—$g_1 \approx 2.08$–2.14, $g_2 \approx g_3 \approx 2.00$–$2.04$. For the all-ferrous cluster in NifH, the balance of six different exchange coupling interactions (one for each possible $Fe^{2+} \cdots Fe^{2+}$ pair) leads to an $S_{TOT} = 4$ ground state with a corresponding parallel-mode EPR feature at $g = 16.4$ [35].

1.5 EPR of [3Fe-4S] Clusters

In biology, [3Fe-4S] are employed as iron sensors [36], O_2 sensors [37], and as evidence that a [4Fe-4S] succumbed to oxidative damage [38]. Two oxidation states are common: $[3Fe\text{-}4S]^{+}$ and $[3Fe\text{-}4S]^{0}$ (Fig. 4c), and only the oxidized form gives rise to a perpendicular-mode EPR spectrum. The EPR spectrum of $[3Fe\text{-}4S]^{+}$ exhibits significantly less g-anisotropy than other [Fe-S], usually giving rise to a narrow spectrum peaking at 2.03 [39, 40]. Somewhat similarly to the oxidized HiPIP clusters, conformational isomers of $[3Fe\text{-}4S]^{+}$ each give rise to unique EPR characteristics [40], but usually, one species is in the great majority. The reduced $[3Fe\text{-}4S]^{0}$ form and an all-ferrous $[3Fe\text{-}4S]^{2-}$ form that can be achieved electrochemically both exhibit parallel-mode EPR signals at low field [41].

1.6 EPR of Greater-Than-4-Iron [Fe-S] Clusters

Here we provide only a summary of valuable EPR-relevant references for clusters composed of more than four iron ions (Table 1). The above treatise has focused on the g-values that are characteristic of [Fe-S] of different nuclearities and in different oxidation states. Other magnetic parameters such as hyperfine interactions that describe the coupling between the electron spin and magnetic nuclei (e.g., ^{1}H, ^{13}C, ^{14}N, etc.) are not often accessible via CW EPR data. However, these interactions as well as inhomogeneity in the g-values contribute to the spectral linewidth (σ). The hyperfine contribution is field-independent, while the peak broadening due to inhomogeneity in g increases with increasing field. This "g-strain" component can only be untangled from the unresolved hyperfine by performing EPR experiments at multiple excitation wavelengths and corresponding magnetic field strengths [42]. Nonetheless, the total linewidth is an important parameter for any simulation of experimental data (*see* Subheading 3.3), and like the set of g-values, it too can be anisotropic. Furthermore, the linewidth can change in response to chemical treatments of [Fe-S]-containing samples. For example, the EPR spectra of samples prepared in deuterated water are often sharper than those of their natural abundance counterparts. The hyperfine interaction of any solvent-exchangeable proton is reduced by a factor of 6.51 (the ratio of the ^{1}H and ^{2}H Larmor frequencies) upon deuteration. The addition of glassing agents (*see* Subheading 3.1) can also sharpen EPR features by reducing heterogeneity and minimizing protein aggregation during freezing. Similarly, the addition of substrate molecules that do not directly interact with the [Fe-S] can lead to signal sharpening [43].

Table 1
High-nuclearity biological [Fe-S]

Number of Fe ions	Example	S_{TOT}	EPR References
5 Fe	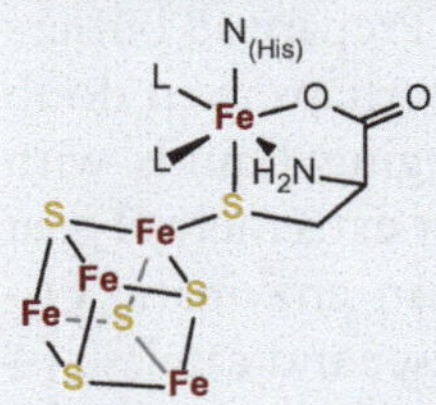	1/2, 3/2, 5/2	[29, 54, 55]
6 Fe	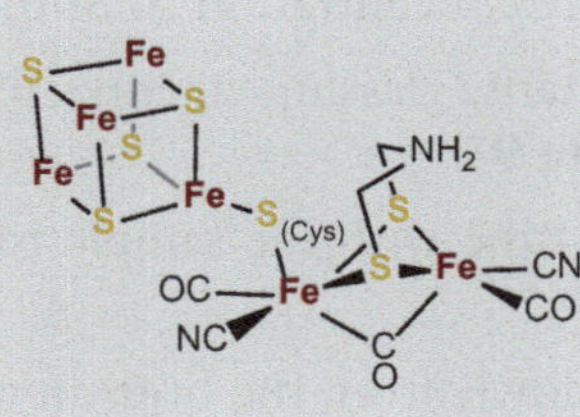	1/2	[56]
7 Fe	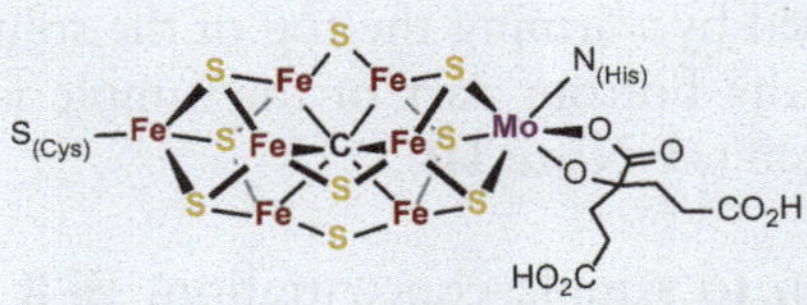	1/2, 3/2	[5, 57, 58]
8 Fe	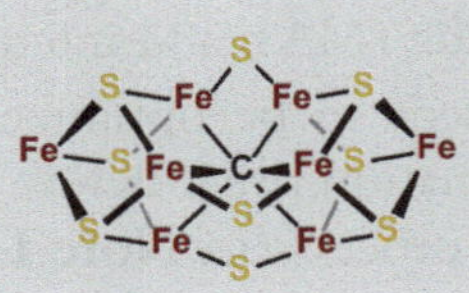	1/2	[59, 60]

2 Materials

1. Quartz tubes: 4 mm outer diameter suitable for most *X*-band CW-EPR spectrometers (*see* **Note 3**).
2. Liquid nitrogen (*see* **Note 4**).
3. Liquid helium (*see* **Note 5**).
4. Isopentane.
5. *X*-band (9–10 GHz) CW EPR spectrometer (*see* **Note 6**).

3 Methods

3.1 Sample Preparation

The most common pitfalls in EPR sample preparation are (1) paramagnetic contamination, (2) insufficient care taken with respect to O_2 sensitivity, and (3) too much or too little paramagnetic analyte.

The following steps in sample preparation are meant to help mitigate these problems:

1. Prepare the sample anaerobically. Samples should be prepared in a glovebox or on the Schlenk line, as [Fe-S] often decompose in the presence of O_2. Prepare all buffers and reagent solutions with high resistivity (perhaps even doubly distilled) water and deoxygenated by sparging liquids with O_2-free nitrogen or argon (*see* **Note 7**). For extremely O_2-sensitive sample preparations, consider using an enzymatic O_2-scrubbing cocktail of glucose, glucose oxidase, and catalase [44].
2. Adjust the sample concentration to 300 μM (*see* **Note 8**).
3. Add cryoprotectant/glassing agents, for example, 10% v/v glycerol (*see* **Note 9**).
4. Fill quartz EPR tube with the sample to a height of no more than 30 mm.
5. Freeze samples slowly from the bottom up while ensuring that the topmost layer of the solution remains liquid. This can be achieved by warming the top of the solution with your fingers while the bottom part of the sample is submerged in liquid nitrogen (*see* **Note 10**).

3.2 Acquisition of CW EPR Spectrum

In addition to sample concentration, EPR signal intensity I is a function of several instrumental settings, including temperature (T), modulation amplitude (MA), and microwave power (P): $I \propto \mathrm{MA}\cdot\sqrt{P}/T$ and each needs to be optimized to maximize signal without sacrificing information [45]. Below are things to consider during this optimization process (*see* **Note 11**):

1. Place sample into sample holder (sometimes called the collet stack) such that bottom of tube will not sit beyond the bottom of the cavity (consult spectrometer manual for specifications, *see* **Note 12**).
2. Wipe off the sample tube with a Kimwipe (*see* **Note 13**) and immediately insert the sample tube into a cooled (at least 100 K) spectrometer.
3. Determine the optimal temperature for data acquisition (*see* **Note 14**).
4. Chose microwave power that maximizes the EPR signal without incurring saturation effects (top of the linear region in Fig. 6b, *see* **Note 15**).
5. Perform the initial scan with the amplitude of field modulation set to 1 mT, then optimize this spectrometer setting (*see* **Note 16**).
6. Chose a B_0 sweep rate that corresponds to "slow-passage" conditions. This is most easily done by collecting the spectrum

twice at a given sweep rate, once sweeping up from the low field and once sweeping down from the high field. Examine the edges of the EPR spectral envelope, and if they appear to be identical regardless of sweep direction, "slow-passage" was achieved.

3.3 Analysis of CW EPR Spectra

As stated in Subheading 1, the magnetic properties that are most commonly extracted from CW EPR data of [Fe-S] clusters are the set of g-values and corresponding linewidth parameters (σ). While these properties can be estimated "by eye" using Eq. 1, for example, more precise and reproducible values can be obtained using EPR spectral fitting programs. There exist several superlative simulation software suites (SpinCount [46], XSophe [47], and more [48]), though we have the most experience using the EasySpin toolbox [49] that runs within MatLab (MathWorks) and so provide a relevant example below:

1. Install MatLab and the EasySpin toolbox following all platform-specific instructions (www.easyspin.org).
2. Copy and paste the example input file (all the text in `Courier` font below) into a new script tab in MatLab and save as 'MySim.m'.
3. Save your experimental data ('Filename.DTA') on to your computer in the same folder as 'MySim.m'.
4. Execute 'MySim.m' in MatLab by clicking 'Run' in the editor tab. Two windows will pop up: One comparing your experimental spectrum to a simulation obtained using the parameters listed in the `Sys` block. And one containing a user-interactive simulation optimization application called "EasySpin Least-Squares Fitting."
5. Manually adjust parameters in the `Sys` block to generate an approximate fit of the experimental spectrum. Repeat as necessary.
6. Shift to the optimization window and click 'Start' to run a root-mean-squared deviation (RMSD) minimization procedure that will optimize each of the parameters listed in the `Vary` block.

EasySpin input file. Text following the percent symbol (`%`) is a comment and not executable. Go to the EasySpin website (www.easyspin.org) for complete definitions of all terms.

```
%sample input file for simulating EPR spectrum of S = 1/2 [Fe-S]
clear all;
%'eprload' reads in experimental data and acquisition parameters
[data_x, data_y, data_param] = eprload('Filename.DTA');
%the 'Sys' block defines your first guess for the magnetic
properties that describe your [Fe-S]
Sys.S = 1/2;
Sys.g = [2.007, 1.96, 1.91];
Sys.lw = 1;
Sys.gStrain = [0.02, 0.02, 0.03];
%the 'Exp' block defines the experimental parameters that you want
your simulated spectrum to mimic
Exp.mwFreq = 9.5; %in units of GHz match to experimental value
Exp.nPoints = 601; %be sure to match to number of data points in
experimental spectrum
Exp.Range = [265 415]; %in units of mT, be sure to match to
experimental field range
[sim_x, sim_y] = pepper(Sys, Exp)
plot(data_x, data_y/max(data_y), sim_x, sim_y/max(sim_y))
%the above line will plot a scaled version of both the simulation
and the experimental spectrum for comparison
%the 'Vary' block designates which of the above magnetic parameters
you wish to float to achieve a best-fit of your data
Vary.g = [0.002 0.002 0.002];
Vary.gStrain = [1 1 1]*0.01;
Vary.lw = .2;
esfit('pepper', data_y, Sys, Vary, Exp)
```

4 Notes

1. Perpendicular-mode ($B_1 \perp B_0$) EPR spectroscopy is the most commonly used form of the technique; however, having $B_1 \parallel B_0$ changes the electron spin transition selection rules from $\Delta m_S = \pm 1$ to $\Delta m_S = 0$, and has been shown to be an effective means to interrogate integer spin systems such as $S = 2$, Fe^{2+}-containing proteins.
2. It is often said that the EPR spectrum is a derivative computed from the instrument-measured absorption spectrum. This is incorrect. The derivative is the raw output from the spectrometer.
3. At low temperatures, quartz is required over borosilicate as the latter contains paramagnetic impurities (most commonly Fe^{2+}, Fe^{3+}, and Cr^{3+}) that can obscure signals from the [Fe-S] analyte. While most spectrometers accept tubes with up to 4 mm, the user should check to see what their specific cavity and probe dimensions dictate for sample tube size.
4. EPR spectra of metalloproteins are intrinsically susceptible to changes in redox state. Therefore, care needs to be taken to limit O_2 exposure during sample preparation (*see* Subheading 3.1). Further, once the desired state of the protein is achieved, it must be maintained by keeping the sample frozen in liquid nitrogen. Reverse electron transfer has been observed for some samples stored in liquid N_2 for long periods of time, so check samples soon after preparation and note if spectral features shift or change intensity over time.
5. For most [Fe-S] samples, liquid helium (or a cryocooler that can obtain temperatures <20 K) is the required cryogen for setting the experimental temperature for EPR spectral acquisition to *see* Subheading 3.2). Be mindful that liquid helium ($T = 4.2$ K at 1 atm) contaminated with air can have substantial amounts of solid water, nitrogen, oxygen, etc. This flocculent material can easily clog the narrow needle valve used to control cryogen flow to the sample. Therefore, source helium from reliable vendors never leave pressurized helium dewars open to the atmosphere for longer than necessary.
6. There are many older brands and models of *X*-band CW-EPR spectrometers still in service as well as newer incarnations that may have a smaller footprint, require less helium, and/or have digital spectral acquisition capabilities. The more common spectrometers include Varian E-109, Varian/IBM 200-D, Bruker ECS-106, Bruker EMX, Bruker EMXmicro, Bruker EMX-nano, Bruker Elexsys-500, MagnetTech (now Bruker) MS-5000, Adani SPINSCAN X, JEOL JES-X3. Each

instrument will have its own idiosyncrasies, so consult the manual and any recent past users for best practices.

7. Protein-containing solutions should NOT be degassed by the freeze–pump–thaw method as this can lead to frothing, which causes protein precipitation, and proteins typically lose some portion of their activity with each free–thaw cycle.

8. Instrument sensitivities are often measured using a narrow organic radicals ("narrow" in this context means all the absorption intensity is centered in a narrow field range, i.e., small *g*-anisotropy) and reported to be in the nano- to the pico-molar range. For [Fe-S], the large *g*-anisotropy causes this intensity to be spread over a wider field range; therefore, concentrations between 50 and 1000 μM are preferred. Even if the protein does not precipitate at higher concentrations, care should be taken going above 1.0 mM as protein molecules could be in such close proximity that their corresponding electron spins interact with one another via a dipolar coupling mechanism that can manifest as peak-broadening in the EPR spectrum. Of course, the final protein concentration may have to be adjusted after initial measurement, given the above considerations.

9. Glassing agents should be added to protein solutions to mitigate the formation of extended water-ice crystals upon freezing. As these pure water networks solidify, solutes are pushed into smaller and smaller volumes, so breaking them up is key to maintaining a homogenous distribution of solute particles throughout the sample volume and preventing dipole-dipole induced broadening of EPR signals. Of course, any such additive will need to be tested to verify that it does not alter the in vitro activity of the protein.

10. Aqueous solutions expand upon freezing and thus have the potential to shatter the tube. Using the method described briefly above, touch the bottom of the tube to the surface of the liquid nitrogen and listen for the "sizzling" of the rapid evaporation of the cryogen to slow or go silent. Then submerge another 2–3 mm of the tube into the liquid nitrogen and again wait for the sizzling to abate. Keep the top of the solution liquid by warming it with your fingers. Repeat until the entire sample is submerged and frozen. Samples can be more rapidly frozen if submerged in liquid nitrogen-cooled slurries of isopentane though tube breakage is somewhat more likely.

11. The Bruker EPR manuals are excellent resources for understanding the different spectrometer components and the effect various settings have on the appearance of the spectrum.

12. The sample material should be centered in the cavity and should not extend significantly—above or below—beyond the dimensions of the cavity. For the most common rectangular *X*-band CW-EPR cavities, this length is about 3 cm and is equal to the wavelength (λ) of the incident radiation ($\lambda = c/\nu$, where c is the speed of light).

13. During storage, water ice can accumulate on the exterior of the tube. This ice and any liquid nitrogen need to be wiped off with a Kimwipe before insertion into the spectrometer as it can bring in with it other paramagnetic contaminants such as liquid $^3O_2$ condensed from the atmosphere (boiling point of O_2 is 90 K versus liquid N_2 temperature of 77 K). O_2 freezes below 50 K, and such O_2 crystalline ice gives rise to sharp EPR signals that can appear anywhere between 0 and 1 T. Simply rotating the sample should cause these spectral features to shift dramatically, allowing for their assignment to solid O_2. Further, by inserting the sample into the spectrometer at ≈100 K, one allows these atmospheric contaminants to boil off and get discarded with the cryostat exhaust.

14. For EPR transitions originating from the lowest energy level, corresponding signal intensity is inversely proportional to temperature, a manifestation of Curie's Law. However, this does not necessarily imply that the lowest achievable temperature is the best for spectral acquisition. At lower temperatures, EPR signals become more easily saturated (meaning there is equivalence in the population of the ground and excited spin levels and there is no longer a net absorption of microwaves) as fewer phonons are available to de-excite spin levels back to their equilibrium populations (Fig. 6). In other words, relaxation times become longer at lower temperatures and signal intensity can be suppressed.

15. The easiest way to see if a signal is partially saturated is to acquire a spectrum at two different powers and divide each by the square root of the power. If the resultant data are superimposable, then higher power is NOT saturating. The power at which one half of the signal intensity is lost to saturation is known at the $P_{1/2}$. We note that a $P_{1/2}$ obtained on one instrument or even one sample cannot be compared to other $P_{1/2}$'s owing to unpredictable differences in the amount of microwave power reaching the sample (sometimes called conversion factor). However, for a single sample, the temperature dependence of its $P_{1/2}$ can reveal thermally accessible low-lying excited states. If multiple species are contributing signals to the EPR spectrum and the relaxation properties are sufficiently different, careful choice of temperature and excitation power may suppress one of the signals [50–52]. This is a powerful tool for deconvolving spectra as illustrated in Fig. 7: the spectral

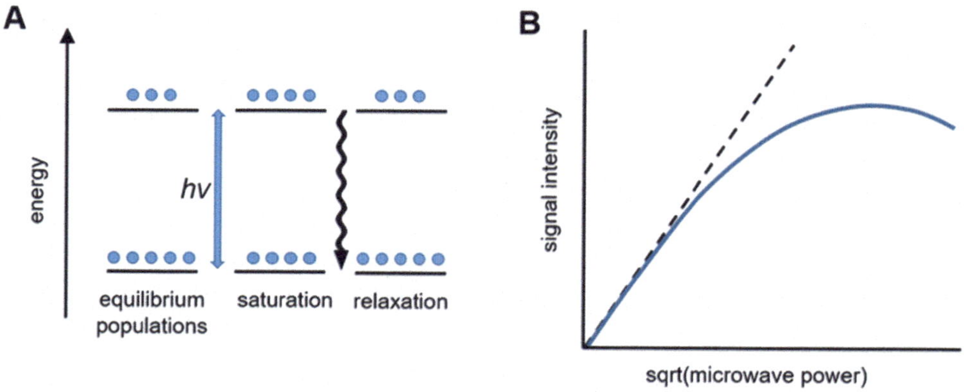

Fig. 6 Saturation/relaxation of a two-level system. (**a**) Boltzmann statistics determine the population of each level at thermal equilibrium (left) as a function of temperature and the energy difference between the two levels. Resonant light excitation (blue arrow) will move the population from the lower level to the upper level. If the flux of resonant photons (in the case of EPR, microwave power) is sufficient to out-pace relaxation (black wavy line) back to the lower level, then population equalization between levels can be achieved—saturation (center). (**b**) EPR signals increase linearly with the square root of the incident power for low powers (the linear region where blue trace and dashed line overlay). As a transition becomes more saturated, photon absorption and emission start to compete, and the net absorption of light decreases, as does the corresponding EPR signal (nonlinear region of blue trace, deflection from the dashed line)

deconvolution and assignment of EPR signals to some of the nine [Fe-S] complexes in respiratory complex I [53].

16. For any EPR spectrum, the magnitude of MA should be chosen to be smaller than the linewidth of the narrowest spectral feature wished to be resolved. [Fe-S] rarely exhibit narrow linewidths owing to considerable in homogenous broadening due to unresolved hyperfine interactions (between the electron spin and nearby magnetic nuclei like ^{1}H, ^{14}N, etc.) and slightly different microenvironments leading to a Gaussian distribution of *g*-values (*g*-strain). Typically, values of MA between 0.5 and 2.0 mT are employed. Always check instrument limits as certain modulation coils may not be rated for electrical currents corresponding to large modulation amplitudes and can be damaged if over-driven.

17. According to Hagen, it is worth "pointing out that the theory of *g*-strain implies continuously varying saturation behavior over the powder envelope of a single species; therefore, differential saturation is not a reliable criterion for deconvolution of these complex spectra" [24]. While this is technically correct, as a practical matter, differential saturation behavior can be used to indicate which EPR features in a spectrum of a complex mixture of paramagnetic species arise from a single species. Additionally, modern single-value decomposition methods should be useful in scoring such deconvolutions.

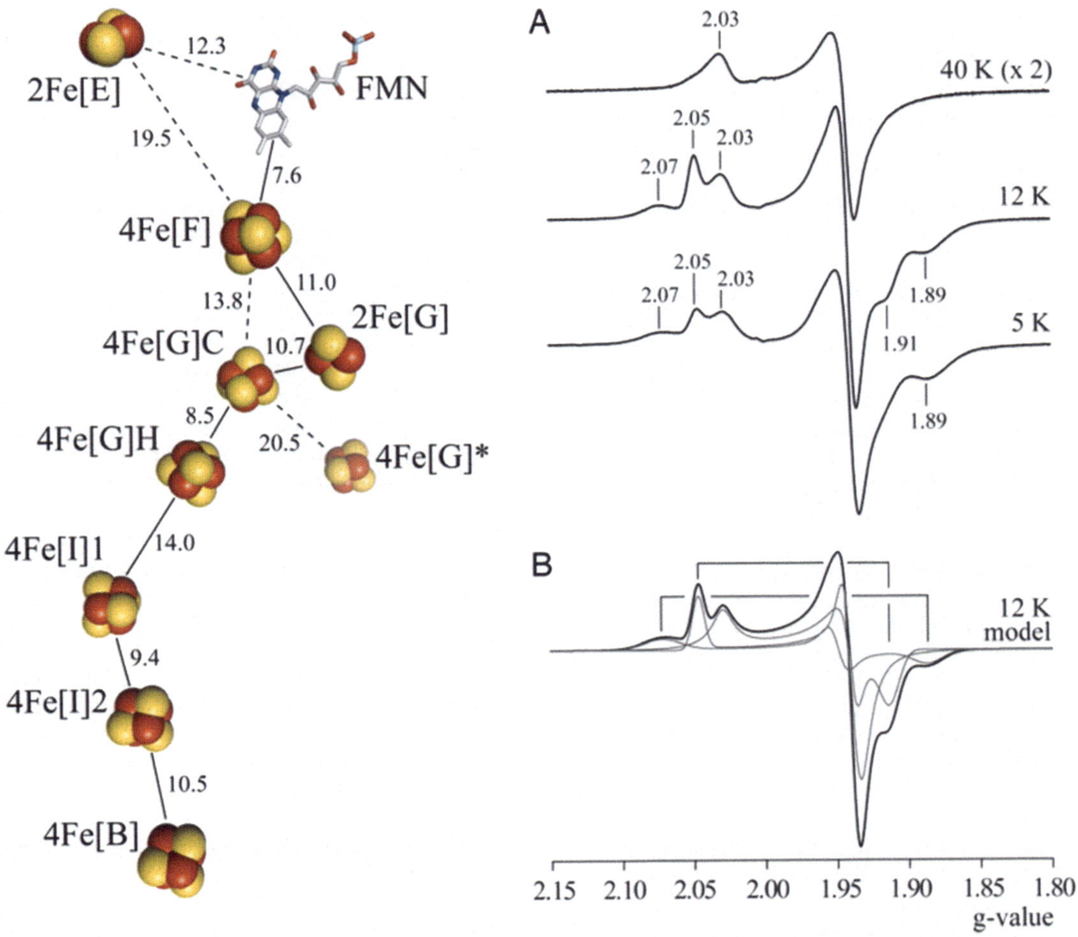

Fig. 7 (Left) Arrangement of the flavin cofactor and the [Fe-S] clusters in *T. thermophilus* complex I. Important edge-to-edge distances are indicated (Å). (Right) EPR spectra of complex I from *E. coli* reduced anaerobically with 1 mM dithionite and frozen immediately. (**a**) The spectrum at 40 K shows predominantly the spectrum of one [2Fe-2S] cluster; the spectra at 12 and 5 K show the spectra of two [4Fe-4S] clusters also. *g*-values for the major features are marked. (**b**) A model of the 12-K spectrum using the following parameters: $g = [2.074, 1.950, 1.887]$, $\sigma = [4.0, 2.5, 4.0]$ mT; $g = [2.048, 1.942, 1.914]$, $\sigma = [1.3, 1.5, 2.5]$ mT; $g = [2.030, 1.939, 1.937]$, $\sigma = [1.5, 2.0, 2.0]$ mT. (Adapted from [53], *see* **Note 17**)

References

1. Palmer G (1975) 1 iron—sulfur proteins. In: Boyer PD (ed) The enzymes. Academic Press, Cambridge, Massachusetts, pp 1–56
2. Beinert H (2003) Iron-sulfur proteins: properties and functions. Imperial College Press, London, pp 47–72
3. Beinert H, Kennedy MC, Stout CD (1996) Aconitase as iron-sulfur protein, enzyme, and iron regulatory protein. Chem Rev 96:2335–2373
4. Evans MCW (1982) Iron-sulfur centers in photosynthetic electron transport. Metal Ions Biol 4:249–284
5. Seefeldt LC, Yang Z-Y, Lukoyanov DA et al (2020) Reduction of substrates by Nitrogenases. Chem Rev 120:5082–5106
6. Jasniewski AJ, Lee CC, Ribbe MW et al (2020) Reactivity, mechanism, and assembly of the alternative Nitrogenases. Chem Rev 120:5107–5157

7. Cammack R (1992) Iron—sulfur clusters in enzymes: themes and variations. In: Cammack R (ed) Adv Inorg Chem. Academic Press, Cambridge, Massachusetts, pp 281–322
8. Hagen WR (2018) EPR spectroscopy of complex biological iron–sulfur systems. J Biol Inorg Chem 23:623–634
9. Abragam A, Bleaney B (1986) In electron paramagnetic resonance of transition ions. Dover Publications, New York
10. Weil JA, Bolton JR (2007) Electron paramagnetic resonance: elementary theory and practical applications. John Wiley & Sons, Hoboken, New Jersey
11. Hendrich MP, Debrunner PG (1989) Integer-spin electron-paramagnetic resonance of iron proteins. Biophys J 56:489–506
12. Hagen WR (1992) EPR spectroscopy of iron—sulfur proteins. In: Cammack R (ed) Adv Inorg Chem. Academic Press, Cambridge, Massachusetts, pp 165–222
13. Dunham WR, Palmer G, Sands RH et al (1971) On the structure of the iron-sulfur complex in the two-iron ferredoxins. Biochim Biophys Acta 253:373–384
14. Sands RH, Dunham WR (1974) Spectroscopic studies on two-iron ferredoxins. Q Rev Biophys 7:443–504
15. Agar JN, Krebs C, Frazzon J et al (2000) IscU as a scaffold for iron-sulfur cluster biosynthesis: sequential assembly of [2Fe-2S] and [4Fe-4S] clusters in IscU. Biochemistry 39:7856–7862
16. Chandramouli K, Unciuleac M-C, Naik S et al (2007) Formation and properties of [4Fe-4S] clusters on the IscU scaffold protein. Biochemistry 46:6804–6811
17. Ribbe M, Tanifuji K, Jasniewski A et al (2020) Nitrogenase M-cluster assembly. American Chemical Society, Washington, p INOR-0100
18. Dicus MM, Conlan A, Nechushtai R et al (2010) Binding of histidine in the $(Cys)_3(his)_1$-coordinated [2Fe−2S] cluster of human mitoNEET. J Am Chem Soc 132:2037–2049
19. Fee JA, Findling KL, Yoshida T et al (1984) Purification and characterization of the Rieske iron-sulfur protein from Thermus Thermophilus - evidence for a 2Fe-2S cluster having non-cysteine ligands. J Biol Chem 259:124–133
20. Gurbiel RJ, Batie CJ, Sivaraja M et al (1989) Electron-nuclear double resonance spectroscopy of nitrogen-15-enriched phthalate dioxygenase from pseudomonas cepacia proves that two histidines are coordinated to the [2Fe-2S] Rieske-type clusters. Biochemistry 28:4861–4871
21. Iwasaki T, Kounosu A, Samoilova RI et al (2006) 15N HYSCORE characterization of the fully deprotonated, reduced form of the archaeal Rieske [2Fe-2S] center. J Am Chem Soc 128:2170–2171
22. Crouse BR, Meyer J, Johnson MK (1995) Spectroscopic evidence for a reduced Fe2S2 cluster with a $S = 9/2$ ground state in mutant forms of clostridium pasteurianum 2Fe ferredoxin. J Am Chem Soc 117:9612–9613
23. Achim C, Golinelli MP, Bominaar EL et al (1996) Mossbauer study of Cys56Ser mutant 2Fe ferredoxin from clostridium pasteurianum: evidence for double exchange in an Fe2S2 (+) cluster. J Am Chem Soc 118:8168–8169
24. Priem AH, Klaassen AK, Reijerse EJ et al (2005) EPR analysis of multiple forms of 4Fe-4S (3+) clusters in HiPIPs. J Biol Inorg Chem 10:417–424
25. Watt GD, Reddy KRN (1994) Formation of an all ferrous Fe4S4 cluster in the iron protein component of *Azotobacter vinelandii* nitrogenase. J Inorg Biochem 53:281–294
26. Pandelia M-E, Lanz ND, Booker SJ et al (2015) Mössbauer spectroscopy of Fe/S proteins. Biochim Biophys Acta 1853:1395–1405
27. Garcia-Serres R, Clémancey M, Latour J-M et al (2018) Contribution of Mössbauer spectroscopy to the investigation of Fe/S biogenesis. J Biol Inorg Chem 23:635–644
28. Telser J, Emptage MH, Merkle H et al (1986) Oxygen-17 electron nuclear double resonance characterization of substrate binding to the 4-iron-4-sulfur ([4Fe-4S]1+) cluster of reduced active aconitase. J Biol Chem 261:4840–4846
29. Rao G, Tao L, Suess DLM et al (2018) A [4Fe–4S]-Fe(CO)(CN)-l-cysteine intermediate is the first organometallic precursor in [FeFe] hydrogenase H-cluster bioassembly. Nat Chem 10:555–560
30. Tao L, Zhu W, Klinman JP et al (2019) Electron paramagnetic resonance spectroscopic identification of the Fe-S clusters in the SPASM domain-containing radical SAM enzyme PqqE. Biochemistry 58:5173–5187
31. Hagen WR, van den Berg WAM, van Dongen WMAM et al (1998) EPR spectroscopy of biological iron-sulfur clusters with spin-admixed S=3-2 ground states. J Chem Soc Faraday Trans 94:2969–2973
32. Lindahl PA, Day EP, Kent TA et al (1985) Moessbauer, EPR, and magnetization studies of the *Azotobacter vinelandii* iron protein. Evidence for a [4-iron-4-sulfur]1+ cluster with spin $S = 3/2$. J Biol Chem 260:11160–11173

33. Conover RC, Kowal AT, Fu WG et al (1990) Spectroscopic characterization of the novel iron-sulfur cluster in Pyrococcus furiosus ferredoxin. J Biol Chem 265:8533–8541
34. Rothery RA, Bertero MG, Cammack R et al (2004) The catalytic subunit of Escherichia coli nitrate reductase a contains a novel [4Fe-4S] cluster with a high-spin ground state. Biochemistry 43:5324–5333
35. Angove HC, Yoo SJ, Burgess BK et al (1997) Mössbauer and EPR evidence for an all-ferrous Fe4S4 cluster with S = 4 in the Fe protein of Nitrogenase. J Am Chem Soc 119:8730–8731
36. Eisenstein RS, Kennedy MC, Beinert H (1998) The iron responsive element (IRE), the iron regulatory protein (IRP), and cytosolic aconitase: posttranscriptional regulation of mammalian iron metabolism. Chapman & Hall, London, pp 157–216
37. Crack JC, Green J, Thomson AJ et al (2014) Iron–sulfur clusters as biological sensors: the chemistry of reactions with molecular oxygen and nitric oxide. Acc Chem Res 47:3196–3205
38. Moura JJG, Moura I, Kent TA et al (1982) Interconversions of [3Fe-3S] and [4Fe-4S] clusters. Moessbauer and electron paramagnetic resonance studies of Desulfovibrio gigas ferredoxin II. J Biol Chem 257:6259–6267
39. Kent TA, Huynh BH, Munck E (1980) Iron-sulfur proteins—spin-coupling model for 3-iron clusters. Proc Natl Acad Sci U S A 77:6574–6576
40. Fan C, Houseman ALP, Doan P et al (1993) Conformational distribution in protein-bound [3Fe-4S]+ clusters: CW and pulsed EPR and iron-57 ENDOR of *D. gigas hydrogenase*. J Phys Chem 97:3017–3021
41. Duff JLC, Breton JLJ, Butt JN et al (1996) Novel redox chemistry of [3Fe–4S] clusters: electrochemical characterization of the all-Fe (II) form of the [3Fe–4S] cluster generated reversibly in various proteins and its spectroscopic investigation in Sulfolobus acidocaldarius ferredoxin. J Am Chem Soc 118:8593–8603
42. Hearshen DO, Hagen WR, Sands RH et al (1986) An analysis of g strain in the EPR of two [2Fe-2S] ferredoxins. Evidence for a protein rigidity model. J Magn Reson 69:440–459
43. Lieder KW, Booker S, Ruzicka FJ et al (1998) S-Adenosylmethionine-dependent reduction of lysine 2,3-aminomutase and observation of the catalytically functional iron-sulfur centers by electron paramagnetic resonance. Biochemistry 37:2578–2585
44. Englander SW, Calhoun DB, Englander JJ (1987) Biochemistry without oxygen. Anal Biochem 161:300–306
45. Eaton GR, Eaton SS, Barr DP et al (2010) Quantitative EPR. Springer, New York
46. Petasis DT, Hendrich MP (2015) Chapter eight - quantitative interpretation of multifrequency multimode EPR spectra of metal containing proteins, enzymes, and biomimetic complexes. In: Qin PZ, Warncke K (eds) Methods enzymol. Academic Press, Cambridge, Massachusetts, pp 171–208
47. Hanson GR, Gates KE, Noble CJ et al (2004) XSophe-Sophe-XeprView®. A computer simulation software suite (v. 1.1.3) for the analysis of continuous wave EPR spectra. J Inorg Biochem 98:903–916
48. Stoll S (2015) Chapter six - CW-EPR spectral simulations: solid state. In: Qin PZ, Warncke K (eds) Methods enzymol. Academic Press, Cambridge, Massachusetts, pp 121–142
49. Stoll S, Schweiger A (2006) EasySpin, a comprehensive software package for spectral simulation and analysis in EPR. J Magn Reson 178:42–55
50. Rupp H, Cammack R, Rao KK et al (1978) Electron-spin relaxation of iron-sulfur proteins studied by microwave-power saturation. Biochim Biophys Acta 537:255–269
51. Beardwood P, Gibson JF, Bertrand P et al (1983) Temperature dependence of the electronic spin-lattice relaxation time in a 2-iron-2-Sulphur model complex. Biochim Biophys Acta 742:426–433
52. Gayda JP, Bertrand P, Deville A et al (1979) Temperature dependence of the electronic spin-lattice relaxation time in a 2-iron-2-sulfur protein. Biochim Biophys Acta 581:15–26
53. Yakovlev G, Reda T, Hirst J (2007) Reevaluating the relationship between EPR spectra and enzyme structure for the iron-sulfur clusters in NADH:quinone oxidoreductase. Proc Natl Acad Sci U S A 104:12720–12725
54. Dinis P, Suess DLM, Fox SJ et al (2015) X-ray crystallographic and EPR spectroscopic analysis of HydG, a maturase in FeFe -hydrogenase H-cluster assembly. Proc Natl Acad Sci U S A 112:1362–1367
55. Suess DLM, Britt RD (2015) EPR spectroscopic studies of FeFe -hydrogenase maturation. Top Catal 58:699–707
56. Lubitz W, Ogata H, Rudiger O et al (2014) Hydrogenases Chem Rev 114:4081–4148
57. Danyal K, Yang Z-Y, Seefeldt LC (2011) Electron paramagnetic resonance spectroscopy. Methods Mol Biol 766:191–205

58. Davis LC, Henzl MT, Burris RH et al (1979) Iron-sulfur clusters in the molybdenum-iron protein component of nitrogenase. Electron paramagnetic resonance of the carbon monoxide inhibited state. Biochemistry 18:4860–4869
59. Rupnik K, Hu Y, Lee CC et al (2012) P+ state of Nitrogenase P-cluster exhibits electronic structure of a [Fe4S4]+ cluster. J Am Chem Soc 134:13749–13754
60. Wiig JA, Hu Y, Ribbe MW (2011) NifEN-B complex of Azotobacter vinelandii is fully functional in nitrogenase FeMo cofactor assembly. Proc Natl Acad Sci U S A 108(8623–8627): S8623/8621–S8623/8622
61. Moebius K, Lubitz W, Cox N et al (2018) Biomolecular EPR meets NMR at high magnetic fields. Magnetochemistry 4:50
62. Beinert H, Holm RH, Munck E (1997) Iron-sulfur clusters: nature's modular, multipurpose structures. Science 277:653–659

Chapter 15

Characterization of Fe-S Clusters in Proteins by Mössbauer Spectroscopy

Chie Ueda, Michelle Langton, and Maria-Eirini Pandelia

Abstract

^{57}Fe Mössbauer spectroscopy is unparalleled in the study of Fe-S cluster-containing proteins because of its unique ability to detect all forms of iron. Enrichment of biological samples with the ^{57}Fe isotope and manipulation of experimental parameters such as temperature and magnetic field allow for elucidation of the number of Fe-S clusters present in a given protein, their nuclearity, oxidation state, geometry, and ligation environment, as well as any transient states relevant to enzyme chemistry. This chapter is arranged in five sections to help navigate an experimentalist to utilize ^{57}Fe Mössbauer spectroscopy for delineating the role and structure of biological Fe-S clusters. The first section lays out the tools and technical considerations for the preparation of ^{57}Fe-labeled samples. The choice of experimental parameters and their effects on the Mössbauer spectra are presented in the following two sections. The last two sections provide a theoretical and practical guide on spectral acquisition and analysis relevant to Fe-S centers.

Key words Mössbauer spectroscopy, Fe-S clusters, Reconstitution, Electronic and chemical structure, Cofactors, Redox state, Isomer shift, Quadrupole splitting, Hyperfine interactions

1 Introduction

^{57}Fe Mössbauer spectroscopy takes advantage of the *recoilless* resonant emission and absorption of γ-rays by ^{57}Fe nuclei in solids [1–3] and is able to resolve electronic and chemical properties of ^{57}Fe nuclei, such as oxidation states, hyperfine interactions, geometry, ligands, etc. [1, 3–5]. ^{57}Fe Mössbauer spectroscopy has thus found a wide application in the study of biological Fe-containing cofactors, especially in the field of Fe-S cluster-containing proteins [1, 6–9]. Fe-S clusters have been extensively studied by ^{57}Fe Mössbauer, and their spectroscopic parameters can serve as "fingerprints" for identifying their type and redox state [1, 3, 6–8]. Mössbauer spectroscopy is also useful for the quantification of Fe-S clusters, identification of changes in their coordination environment and nuclear configuration, as well as trapping of intermediates or short-lived states [10]. Combined, these factors make ^{57}Fe Mössbauer the

Patricia C. Dos Santos (ed.), *Fe-S Proteins: Methods and Protocols*, Methods in Molecular Biology, vol. 2353, https://doi.org/10.1007/978-1-0716-1605-5_15,

method of choice for the characterization of Fe-S clusters, yielding invaluable information about the structure and chemistry of these cofactors.

2 Materials

2.1 Basic Mössbauer Spectrometer Setup

1. Mössbauer spectrometer: A typical Mössbauer spectrometer for studies on Fe-S cluster-containing proteins consists of a ^{57}Co γ-ray radiation source (10–100 mCi) mounted on a velocity transducer (see below), a cryostat for performing measurements in a temperature range between 1.5 and 220 K, and a γ-ray counting system. A weak permanent magnet (30–80 mT) can be preferentially mounted, which helps with the identification of biological Fe-S clusters. A spectrometer equipped with a superconducting system can be alternatively used to carry out field-dependent measurements at higher magnetic field strengths (i.e., 0.1–10 T).

2.2 Solutions for In Vivo and In Vitro Fe-S Enrichment

1. ^{57}Fe solution: 500 mM Fe^{2+} in 2 N H_2SO_4. Dissolve 57 mg of the ^{57}Fe metal in 2 mL of 2 N H_2SO_4 (*see* **Note 1**). Allow solvation to proceed overnight at room temperature or for 1–2 h at 70 °C. At high temperatures, the solution will bubble due to the release of H_2, so keep the lid loose. Dilute the ^{57}Fe stock with water to 125 mM and store indefinitely at −20 °C. Confirm Fe concentration using the ferrozine assay [11]. For in vivo assembly, add the ^{57}Fe solution directly to the cell cultures. For reconstitution reactions, prepare and store the ^{57}Fe stock solution anaerobically. Prior to addition to the reconstitution mixture, dilute the Fe solution with a buffer compatible with the protein of interest to achieve working concentrations of Fe between 10–50 mM and check the pH with a pH strip. If very acidic, adjust the buffer concentration or re-pH as necessary. Care should be taken to ensure that solutions are strictly O_2-free; otherwise, the $^{57}Fe^{2+}$ will readily oxidize to $^{57}Fe^{3+}$.
2. Na_2S solution: 100 mM sodium sulfide solution. Dissolve 24 mg of solid $Na_2S{\cdot}9H_2O$ in 1 mL buffer for a 100 mM solution. The pH of the buffer should be kept basic (i.e., above pH 7.5) to prevent the formation of volatile H_2S. Na_2S solutions should always be freshly prepared.
3. Pyridoxal phosphate (PLP) solution: 100 mM pyridoxal-5-′-phosphate solution. Dissolve 24 mg of solid PLP in 1 mL buffer. If very acidic, adjust the buffer concentration or re-pH as necessary.

4. L-cysteine solution: 100 mM cysteine solution. Dissolve 12.1 mg of solid L-cysteine in 1 mL buffer. The L-cysteine solution should always be freshly prepared.

3 Methods

3.1 Preparing the Optimal Biological Mössbauer Sample

The quality of Mössbauer spectra is reflected in the signal-to-noise (S/N) ratio and depends on both the sample (i.e., extent of ^{57}Fe labeling, purity, and homogeneity) and spectroscopy-related (experimental and theoretical) factors [3, 4]. When designing an experiment, the aim is to maximize the S/N ratio and minimize acquisition time. In this respect, the most important considerations are usually sample thickness, ^{57}Fe concentration, and sample purity.

3.1.1 Sample Size, Thickness, and ^{57}Fe Concentration

1. Optimize the thickness and size of a Mössbauer sample. These factors greatly influence the S/N ratio. Optimal samples have a thickness of ~6 mm and volumes ranging between 300 and 500 μL depending on the Mössbauer cell type (*see* **Note 2**). Thicker samples (higher effective ^{57}Fe concentration) do not increase the S/N ratio further because nonresonant absorption attenuates the Mössbauer effect [3, 12].
2. Optimize the ^{57}Fe sample concentration. The natural abundance of ^{57}Fe is 2.2%, and samples for Mössbauer spectroscopy need to be isotopically enriched. As a rule of thumb, ^{57}Fe concentrations of 1–5 mM [3] provide spectra in relatively short acquisition times (1 mM ^{57}Fe typically yields high-quality statistics between 12 and 24 h, Fig. 1). Because spectral quality is also dependent on the spin state (diamagnetic, paramagnetic) and sample purity (distinct Fe species present), longer times or higher isotope concentrations may be needed.

3.1.2 In Vivo ^{57}Fe Enrichment of Fe-S Cluster-Containing Proteins

The endogenous Fe-S cluster biosynthetic machinery of *Escherichia (E.) coli* is often outpaced by the demand for cluster assembly during heterologous expression. Therefore, a large fraction of the isolated protein is devoid of the cofactor (apo) or incorporates nonnative Fe-S cluster fragments. This limitation can be overcome by stimulating endogenous or exogenous expression of Fe-S biosynthetic pathways.

1. For upregulation of the endogenous de novo Fe-S cluster biosynthetic pathways ISC and SUF [13, 14] in *E. coli*, use the derivative strains ΔiscR and SufFeScient, which lack the repressor *iscR* [9, 15] and Fur binding site at the *sufA* promoter [14, 16], respectively (see also Chapter 4 of this edition).
2. Alternatively, coexpress the protein of interest with the plasmid encoding for the *Azotobacter (A.) vinelandii isc* operon

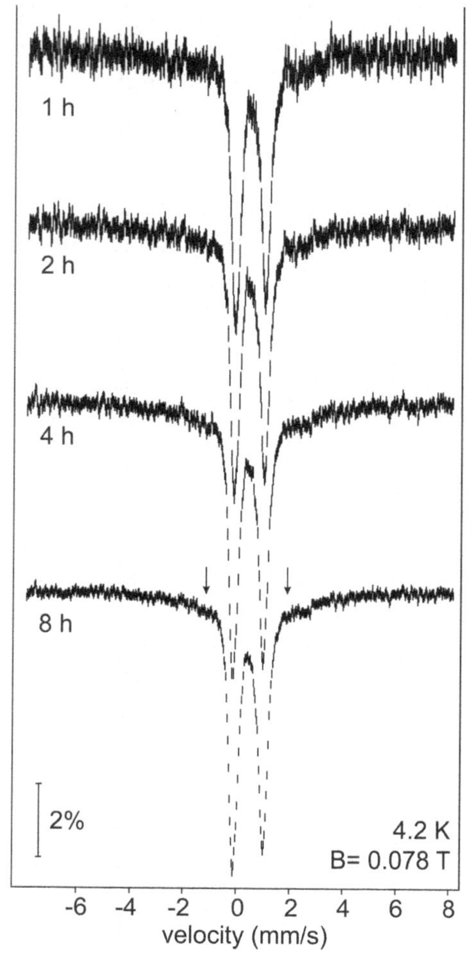

Fig. 1 Effect of acquisition time on the S/N ratio of Mössbauer spectra and the importance of good statistics. The Mössbauer spectrum shown corresponds to that of the chemically reconstituted MftC [47], a radical-SAM enzyme that contains three [4Fe-4S] clusters, one that binds SAM and two auxiliary clusters. The sample has an Fe concentration of 2.3 mM. The spectra were acquired at 4.2 K and in the presence of a magnetic field applied parallel to the γ beam. Although good statistics can be obtained after a two-hour acquisition time, only after 8 h of accumulation are the broad paramagnetic contributions of adventitious Fe (indicated by arrows) reliably detectable

(pDB1282) or the *E. coli suf* operon (pGSO164 or pPH151) [17, 18]. Protocols for overexpression of ^{57}Fe-labeled Fe-S-containing proteins are summarized in Fig. 2 and elaborated in the literature [19–22]. These can be modified to include the addition of ^{57}Fe to final concentrations of 25–125 μM upon induction of protein expression.

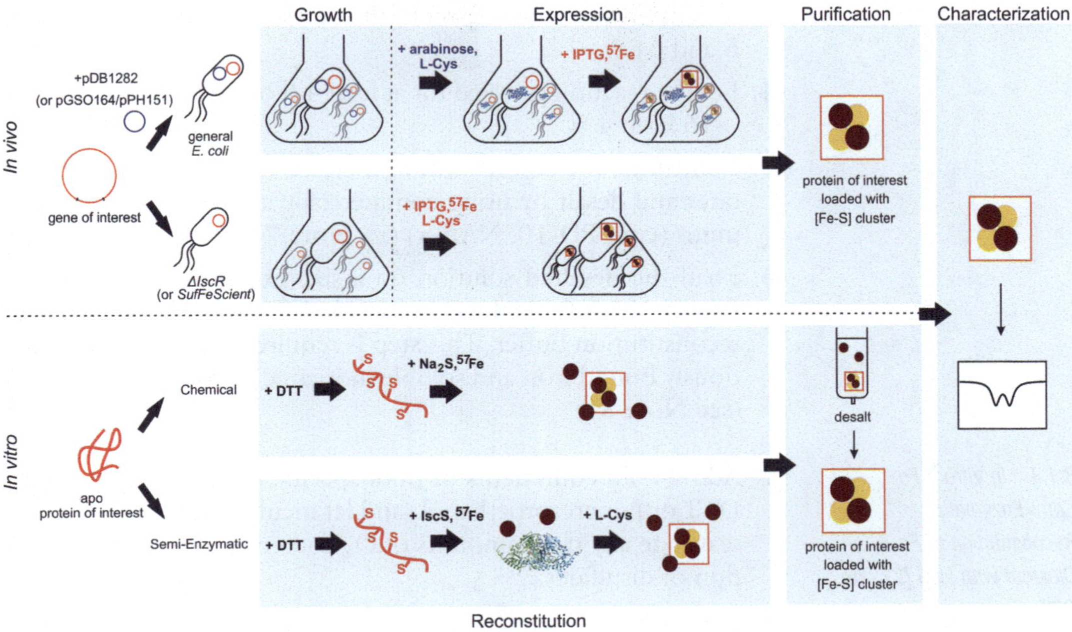

Fig. 2 ^{57}Fe enrichment of Fe-S clusters in proteins. In vivo assembly of Fe-S clusters with the Mössbauer isotope can be enhanced by coexpression with either the plasmids expressing biological cluster assembly machineries (pDB1282 or pGSO164/pPH151) or in cell strains lacking the repressors IscR or SufR. In both cases, cells should be grown in minimal media and with addition of ^{57}Fe and L-cysteine. In vivo methods afford protein loaded with the cofactor after purification. Alternatively, protein can be purified in its apo form and the cofactor assembled in vitro via reconstitution procedures. This can be performed chemically by addition of a reducing agent followed by the slow addition of sulfide and ^{57}Fe, or semi-enzymatically by addition of a reducing agent followed by the desulfurase IscS, L-cysteine, and ^{57}Fe. Following all reconstitution procedures, it is important to extensively desalt the protein to remove any adventitiously bound iron species

3.1.3 In Vitro ^{57}Fe Chemical Reconstitution of Fe-S Clusters

Fe-S clusters can be reconstituted into apoproteins via chemical or enzymatic methods [21–25]. General protocols are outlined below. Reactions should be performed under strictly O_2-free conditions and preferably on ice.

1. Add 20–50 equivalents of thiol-specific reducing agents (e.g., dithiolthreitol (DTT) or β-mercaptoethanol) to the protein (*see* **Notes 3** and **4**) and let incubate for 15–20 min to scavenge any trace amounts of O_2 and ensure complete reduction of disulfides.
2. Add ^{57}Fe in 4 aliquots over the course of an hour. Protein commonly precipitates upon the addition of iron salts, so the reconstitution mixture should be thoroughly mixed upon every aliquot addition. The ^{57}Fe amount added should correspond to that expected for the number and type of Fe-S clusters bound to the protein of interest.
3. Add Na_2S in 4 aliquots over the course of an hour with thorough mixing. The final concentration of Na_2S should be

stoichiometric with respect to the Fe added in **step 2** (*see* **Notes 5** and **6**).

4. Let the reaction proceed for at least 4 h or preferably overnight (~16 h) at 4 °C.
5. Centrifuge the reconstitution mixture to remove any precipitates and desalt by using commercially available desalting columns (e.g., PD-10, NAP-5) (*see* **Note 7**).
6. Load the desalted solution on a size-exclusion column (e.g., Hiprep 16/60 Sephracryl HR S-200) equilibrated in the reconstitution buffer. This step is required to remove adventitiously bound iron and soluble inorganic or protein aggregates (see **Note 8**).

3.1.4 In Vitro ^{57}Fe Semi-Enzymatic Reconstitution of Fe-S Clusters with IscS [22, 25, 26]

1. Add 20–50 equivalents of thiol-specific reducing agents (e.g., DTT or β-mercaptoethanol) and let incubate for 15–20 min to scavenge any trace amounts of O_2 and ensure complete reduction of disulfides.
2. Add IscS (typically in a molar ratio IscS: protein of 1:50–200). Estimate IscS concentration based on the PLP-cofactor bound ($\varepsilon_{388} = 5305\ M^{-1}\ cm^{-1}$) or supplement with PLP at an equimolar concentration of IscS.
3. Add ^{57}Fe in 4 aliquots over the course of an hour.
4. Initiate reaction with the addition of L-cysteine and incubate overnight (~16 h) at 4 °C.
5. Follow **steps 5** and **6** of Subheading 3.1.3 for the cleanup of the reconstitution reaction. If chemical or semi-enzymatic Fe-S reconstitutions are not successful, there are other avenues that one can resort to (*see* **Note 9**).

3.2 Spectroscopic Parameters and Chemical and Electronic Information Obtained from Mössbauer

For a ^{57}Fe nucleus in the sample to resonantly absorb the γ irradiation from the source, there must be an overlap between their respective absorption and emission spectra. The emission spectrum of the source and the absorption spectrum of the sample occur at slightly different energies as a result of the different electric and magnetic properties experienced by the respective Mössbauer atoms that perturb or split the ^{57}Fe energy levels differently (Fig. 3). The energy of the source γ-photons can be modulated to match that of the sample via the Doppler effect by attaching the source to a velocity transducer [3, 5]. Thus, the *x*-axis and parameters of a Mössbauer spectrum are cited in a velocity scale (mm/s). The fundamental ^{57}Fe Mössbauer spectrum in the absence of magnetic interactions consists of a "quadrupole doublet," with two resonance lines of equal integrated intensities [3, 4]. The quadrupole doublet is described by the two principal Mössbauer parameters: isomer shift (δ) and quadrupole splitting (ΔE_Q) [1, 2, 5].

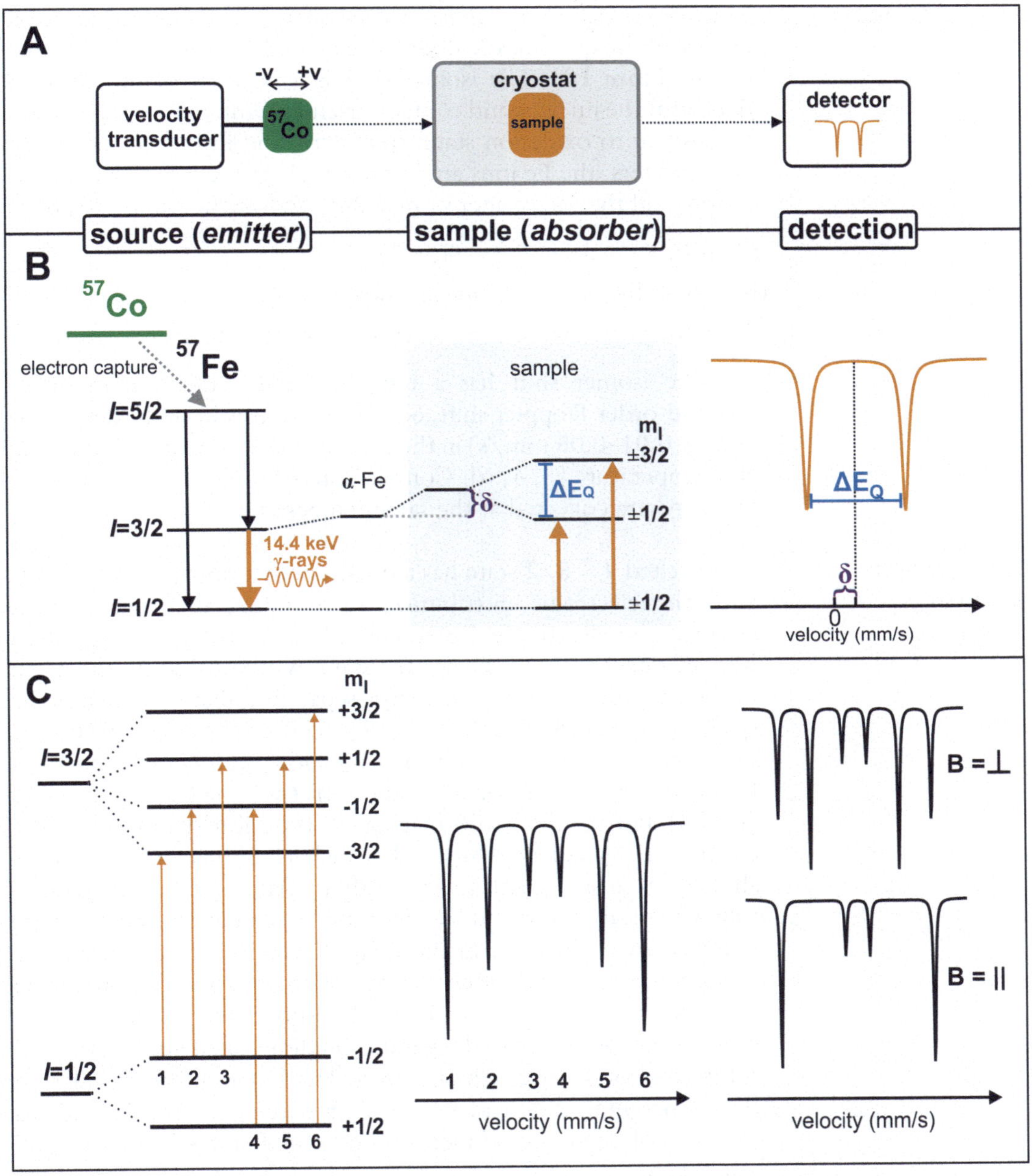

Fig. 3 Experimental setup and chemical and electronic information obtained from ^{57}Fe Mössbauer spectroscopy. (**a**) The ^{57}Co source (mounted on a velocity transducer), the sample (housed in a cryostat), and detector (proportional gas flow counter) are all aligned in the spectrometer. (**b**) Decay scheme of ^{57}Co to ^{57}Fe emitting the 14.4 keV γ-photon and the relevant transitions in the sample that give rise to a quadrupole doublet. Zero velocity on the velocity scale is calibrated with respect to α-iron (**c**) Mössbauer transitions in the presence of an external magnetic field for an effective $S = 1/2$ compound. The number of allowed transitions and their relative intensities depend on the orientation of the external magnetic field with respect to the γ-beam

3.2.1 Isomer Shift (δ)

The isomer shift (δ) is determined by taking the average absorption energy of the quadrupole doublet (centroid) relative to zero velocity (*see* **Note 10**). The isomer shift is a measure of the electron density at the nucleus and contains useful chemical information as it is sensitive to oxidation state, spin state, and ligand covalency. In Fe-S clusters, the Fe ions are in their high-spin electronic configuration, and the Mössbauer isomer shifts follow the trends: [1, 6, 7]

(i) $\delta(Fe^{3+}) < \delta(Fe^{2.5+}) < \delta(Fe^{2+})$

(ii) δ(soft ligands) < δ(ionic ligands)

(iii) δ(four-coordinate complexes) < δ(six-coordinate complexes)

The isomer shift has a temperature-dependent component (second-order Doppler shift, δ_{SOD}) that results in slight deviations (i.e., ± 0.01–0.05 mm/s) in the isomer shift values upon increasing the temperature [3, 4, 6]. Comparisons of δ should thus be made from spectra collected at the same temperature.

3.2.2 Quadrupole Splitting (ΔE_Q)

The excited $I = 3/2$ state has a quadrupole moment that interacts with the electronic distribution surrounding the ^{57}Fe nucleus (termed the electric field gradient) [1, 3, 4, 6]. This interaction lifts the degeneracy of the excited state (without shifting the mean of the energy), leading to an energy separation that is defined as the quadrupole splitting, ΔE_Q (Fig. 3). The quadrupole splitting is determined from the spectrum by taking the energy separation between the two resonance lines of the quadrupole doublet (Fig. 3) and provides information on the local symmetry of the electric field gradient of the ^{57}Fe nucleus. Similar to the isomer shift, Fe-S clusters, with their mostly thiolate-based coordination, follow typical trends, and a decrease in oxidation state can be followed by an increase in the ΔE_Q. Exceptions in this trend are not uncommon, and often deviations from commonly reported ΔE_Q values are suggestive of the Fe ions being coordinated by more ionic protein-based ligands, binding small molecules and substrates, or having a higher coordination number. Similar to the isomer shift, ΔE_Q can also be slightly temperature-dependent, especially if there are admixtures of excited states in the ground state [1, 6].

3.2.3 Magnetic Hyperfine Interactions (A)

Fe compounds with a paramagnetic ground state ($S > 0$) exhibit more complex spectra than a quadrupole doublet because the interaction between the unpaired electron and the ^{57}Fe nucleus perturbs the nuclear energy levels. In the presence of an external magnetic field for a system with an effective $S = 1/2$ and as described in detail elsewhere (*see* **Note 11**) [1, 3, 4, 6, 27], the excited $I = 3/2$ state splits into four, and the ground $I = 1/2$ state splits into two energy levels. This results in a total of four or six allowed transitions depending on whether the applied field is parallel or perpendicular with respect to the γ-beam, Fig. 3. Hamiltonian-based simulations of spectra recorded at different applied magnetic fields provide the hyperfine constants (A), which are tensorial values commonly expressed in units of MHz, Tesla, or Gauss (*see* **Note 12**). The magnitude and sign of hyperfine constants inform on the spin coupling scheme in polynuclear Fe-S clusters and their electronic configuration [1, 6]. The observed hyperfine constants between Fe-S clusters of the same nuclearity will be very similar and only deviate if local geometry changes perturb their electronic structure. Hyperfine constants are therefore useful for deducing the electronic structure, especially for Fe-S clusters with unusual ligation or chemistry.

3.3 Experimental Considerations for Recording Mössbauer Spectra of Fe-S Proteins

Although the measurement conditions are dependent on the Fe-S type under interrogation, the experimental approaches for their characterization share common features that are broadly applicable for different Fe-S cluster types.

3.3.1 Temperature

The choice of temperature has two major experimental implications in the Mössbauer spectra: (1) the S/N ratio and (2) the spectral complexity of paramagnetic Fe-S clusters.

1. The probability for recoil-free emission and absorption is known as the f (or Lamb-Mössbauer) factor and is a measure of the Mössbauer effect. The f factor is strongly temperature-dependent because it is inversely proportional to the mean-square displacement of the ^{57}Fe nucleus. Thus, lower experimental temperatures will increase the f factor and consequently, the S/N ratio. The f factor can vary after a change in redox state or addition of a ligand, but at liquid helium temperatures (4.2 K), it has been semi-empirically shown to be equal to ~0.8 for all ^{57}Fe species, allowing for their reliable quantification (*see* **Note 13**) [3, 4].
2. At 4.2 K, electronic spin relaxation is slower than the Larmor precession frequency of the ^{57}Fe nucleus (slow relaxation), leading to magnetically split Mössbauer spectra (Fig. 4) [1, 3, 4, 6, 7, 28]. At higher temperatures, typically > 80 K, electronic spin relaxation is faster than the ^{57}Fe nucleus Larmor frequency (fast relaxation), and paramagnetic spectra collapse into quadrupole doublets.

3.3.2 Applied Magnetic Field

The presence and strength of the applied magnetic field have different implications in the Mössbauer spectra of diamagnetic and paramagnetic Fe-S clusters, respectively. At 4.2 K, small applied fields (30–80 mT) will, in most cases, yield magnetically split spectra for paramagnetic Fe-S clusters and quadrupole doublets for

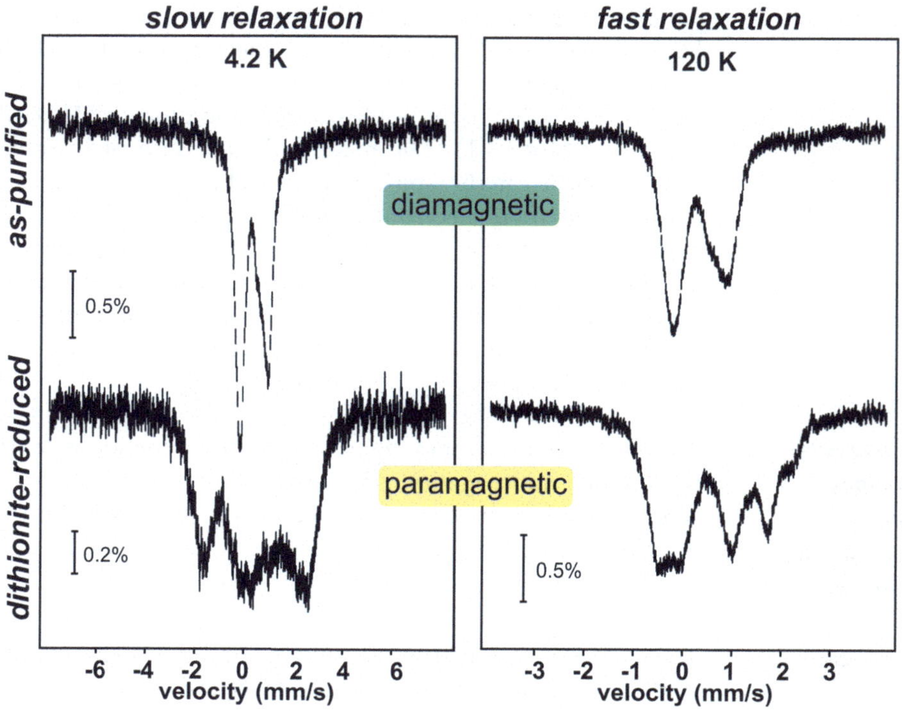

Fig. 4 Mössbauer spectra of the as-purified and dithionite-reduced carbon monoxide dehydrogenase (CODH) from *Desulfovibrio vulgaris* [28] demonstrating the effects of slow and fast relaxation on diamagnetic and paramagnetic states of Fe-S clusters. (Left) Recorded in the slow relaxation limit (4.2 K) in a $\pm$ 8 mm/s velocity scale, and in the presence of a small magnetic field (78 mT) applied parallel to the γ-beam. (Right) Recorded in the fast relaxation limit (120 K) in a $\pm$ 4 mm/s velocity scale, and in the absence of an applied magnetic field. The as-isolated form of CODH contains four [4Fe]-containing clusters and one [2Fe-2S] clusters in their 2+ states (all are $S = 0$, diamagnetic), while the dithionite-reduced form contains the same number of clusters in their one-electron-reduced forms (all are $S = 1/2$, paramagnetic)

diamagnetic Fe-S clusters. Application of high magnetic fields (> 0.1 T) is mostly employed to verify the diamagnetism of the ground state and obtain the complete set of magnetic parameters. However, measurements at strong magnetic fields require specialized Mössbauer cryostats equipped with superconducting magnets and are often not required for the basic characterization of common Fe-S clusters. Instead, the use of a weak permanent magnet is less experimentally demanding and sufficient to lift the degeneracy of nuclear states and help distinguish between paramagnetic and diamagnetic Fe-S clusters. The applied field can also help identify adventitiously bound Fe that is not assembled in Fe-S clusters (Fig. 1, *see* **Note 14**).

3.3.3 Velocity Scale

The velocity scale represents the range of energies scanned by the spectrometer. Generally, transitions of ^{57}Fe atoms have energies within the $\pm$ 12 mm/s range, but those typically observed for Fe-S clusters are within $\pm$ 8 mm/s [1, 6, 7, 29]. Larger velocity scales will increase acquisition times and decrease resolution; therefore, velocity scales are chosen on the basis of maintaining spectral resolution and capturing all signals in the minimum amount of time (Fig. 4). Typically, if the type of Fe-S clusters in the sample is unknown or paramagnetic Fe-S clusters are present, scales of ± 8 mm/s (or greater) are initially chosen. A smaller scale (± 4 mm/s) is employed for diamagnetic compounds to obtain the best spectral resolution in the shortest acquisition time.

3.4 Types of Fe-S Clusters, Their Electronic Configurations, and Representative Mössbauer Fingerprints

Fe-S clusters have unique ^{57}Fe Mössbauer spectroscopic signatures that allow for their identification as well as characterization of their electronic and geometric structures, irrespective of their oxidation and spin state. The characteristic Mössbauer parameters of the most commonly occurring Fe-S centers are listed in Table 1. Fe-S clusters are assemblies of elemental iron and sulfide and can adopt a vast range of structures of variable complexity (Fig. 5). Only the inorganic part of their core is inserted in their notation and formal charge assignment. Their nomenclature can be simply viewed as $[x\text{Fe-}y\text{S}]^{z+}$, in which x is the number of Fe ions, y is the number of sulfide ions, and z is the overall charge accounting only for the Fe and S^{2-} constituents (and not the protein ligands). The individual Fe ions attain the formal valences Fe^{2+} and Fe^{3+} and adopt a high-spin electronic configuration due to coordination to “soft” ligands (e.g., S-, O-, and N-based).

3.4.1 [1Fe] Centers

The simplest form of an Fe-S cluster occurs in the so-called rubredoxins and consists of a single Fe ion in a tetrahedral geometry bound to the protein polypeptide by the thiols of four cysteine residues (Fig. 5). This is the building block of all higher-order Fe-S cluster structures. The as-isolated form of the rubredoxin-like [Fe] center is in the ferric form (Fe^{3+}, $S = 5/2$), and one-electron reduction results in the formation of the ferrous form (Fe^{2+}, $S = 2$). Both states are paramagnetic, and their spectra have been described in detail elsewhere [66, 67]. Typical Mössbauer parameters are listed in Table 1.

3.4.2 [2Fe-2S] Clusters

[2Fe-2S] clusters have a diamond-like core (Fig. 5) and are bound to the protein polypeptide by four amino acid residues. To date, three formal redox states of [2Fe-2S] clusters have been observed (i.e., 2+, 1+, and 0). On the basis of the chemical nature of their protein ligands, [2Fe-2S] clusters are classified into three subfamilies: Thioredoxin-like clusters (ligated by four cysteine residues), NEET-like clusters (ligated by three cysteines and one histidine), and Rieske clusters (one Fe is ligated to two cysteines and one Fe is ligated to two histidines).

The oxidized $[2\text{Fe-}2\text{S}]^{2+}$ form consists of an antiferromagnetically (AF) coupled high-spin diferric pair with a diamagnetic ($S = 0$) ground state. For a $[2\text{Fe-}2\text{S}]^{2+}$ cluster, two quadrupole doublets

Table 1
Mössbauer parameters of common Fe-S clusters

Type	Formal valences	S_{tot}	δ (mm/s)[a]	$\lvert\Delta E_Q\rvert$ (mm/s)[a]	*A*-values (A_x, A_y, A_z) (MHz)
$[1Fe]^{3+}_{rubredoxin}$ [66]	Fe^{3+}	5/2	0.32	0.5	−22.6, −21.8, −23.2
$[1Fe]^{3+}_{desulforedoxin}$ [67]	Fe^{3+}	5/2	0.25	0.75	−10.4, −10.4, −10.4
$[1Fe]^{2+}_{rubredoxin}$ [66]	Fe^{2+}	2	0.70	3.25	−27.6, −11.4, −41.3
$[1Fe]^{2+}_{desulforedoxin}$ [67]	Fe^{2+}	2	0.70	3.55	−27.6, −27.6, −9.3
$[2Fe\text{-}2S]^{2+}_{4Cys}$ [30]	2 Fe^{3+}	0	0.27	0.60	
$[2Fe\text{-}2S]^{2+}_{NEET}$ [68]	Fe^{3+} (Cys) Fe^{3+} (His)	0	0.26 0.30	0.47 0.96	
$[2Fe\text{-}2S]^{2+}_{Rieske}$ [31]	Fe^{3+} (Cys) Fe^{3+} (His)	0	0.24 0.32	0.52 0.91	
$[2Fe\text{-}2S]^{1+}_{4Cys}$ [30]	Fe^{3+} Fe^{2+}	1/2	0.35 0.60	0.65 2.70	−56, −50, −43 +14, +21, +35
$[2Fe\text{-}2S]^{1+}_{NEET}$ [32]	Fe^{3+} Fe^{2+}	1/2	0.32 0.68	1.07 3.15	−55.7, −46.7, −41.6 +29.5, +28.5, +9.1
$[2Fe\text{-}2S]^{1+}_{Rieske}$ [31]	Fe^{3+} Fe^{2+}	1/2	0.31 0.74	0.63 3.05	−55, −50, −43 +11, +14, +33
$[2Fe\text{-}2S]^{0}$ [33]	Fe^{2+} (Cys) Fe^{2+} (His)	0	0.70 0.81	2.76 2.32	
$[3Fe\text{-}4S]^{1+}_{cube}$ [36]	3 Fe^{3+}	1/2	0.27	0.63	−45, +20, +6
$[3Fe\text{-}4S]^{1+}_{linear}$ [69]	3 Fe^{3+}	5/2	0.3	n.d.	−19.2; −18.2; +13.9[b]
$[3Fe\text{-}4S]^{0}$ [35]	1 Fe^{3+} 2 $Fe^{2.5+}$	2	0.32 0.46	0.52 1.47	+13.7, +15.8, +17.3 −20.5, −20.5, −16.4
$[4Fe\text{-}4S]^{3+}$ [42]	2 Fe^{3+} 2 $Fe^{2.5+}$	1/2	0.29 0.40	0.88 1.03	+19.2, +22.4, +19.3 −28.2, −30.6, −32.6
$[4Fe\text{-}4S]^{2+}$ [43]	4 $Fe^{2.5+}$	0	0.42	1.12[c]	
$[4Fe\text{-}4S]^{2+}\text{-}_{SAM}$ [39, 40, 41]	2 $Fe^{2.5+}$ 1 Fe^{3+}-like 1 Fe^{2+}-like	0	0.42–0.48 0.34–0.40 0.64–0.80	0.97–1.21 0.86–1.10 1.15–1.62	
$[4Fe\text{-}4S]^{1+}$ [43]	2 Fe^{2+} 2 $Fe^{2.5+}$	1/2[c]	0.58 0.49	1.89 1.32	+26.5, +13.5, +8.6 −31.7, −32.6, −27.9
$[4Fe\text{-}4S]^{0}$ [70]	3 Fe^{2+} 1 Fe^{2+}	4	0.65 0.65	1.51 2.19	−14.2, −8.8, −10.0 +6.9, +10.7, +10.8

[a]Mössbauer parameters for spectra acquired at 4.2 K
[b]These values are represented as A_{iso}, the average of A_x, A_y, and A_z, for each Fe site
[c]Higher spin ground states, such as $S = 3/2$, 5/2, and 7/2, have been also observed, but are not as common

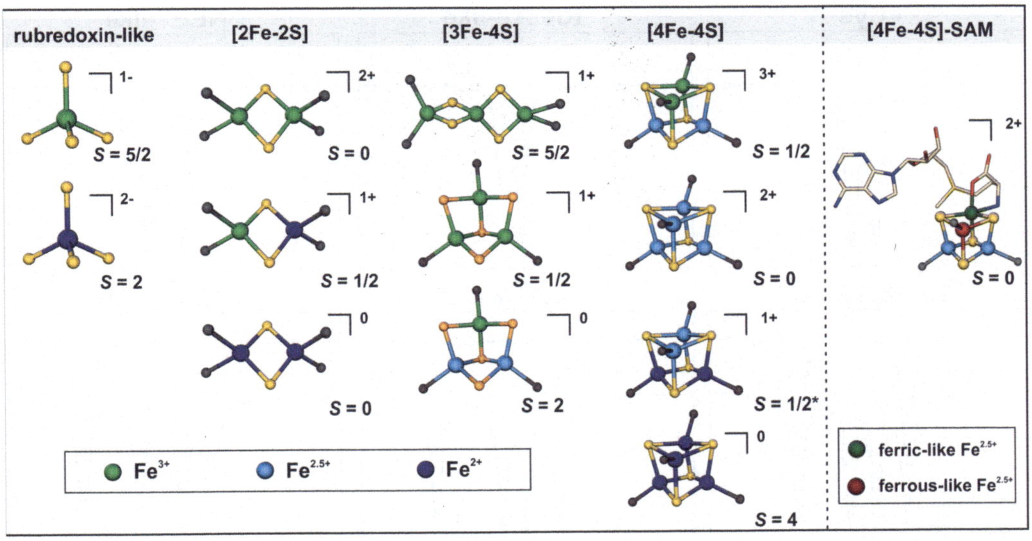

Fig. 5 Structures, oxidation states, Fe valences, and spin states of biologically relevant Fe-S clusters. *Higher spin ground states of $[4Fe\text{-}4S]^{1+}$ clusters, such as S = 3/2, 5/2, and 7/2 have been also observed, but are not as common

are expected (one for each Fe^{3+}), which may overlap or differ depending on the local environment of the Fe ions (Fig. 6). Thus, the Mössbauer spectra of thioredoxin-like $[2Fe\text{-}2S]^{2+}$ clusters with two chemically equivalent Fe ions exhibit one quadrupole doublet (superposition of two doublets) [30], whereas Mössbauer spectra of NEET-like and Rieske clusters with two Fe ions in unique chemical environments exhibit two distinct quadrupole doublets [31, 32].

The one-electron-reduced $[2Fe\text{-}2S]^{1+}$ form consists of a ferric and a ferrous Fe ion that is AF-coupled to yield a paramagnetic ground state ($S = 1/2$) [30]. At low temperatures (slow relaxation) and in the presence of a magnetic field, each Fe ion produces complex spectra due to hyperfine interactions [7]. The Mössbauer spectrum at high temperatures (fast relaxation) collapses into two quadrupole doublets, one for the Fe^{2+} and one for the Fe^{3+} center (*see* **Note 15**, Fig. 6).

The two-electron-reduced form, $[2Fe\text{-}2S]^0$, is not accessible in biological systems but has been demonstrated in Rieske clusters by reduction with a europium salt [33]. The two Fe^{2+} ions are AF-coupled, yielding a diamagnetic ($S = 0$) ground state (Table 1).

3.4.3 [3Fe-4S] Clusters

[3Fe-4S] clusters typically have a cuboidal structure (Fig. 5, *see* **Note 16**). To date, two formal redox states of [3Fe-4S] clusters have been observed (i.e., 1+ and 0). The oxidized $[3Fe\text{-}4S]^{1+}$ state consists of three high-spin Fe^{3+} ions coupled to yield a ground state of $S = 1/2$ as a result of "spin frustration" [34, 35]. In the slow

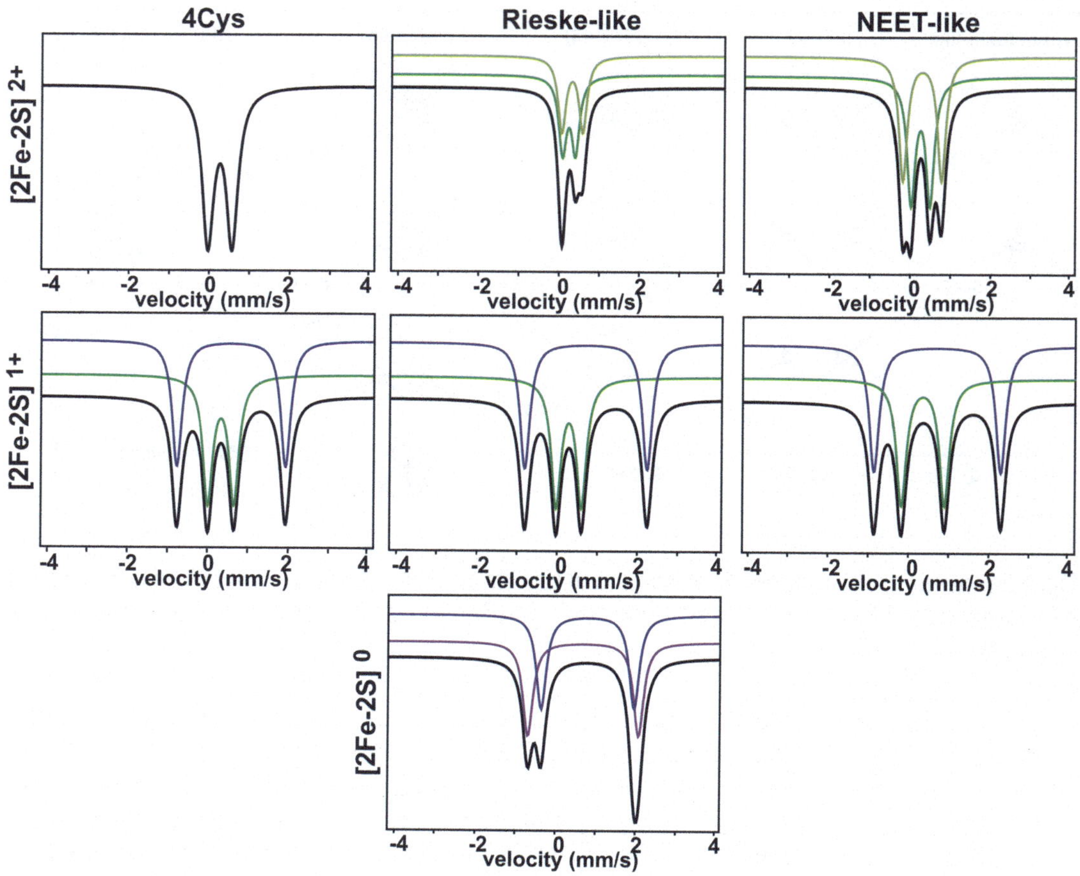

Fig. 6 Representative simulated ^{57}Fe Mössbauer spectra of [2Fe-2S] clusters in all three oxidation states and under conditions of fast relaxation (quadrupole doublet spectra) in a ± 4 mm/s velocity scale. In [2Fe-2S] clusters, differences in ligation state are reflected in site differentiation and distinct quadrupole doublets. Black lines represent the total (observed) ^{57}Fe Mössbauer spectrum. The individual quadrupole doublets corresponding to the ferric sites are shown with green lines, and those corresponding to the ferrous sites are shown with purple lines

relaxation limit, the Mössbauer spectrum of a $[3Fe\text{-}4S]^{1+}$ cluster exhibits a complex hyperfine structure [36], whereas in the fast relaxation limit, the spectrum collapses into a single quadrupole doublet (three overlapping ones) with parameters $\delta \sim 0.3$ mm/s and $\Delta E_Q \sim 0.53$ mm/s (Fig. 7). One-electron reduction yields the $[3Fe\text{-}4S]^0$ state that contains one Fe^{3+} ion and a valence-delocalized "ferrous-ferric" pair (*see* **Note 17**) [6, 8, 35, 37]. The valence-delocalized pair of the $[3Fe\text{-}4S]^0$ cluster can be assigned the "observed" valences $Fe^{2.5+}\text{-}Fe^{2.5+}$ ($S = 9/2$) and is AF-coupled to the third Fe^{3+} ion ($S = 5/2$), resulting in a $S = 2$ ground state. At 4.2 K and in the absence of a magnetic field, the Mössbauer spectrum is a combination of two quadrupole doublets with intensity ratios of 2:1 for the $Fe^{2.5+}\text{-}Fe^{2.5+}$ pair and Fe^{3+} ion, respectively. The quadrupole doublet of the $Fe^{2.5+}\text{-}Fe^{2.5+}$ pair has

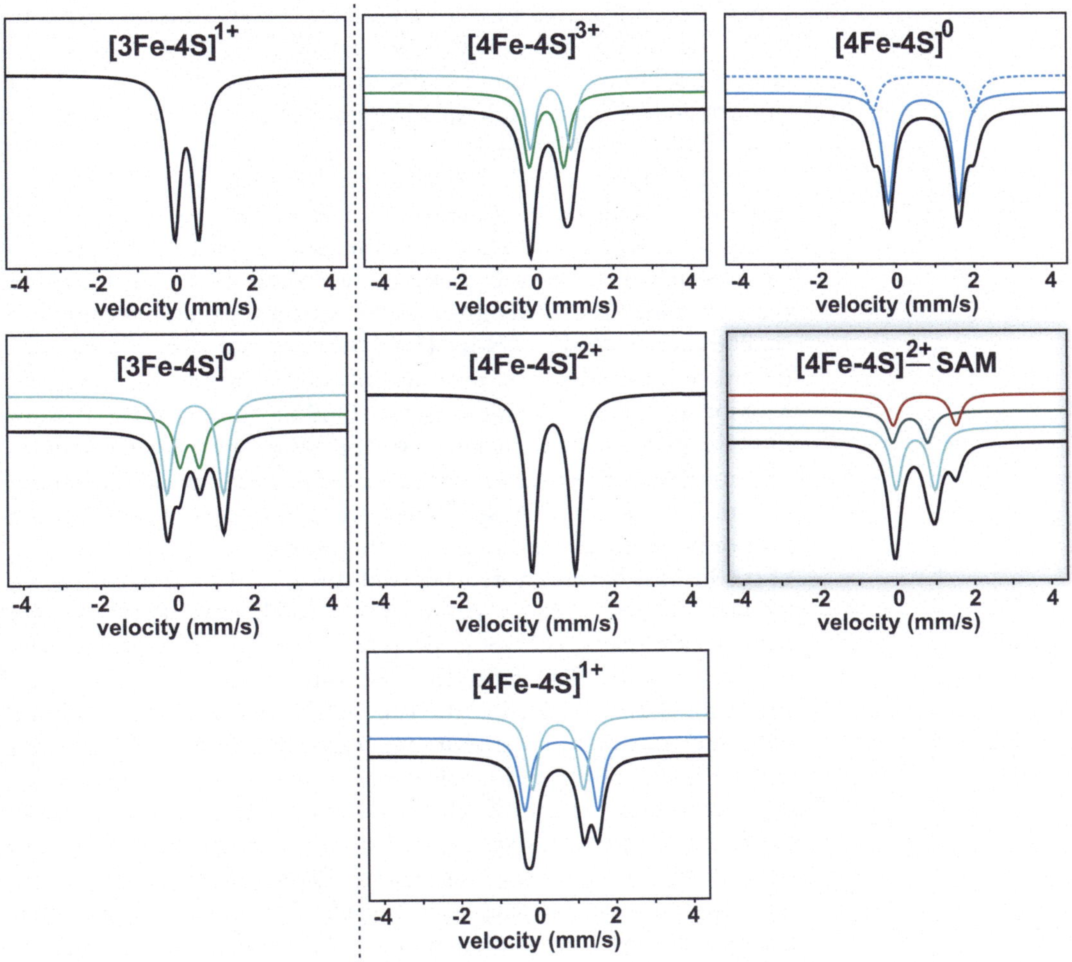

Fig. 7 Representative simulated Mössbauer spectra of [3Fe-4S] and [4Fe-4S] clusters at their various oxidation states. Spectra are under conditions of fast relaxation and in a $\pm$ 4 mm/s velocity scale. Black lines represent the total (observed) spectra. The subspectra from ferric sites are shown in green, ferrous sites shown in purple, and the mixed-valent $Fe^{2.5+}$ are shown in light blue. The ferric-like and ferrous-like sites observed in radical SAM clusters upon SAM binding are shown in dark green and red, respectively

$\delta \sim 0.46$ mm/s and $\Delta E_Q \sim 1.47$ mm/s, whereas the doublet of the Fe^{3+} site has $\delta \sim 0.32$ mm/s and $\Delta E_Q \sim 0.52$ mm/s (Table 1).

3.4.4 [4Fe-4S] Clusters

[4Fe-4S] clusters have four Fe ions arranged in a distorted tetrahedral cubane structure and attain four possible formal redox states (i.e., 3+, 2+, 1+, and 0) (Fig. 5). [4Fe-4S] clusters are classified into high-potential and low-potential centers depending on whether they operate in the 3+/2+ or the 2+/1+ redox couple, respectively. The central redox state for both types, $[4Fe\text{-}4S]^{2+}$, consists of two valence-delocalized $Fe^{2.5+}$-$Fe^{2.5+}$ pairs that are AF-coupled leading to a diamagnetic ($S = 0$) ground state [1, 6, 7]. The four $Fe^{2.5+}$ ions experience similar chemical environments and are thus spectroscopically indistinguishable with parameters $\delta \sim 0.42$ mm/s and $\Delta E_Q \sim 1.12$ mm/s. Therefore, the spectrum of a $[4Fe\text{-}4S]^{2+}$ cluster often manifests as a single quadrupole doublet instead of two or four (Table 1, *see* **Note 18**).

In contrast to [2Fe-2S] clusters, for [4Fe-4S] clusters it is sometimes challenging (and often impossible) to resolve small changes in the coordination environment. Typically, in [4Fe-4S] clusters in which the fourth Fe has a protein-based ligand other than cysteine, the differences in Mössbauer parameters are too small to be resolved (*see* **Note 19**). However, addition of substrates or small molecules can produce distinct features that are not Mössbauer "silent." Small molecule binding to one of the Fe sites of $[4Fe\text{-}4S]^{2+}$ clusters causes a soft "breaking" of the valence delocalization in one of the two mixed-valent pairs as a result of changes in coordination number, symmetry, etc. The observed outcome is a pronounced site differentiation in the Mössbauer spectra with one of the Fe ions having more ferric and the other Fe ion having more ferrous character [38]. The spectra of such $[4Fe\text{-}4S]^{2+}$ clusters consist of three quadrupole doublets (instead of one), and the relative contribution of the four Fe ions to the total area of the spectrum is 2:1:1 reflecting the two $Fe^{2.5+}$, one Fe^{2+}-like, and one Fe^{3+}-like sites. The most common example of site-differentiated clusters is that observed in radical *S*-adenosyl-L-methionine (SAM)-dependent enzymes, in which the bidentate coordination of SAM via the amide and carboxylic groups to one of the Fe sites produces these distinct spectral features (Fig. 7, Table 1) [38–41].

The one-electron-oxidized $[4Fe\text{-}4S]^{3+}$ form of high-potential clusters consists of a diferric pair AF-coupled to a valence-delocalized $Fe^{2.5+}$-$Fe^{2.5+}$ pair, resulting in an $S = 1/2$ ground state [42]. The presence of the diferric subcore in the $[4Fe\text{-}4S]^{3+}$ leads to a decrease in the average isomer shift and quadrupole coupling parameters, which can be seen in the high-temperature spectra as a shift toward lower energies (i.e., velocities) with respect to those of $[4Fe\text{-}4S]^{2+}$ clusters (Fig. 7). At 4.2 K and in the presence of a magnetic field, a complex magnetically split spectrum is observed. Characteristic parameters for the Fe^{3+}-Fe^{3+} pair are $\delta \sim 0.29$ mm/s and $\Delta E_Q \sim 0.88$ mm/s and $\delta \sim 0.40$ mm/s and $\Delta E_Q \sim 1.03$ mm/s for the mixed-valent $Fe^{2.5+}$-$Fe^{2.5+}$ pair (Table 1).

The one-electron-reduced $[4Fe\text{-}4S]^{1+}$ form of low-potential clusters consists of one diferrous pair AF-coupled to a valence-delocalized $Fe^{2.5+}$-$Fe^{2.5+}$ pair, yielding an $S = 1/2$ ground state (*see* **Note 20**). The presence of the diferrous subcore in $[4Fe\text{-}4S]^{1+}$ clusters is seen in the high-temperature spectra as a shift toward higher energies (i.e., velocities) as a result of an increase in the average isomer shift and quadrupole coupling parameters for the diferrous pair (with respect to those of the mixed-valent pair). At 4.2 K and in the presence of a magnetic field, the spectrum is a convolution of two magnetically split species of equal intensity ratios, one for the ferrous pair and one for the mixed-valent pair [43]. Representative isomer shift values are $\delta \sim 0.58$ mm/s for the Fe^{2+}-Fe^{2+} pair and $\delta \sim 0.49$ mm/s for the $Fe^{2.5+}$-$Fe^{2.5+}$ pair (Table 1, *see* **Note 21**).

In some proteins, the two-electron-reduced all ferrous $[4Fe\text{-}4S]^0$ state of low-potential clusters is attainable by treatment with strong reducing agents [44, 45]. In this form, the $S = 4$ ground state results from AF-coupling of one Fe^{2+} ion to the remaining three Fe^{2+} ions (Table 1). The Mössbauer spectrum consists of two components in a 3:1 intensity ratio with parameters $\delta = 0.71$ mm/s and $\Delta E_Q = 1.80$ mm/s (75%) and $\delta = 0.71$ mm/s and $\Delta E_Q = 2.60$ mm/s (25%), which are representative of high-spin $Fe^{2+}S_4$ sites [29].

3.4.5 Other Clusters

Fe-S clusters are cofactors of extraordinary structural plasticity and can incorporate additional protein-based elements to extend their redox capabilities or adopt more complex structures of higher nuclearities beyond the four classical types discussed in Subheadings 3.4.1–3.4.4. Despite their unusual structures or higher nuclearity, their Mössbauer parameters resemble those of the simpler clusters (*see* **Note 22**).

3.5 Applications of Mössbauer Spectroscopy

3.5.1 Determination of Fe-S Cluster Stoichiometry—"Counting" Fe-S Clusters

Mössbauer spectroscopy, when combined with elemental analysis such as quantification of elemental iron and sulfide, allows for the absolute quantification of the Fe-S clusters. The amount of inorganic sulfide can easily be measured via a colorimetric assay by monitoring methylene blue formation and has been described in detail elsewhere [22]. Iron can be quantified by two common methods: the ferrozine assay and inductively coupled plasma atomic emission spectrometry (ICP-AES). Although ICP-AES detects all metal ions independent of oxidation state, quantification using the colorimetric ferrozine assay does not rely on expensive instrument time and results can be quickly obtained (~20 min) [11, 46].

The presence, type, and number of Fe-S clusters in a protein cannot be solely inferred on the basis of amino acid sequence. In this respect, pairing elemental analysis with Mössbauer spectroscopy is an unambiguous method for assigning stoichiometry and type of Fe-S centers in a protein [7]. This approach has been successfully applied for radical-SAM enzymes that often coordinate extra auxiliary clusters in addition to the SAM-binding cluster (e.g., LipA, MftC, etc.) [23, 47]. A typical approach for "counting" the number of Fe-S clusters is outlined below (*see* **Note 23**).

1. Estimate the concentration of elemental Fe and sulfide by employing the colorimetric assays referenced above.

2. Ensure that your protein is > 95% homogeneous and pure. Determine protein concentration (*see* **Note 24**).
3. The Fe and sulfide should ideally be in 1:1 stoichiometry except in the case of [3Fe-4S] or more complex clusters. Excess Fe with respect to sulfide may suggest that not all Fe is assembled in Fe-S clusters (*see* **Notes 8** and **13**).
4. The following formula can be used to calculate the number of clusters per protein:

$$\# \text{ of clusters} = ([\text{Fe}] \times p/[\text{protein}])/n,$$

in which p is the fractional percentage (1 for 100%), n is the cluster nuclearity (e.g., $n = 4$ for a [4Fe-4S] cluster).

Numerical example 1. Assume [protein] = 100 μM, [Fe] = 600 μM, and the Mössbauer spectrum is a "clean" quadrupole doublet with parameters characteristic of [4Fe-4S] clusters, which means that the spectrum corresponds to 100% of the absorption. The number of [4Fe-4S] clusters is therefore $(600 \times 1/100)/4 = 1.5$ [4Fe-4S] clusters.

Numerical example 2. Assume the Mössbauer spectrum consists of two doublets, one with characteristic values for a $[2Fe\text{-}2S]^{2+}$ and one for a $[4Fe\text{-}4S]^{2+}$ cluster in 1:2 ratio. Assume [protein] = 100 μM, [Fe] = 600 μM. Employing the formula from 4 and assuming a fractional percentage for [2Fe-2S] 0.33 and [4Fe-4S] 0.66, we can calculate the following number of clusters per protein:

$$\begin{aligned}\# \text{ of } [2\text{Fe} - 2\text{S}] \text{ clusters} &= (600 \times 0.33)/100/2 \\ &= 0.99\ [2\text{Fe} - 2\text{S}] \text{ clusters}\end{aligned}$$

and

$$\begin{aligned}\# \text{ of } [4\text{F e} - 4\text{S}] \text{ clusters} &= (600 \times 0.66)/100/4 \\ &= 0.99\ [4\text{Fe} - 4\text{S}] \text{ clusters}\end{aligned}$$

Therefore, the protein contains one [2Fe-2S] and one [4Fe-4S] cluster. In these examples, the spectra were free from any adventitiously bound Fe or any other contaminating species, which may complicate analysis and lower the confidence in Fe-S cluster "counting" (*see* **Notes 5** and **8**).

3.5.2 Fe-S Cluster Interconversions and Identification of Multiple Cluster Binding Sites

Fe-S clusters in proteins can be modular and adopt different nuclearities upon oxidative, reductive, and reconstitution processes, and such interconversions in their structure can be uniquely identified by Mössbauer spectroscopy [8, 10, 48]. This is particularly important if both forms are EPR-silent and have overlapping spectroscopic signals in UV/VIS or other techniques. Some seminal examples consist of (1) the global transcriptional regulator FNR, in which sensing of O_2 involves the transformation of a $[4Fe\text{-}4S]^{2+}$ cluster to a $[2Fe\text{-}2S]^{2+}$ cluster [49], and (2) the scaffold proteins

IscU and IscA, in which treatment with a reductant leads to conversion of the $[2Fe\text{-}2S]^{2+}$ cluster to a $[4Fe\text{-}4S]^{2+}$ cluster [24, 26]. Other proteins can coordinate multiple clusters in distinct binding sites. For example, Mössbauer spectroscopy revealed two distinct cluster binding sites in biotin synthase (BS) and CIAPIN1 [50, 51].

3.5.3 Trapping of Reaction "Intermediates" and Selectively Labeling Fe Sites of Interest

Mössbauer can also provide insight into substrate binding, redox-state changes, and Fe-S cluster interconversions occurring within a subsecond timescale. Intermediate states in a reaction pathway are "trapped" in frozen solutions by utilizing freeze-quench techniques. One such example is the elucidation of the role of the auxiliary cluster in LipA, which serves as a two sulfur atom donor and leads to its sacrificial disassembly after substrate turnover [39]. In other cases, one can specifically label the Fe site of interest with ^{57}Fe, whereas all other Fe ions are ^{56}Fe-labeled, rendering them Mössbauer-silent [38]. This allows for elucidation of the electronic structure of labile Fe sites in Fe-S clusters that participate in chemistry.

3.5.4 Mössbauer on ^{57}Fe-Enriched Whole Cells

In some cases, the chemical nature of Fe-S cofactors must be interrogated in vivo to assess the biological relevance of in vitro studies, obviate any artifactual states introduced during purification, or if the protein of interest is insoluble. Under these circumstances, Mössbauer can be performed on intact (whole) cells using the following protocol [52–54].

1. Prepare ^{57}Fe-enriched whole cells that express the empty parent vector and the vector containing the gene of interest (*see* **Note 25**).
2. Directly transfer the same quantity of cell pellet to the Mössbauer cup for both samples (*see* **Note 26**).
3. Record Mössbauer spectra on both samples and subtract the spectrum of the parent vector to obtain the contribution of the protein of interest.

4 Notes

1. ^{57}Fe is often obtained as fine metal powder and needs to be dissolved either in HCl or H_2SO_4 to yield the respective ferrous salt. The reaction of elemental Fe with H_2SO_4 produces $FeSO_4$ and H_2. Fe dissolution should be performed with dilute amounts of acid to prevent the formation of ferric salts and alteration of the pH of the growth media or reaction solutions [55].

2. A typical Mössbauer cell (also referred to as holder or cup) is cylindrical (12.3 mm OD, 10 mm height), made of delrin or nylon, and requires a volume of approximately 500 μL. However, it can also be tapered on the inside to afford the same sample thickness with less volume (300–400 μL). After the transfer of the sample to the cell, the solution should be frozen by *slowly* submerging the cell into liquid nitrogen to avoid formation of a "peak" at the center of the sample.
3. There is no restriction in the buffer composition of the samples (salt, pH, glycerol, additives).
4. Reconstitutions are typically performed with dilute protein solutions (10–100 μM); however, some are more successful at higher protein concentrations. At high sample concentrations, care should be taken that there is no aggregation leading to the formation of biologically irrelevant states. Centrifuge the sample to remove any precipitated protein or particles. Protein precipitates will affect line shape (in addition to introducing nonrelevant Fe species in the spectra).
5. Assembly of Fe-S clusters is in direct competition with undesirable formation of polysulfide-iron aggregates that often appear as black precipitates. Chemical methods of reconstitution circumvent this issue by gradually adding iron and sulfide. For a slower and more controlled release of sulfide, reconstitution is performed utilizing IscS desulfurase, which produces sulfide from L-cysteine.
6. If there is precipitation upon addition of iron or sulfide, add less concentrated solutions (1–10 mM) or try modifying the order in which the iron and sulfide are added (e.g., add sulfide first or alternate adding iron and sulfide).
7. This step is suggested to achieve a crude cleanup of the sample and remove any particulates that decrease resolution of the next chromatographic step and overall performance of the analytical column.
8. Iron can also be adventitiously bound to the protein of interest and not in the form of Fe-S clusters. Non-Fe-S cluster-bound Fe can exist either in the high-spin ferrous or ferric forms, which often complicate analyses and cluster counting. Typical parameters for adventitious Fe^{2+} are $\delta \sim 1.15$ mm/s and $\Delta E_Q \sim 3.2$ mm/s and for adventitious Fe^{3+} are $\delta \sim 0.45$ mm/s and $\Delta E_Q \sim 0.8$ mm/s [56].
9. Enzymatic transfer of an intact Fe-S cluster by the Fe-S scaffold protein IscU has been successful for [2Fe-2S] and [4Fe-4S] clusters [57, 58]. For more complex cases, such as NifB, reconstitution with preassembled, chemically synthesized [4Fe-4S] clusters has been successful [21, 59, 60].

10. The Mössbauer velocity scale is calibrated relative to the centroid of the magnetically spit spectrum of α-iron, which is considered zero velocity [3].
11. Removal of the external magnetic field often simplifies paramagnetic spectra; however, some Fe compounds will exhibit paramagnetic splitting even in the absence of an external field depending on the strength of the local magnetic field exerted by the unpaired electron density.
12. Conversion factors for hyperfine constant (A) units for the ground state ($I = 1/2$) of ^{57}Fe.
    (a) 1 Tesla = 10,000 Gauss (G) or 10 kG
    (b) 1 Tesla = 0.7238 MHz
13. The Mössbauer spectrum is a weighted superposition of the individual subspectra of the different Fe sites in the sample. The relative area of each subspectrum is proportional to the relative concentration of the corresponding Fe center. At 4.2 K, Mössbauer spectroscopy can be used more reliably in a quantitative manner due to the similarity of f factors at this temperature.
14. The presence of high-spin Fe^{3+} manifests itself as broad paramagnetic features and lowers the confidence of Fe-S cluster quantifications.
15. The caveat of high-temperature measurements is that they cannot distinguish between an Fe^{3+} in a mononuclear center or an Fe^{3+} in a polynuclear cluster. Therefore, an unequivocal determination of the chemical nature of the distinct Fe species contributing to the spectra demands that spectra are also collected at 4.2 K and in the absence and presence of a magnetic field.
16. "Linear" [3Fe-4S] cluster forms are also possible and are afforded by a rearrangement allowing for the central iron to be ligated to four sulfides (Fig. 5). Linear $[3Fe\text{-}4S]^{1+}$ clusters have an $S = 5/2$ ground state.
17. Delocalization can be simply viewed as an extended sharing of the extra electron between the two Fe ions, resulting in the observed $Fe^{2.5+}$-$Fe^{2.5+}$ valences and is a common feature in [3Fe-4S] and [4Fe-4S] clusters. Parameters for the $Fe^{2.5+}$ are intermediary between those of Fe^{3+} and Fe^{2+}.
18. In practice, each of the Fe ions of the $[4Fe\text{-}4S]^{2+}$ experiences a slightly different chemical environment (second sphere interactions, hydrogen bonding, etc.). Thus, the two resonance lines of the single doublet observed are often asymmetric both in intensity and linewidth.
19. Greater resolution and spectral separation may be attained by recording spectra at higher magnetic fields (i.e., 5 T), which

may allow for distinguishing the Fe site that has a protein-based ligand other than a cysteine residue. Some examples of such [4Fe-4S] clusters are (1) the distal [4Fe-4S] cluster in [Ni-Fe] hydrogenases in which the fourth Fe is coordinated by a histidine [61] and (2) lipoyl synthase (LipA) in which the fourth Fe is coordinated by an aspartate [62].

20. Higher-spin ground states (i.e., $S = 3/2$, $5/2$, and $7/2$) for $[4Fe\text{-}4S]^{1+}$ clusters are possible as a result of perturbations in cluster coordination or distortions in local geometry [63, 64]. If experimentally available, one should carry out measurements at higher magnetic fields (i.e., 5 T) to assess the exact electronic configuration and spin state.

21. In the absence of a magnetic field, a quadrupole doublet spectrum is usually expected for a $[4Fe\text{-}4S]^{1+}$ cluster. Measurements both in the presence and absence of a small applied magnetic field help ascertain the paramagnetism, while taking the difference spectrum allows analysis of only the paramagnetic components.

22. [4Fe-4S] clusters have electronic properties majorly influenced by valence delocalization (double exchange) and exhibit enormous plasticity, such that in the presence of structural changes, the Fe ions rearrange themselves to preserve this electronic configuration. Therefore, [4Fe-4S] clusters incorporating additional metal ions or adopting distorted or higher nuclearity structures greatly maintain the Mössbauer and hyperfine parameters of prototypical [4Fe-4S] clusters.

23. A widely used program for fitting and simulating Mössbauer spectra is the WMOSS software (http://www.wmoss.org), which is freely available. The parameters collected in Table 1 can be employed as an initial set of parameters for analyzing and interpreting Mössbauer data.

24. Fe-S clusters in one or more oxidation states can have distinct optical spectra, originating from S-*to*-Fe charge transfer transitions that can aid in their identification [65]. Most clusters have a broad absorption peak around ~310 nm that can interfere with the protein absorbance at 280 nm. Thus, concentrations of Fe-S cluster-containing protein samples may be estimated by the Bradford assay or amino acid analysis.

25. Mössbauer can detect all endogenous ^{57}Fe-containing proteins in addition to the overexpressed protein of interest.

26. Typically, the whole cell sample should fill less than half of the Mössbauer cup due to the high concentration of ^{57}Fe, and high-quality Mössbauer spectra can be recorded within a few hours [54].

References

1. Münck E (2000) In: Que LJ (ed) Physical methods in bioinorganic chemistry. Spectroscopy and magnetism. University Science Books, Mill Valley, pp 287–319
2. Martinho M, Münck E (2010) ^{57}Fe Mössbauer spectroscopy in chemistry and biology. In: Bakac A (ed) Physical inorganic chemistry: Principles, methods, and models, John Wiley & Sons, Ltd, Hoboken, New Jersey, pp 39–67
3. Gütlich P, Bill E, Trautwein A, X. (2011) Mössbauer spectroscopy and transition metal chemistry. Fundamentals and Applications. John Wiley & Sons, Ltd, Hoboken, New Jersey
4. Greenwood NN, Gibb TC (1971) Moessbauer spectroscopy. Chapman and Hall, London
5. Bill E (2013) Chapter 6—57Fe-Mössbauer spectroscopy and basic interpretation of Mössbauer parameters. In: Crichton RR, Louro RO (eds) Practical approaches to biological inorganic chemistry, 2nd edn. Elsevier, Amsterdam, pp 201–228
6. Schünemann V, Winkler H (2000) Structure and dynamics of biomolecules studied by Mössbauer spectroscopy. Rep Prog Phys 63:263–353
7. Pandelia M-E, Lanz ND, Booker SJ et al (2015) Mössbauer spectroscopy of Fe/S proteins. Biochim Biophys Acta 1853:1395–1405
8. Beinert H, Holm RH, Münck E (1997) Iron-Sulfur Clusters: Nature's Modular, Multipurpose Structures. Science 277:653
9. Johnson DC, Dean DR, Smith AD et al (2005) Structure, FUNCTION, and formation of biological iron-sulfur clusters. Annu Rev Biochem 74:247–281
10. Huynh BH (2011) Mossbauer spectroscopy. In: Ribbe MW (ed) Meth Enzymol. Springer Science and Business Media, Berlin/Heidelberg
11. Bollinger J (1993) On the chemical mechanism of assembly of the tyrosyl radical- dinuclear iron cluster cofactor of *E. coli* ribonucleotide reductase. Massachusetts Institute of Technology, Cambridge, MA
12. Fultz B, Morris JW (1981) The thickness distortion of ^{57}Fe backscatter Mössbauer spectra. Nuc Inst Meth Phys Res 188:197–201
13. Albrecht AG, Netz DJA, Miethke M et al (2010) SufU is an essential iron-sulfur cluster scaffold protein in Bacillus subtilis. J Bacteriol 192:1643–1651
14. Corless EI, Mettert EL, Kiley PJ et al (2020) Elevated expression of a functional Suf pathway in Escherichia coli BL21(DE3) enhances recombinant production of an iron-sulfur cluster-containing protein. J Bacteriol 202: e00496–e00419
15. Johnson DC, Unciuleac M-C, Dean DR (2006) Controlled expression and functional analysis of iron-sulfur cluster biosynthetic components within Azotobacter vinelandii. J Bacteriol 188:7551
16. Akhtar MK, Jones PR (2008) Deletion of iscR stimulates recombinant clostridial Fe-Fe hydrogenase activity and H2-accumulation in Escherichia coli BL21(DE3). Appl Microbiol Biotechnol 78:853–862
17. Outten FW, Wood MJ, Muñoz FM et al (2003) The SufE protein and the SufBCD complex enhance SufS cysteine Desulfurase activity as part of a sulfur transfer pathway for Fe-S cluster assembly in *Escherichia coli*. J Biol Chem 278:45713–45719
18. Hänzelmann P, Hernández HL, Menzel C et al (2004) Characterization of MOCS1A, an oxygen-sensitive iron-sulfur protein involved in human molybdenum cofactor biosynthesis. J Biol Chem 279:34721–34732
19. Kuchenreuther JM, Grady-Smith CS, Bingham AS et al (2010) High-yield expression of heterologous [FeFe] hydrogenases in *Escherichia coli*. PLoS One 5:e15491
20. Lanz ND, Grove TL, Gogonea CB, Lee K-H, Krebs C, Booker SJ (2012) RlmN and AtsB as models for the overproduction and characterization of radical SAM proteins. Methods Enzymol 516:125–152
21. Mccarthy EL, Booker SJ (2017) Destruction and reformation of an iron-sulfur cluster during catalysis by lipoyl synthase. Science 358:373–377
22. Mccarthy EL, Booker SJ (2018) Biochemical approaches for understanding iron-sulfur cluster regeneration in *Escherichia coli* Lipoyl synthase during catalysis. Methods Enzymol 606:217–239
23. Cicchillo RM, Lee K-H, Baleanu-Gogonea C et al (2004) *Escherichia coli* Lipoyl synthase binds two distinct [4Fe–4S] clusters per polypeptide. Biochemistry 43:11770–11781
24. Agar JN, Krebs C, Frazzon J et al (2000) IscU as a scaffold for iron–sulfur cluster biosynthesis: sequential assembly of [2Fe-2S] and [4Fe-4S] clusters in IscU. Biochemistry 39:7856–7862
25. Freibert S-A, Weiler BD, Bill E et al (2018) Chapter seven—biochemical reconstitution and spectroscopic analysis of iron–sulfur proteins. In: David SS (ed) Methods enzymol.

Academic Press, Cambridge, Massachusetts, pp 197–226

26. Mapolelo DT, Zhang B, Naik SG et al (2012) Spectroscopic and functional characterization of iron–sulfur cluster-bound forms of *Azotobacter vinelandii* NifIscA. Biochemistry 51:8071–8084
27. Münck E (1978) Mössbauer spectroscopy of proteins: electron carriers. Methods Enzymol 20:346–379
28. Wittenborn EC, Merrouch M, Ueda C et al (2018) Redox-dependent rearrangements of the NiFeS cluster of carbon monoxide dehydrogenase. Elife 7:e39451
29. Bill E (2012) Iron-sulfur clusters—new features in enzymes and synthetic models. Hyperfine Interact 205:139–147
30. Muenck E, Debrunner PG, Tsibris JCM et al (1972) Moessbauer parameters of putidaredoxin and its selenium analog. Biochemistry 11:855–863
31. Fee JA, Findling KL, Yoshida T et al (1984) Purification and characterization of the Rieske iron-sulfur protein from Thermus thermophilus. Evidence for a [2Fe-2S] cluster having non-cysteine ligands. J Biol Chem 259:124–133
32. Golinelli-Cohen M-P, Lescop E, Mons C et al (2016) Redox control of the human iron-sulfur repair protein MitoNEET activity via its iron-sulfur cluster. J Biol Chem 291:7583–7593
33. Leggate EJ, Bill E, Essigke T et al (2004) Formation and characterization of an all-ferrous Rieske cluster and stabilization of the [2Fe-2S] core by protonation. Proc Natl Acad Sci U S A 101:10913
34. Huynh BH, Moura JJ, Moura I et al (1980) Evidence for a three-iron center in a ferredoxin from *Desulfovibrio gigas*. Mössbauer and EPR studies. J Biol Chem 255:3242–3244
35. Papaefthymiou V, Girerd JJ, Moura I et al (1987) Moessbauer study of *D. gigas* ferredoxin II and spin-coupling model for Fe3S4 cluster with valence delocalization. J Am Chem Soc 109:4703–4710
36. Kent TA, Huynh BH, Münck E (1980) Iron-sulfur proteins: spin-coupling model for three-iron clusters. Proc Natl Acad Sci U S A 77:6574–6576
37. Noodleman L, Case DA, Mouesca JM et al (1996) Valence electron delocalization in polynuclear iron-sulfur clusters. J Biol Inorg Chem 1:177–182
38. Krebs C, Broderick WE, Henshaw TF et al (2002) Coordination of Adenosylmethionine to a unique iron site of the [4Fe-4S] of pyruvate Formate-Lyase activating enzyme: a Mössbauer spectroscopic study. J Am Chem Soc 124:912–913
39. Lanz ND, Pandelia M-E, Kakar ES et al (2014) Evidence for a catalytically and kinetically competent enzyme-substrate cross-linked intermediate in catalysis by lipoyl synthase. Biochemistry 53:4557–4572
40. Molle T, Moreau Y, Clemancey M et al (2016) Redox behavior of the S-Adenosylmethionine (SAM)-binding Fe–S cluster in Methylthiotransferase RimO, toward understanding dual SAM activity. Biochemistry 55:5798–5808
41. Cosper MM, Jameson GNL, Davydov R et al (2002) The [4Fe−4S]2+ cluster in reconstituted biotin synthase binds S-Adenosyl-l-methionine. J Am Chem Soc 124:14006–14007
42. Middleton P, Dickson DPE, Johnson CE et al (1980) Interpretation of the Mössbauer spectra of the high-potential iron protein from Chromatium. Eur J Biochem 104:289–296
43. Middleton P, Dickson DPE, Johnson CE et al (1978) Interpretation of the Mössbauer spectra of the four-iron ferredoxin from *Bacillus stearothermophilus*. Eur J Biochem 88:135–141
44. Angove HC, Yoo SJ, Burgess BK et al (1997) Mössbauer and EPR evidence for an all-ferrous Fe4S4 cluster with $S = 4$ in the Fe protein of Nitrogenase. J Am Chem Soc 119:8730–8731
45. Yoo SJ, Angove HC, Burgess BK et al (1999) Mössbauer and integer-spin EPR studies and spin-coupling analysis of the [4Fe-4S]0 cluster of the Fe protein from *Azotobacter vinelandii* Nitrogenase. J Am Chem Soc 121:2534–2545
46. Stookey LL (1970) Ferrozine---a new spectrophotometric reagent for iron. Anal Chem 42:779–781
47. Ayikpoe R, Ngendahimana T, Langton M et al (2019) Spectroscopic and electrochemical characterization of the Mycofactocin biosynthetic protein, MftC, provides insight into its redox flipping mechanism. Biochemistry 58:940–950
48. Moura JJ, Moura I, Kent TA et al (1982) Interconversions of [3Fe-3S] and [4Fe-4S] clusters. Mössbauer and electron paramagnetic resonance studies of Desulfovibrio gigas ferredoxin II. J Biol Chem 257:6259–6267
49. Khoroshilova N, Popescu C, Münck E et al (1997) Iron-sulfur cluster disassembly in the FNR protein of Escherichia coli [4Fe-4S] to [2Fe-2S] conversion with loss of biological activity. Proc Natl Acad Sci U S A 94:6087
50. Ugulava NB, Surerus KK, Jarrett JT (2002) Evidence from Mössbauer spectroscopy for distinct [2Fe-2S]2+ and [4Fe-4S]2+ cluster

binding sites in biotin synthase from *Escherichia coli*. J Am Chem Soc 124:9050–9051

51. Banci L, Ciofi-Baffoni S, Mikolajczyk M, Winkelmann J, Bill E, Pandelia M-E (2013) Human anamorsin binds [2Fe-2S] clusters with unique electronic properties. J Biol Inorg Chem 18(8):883–893
52. Cosper MM, Jameson GNL, Eidsness MK et al (2002) Recombinant *Escherichia coli* biotin synthase is a $[2Fe–2S]^{2+}$ protein in whole cells. FEBS Lett 529:332–336
53. Popescu CV, Bates DM, Beinert H et al (1998) Mössbauer spectroscopy as a tool for the study of activation/inactivation of the transcription regulator FNR in whole cells of *Escherichia coli*. Proc Natl Acad Sci U S A 95:13431
54. Vrajmasu VV, Bominaar EL, Meyer J et al (2002) Mössbauer study of reduced Rubredoxin as purified and in whole cells. Structural correlation analysis of spin Hamiltonian parameters. Inorg Chem 41:6358–6371
55. Huffman RE, Davidson N (1956) Kinetics of the ferrous iron-oxygen reaction in sulfuric acid solution. J Am Chem Soc 78:4836–4842
56. Emptage MH, Kent TA, Kennedy MC et al (1983) Mössbauer and EPR studies of activated aconitase: development of a localized valence state at a subsite of the [4Fe-4S] cluster on binding of citrate. Proc Natl Acad Sci U S A 80:4674–4678
57. Unciuleac M-C, Chandramouli K, Naik S et al (2007) In vitro activation of Apo-Aconitase using a [4Fe-4S] cluster-loaded form of the IscU [Fe–S] cluster scaffolding protein. Biochemistry 46:6812–6821
58. Wachnowsky C, Fidai I, Cowan JA (2016) Cytosolic iron-sulfur cluster transfer-a proposed kinetic pathway for reconstitution of glutaredoxin 3. FEBS Lett 590:4531–4540
59. Tanifuji K, Lee CC, Sickerman NS et al (2018) Tracing the 'ninth sulfur' of the nitrogenase cofactor via a semi-synthetic approach. Nat Chem 10:568–572
60. Barclay JE, Davies SC, Evans DJ et al (1999) Lattice effects in the Mössbauer spectra of salts of [Fe4S4{S(CH2)nOH}4]2−. Crystal structures of [PPh4]2[Fe4S4{S(CH2)nOH}4] (n=2, 3 and 4). Inorg Chim Acta 291:101–108
61. Dementin S, Belle V, Bertrand P et al (2006) Changing the ligation of the distal [4Fe4S] cluster in NiFe hydrogenase impairs inter- and intramolecular electron transfers. J Am Chem Soc 128:5209–5218
62. Mclaughlin MI, Lanz ND, Goldman PJ et al (2016) Crystallographic snapshots of sulfur insertion by lipoyl synthase. Proc Natl Acad Sci U S A 113:9446–9450
63. Duderstadt RE, Brereton PS, Adams MWW et al (1999) A pure S=3/2 [Fe4S4]+ cluster in the A33Y variant of *Pyrococcus furiosus* ferredoxin. FEBS Lett 454:21–26
64. Boll M, Fuchs G, Tilley G et al (2000) Unusual spectroscopic and electrochemical properties of the 2[4Fe-4S] ferredoxin of Thauera aromatica. Biochemistry 39:4929–4938
65. Lippard S, Berg J (1994) Principles of bioinorganic chemistry. University Science Books, Mill Valley
66. Schulz C, Debrunner PG (1976) Rubredoxin, a simple iron-sulfur protein : its spin hamiltoman and hyperfine parameters. J Phys Colloques 37:C6-153–C156-158
67. Moura I, Huynh BH, Hausinger RP et al (1980) Mössbauer and EPR studies of desulforedoxin from *Desulfovibrio gigas*. J Biol Chem 255:2493–2498
68. Ferecatu I, Gonçalves S, Golinelli-Cohen M-P et al (2014) The diabetes drug target MitoNEET governs a novel trafficking pathway to rebuild an Fe-S cluster into cytosolic Aconitase/iron regulatory protein 1. J Biol Chem 289:28070–28086
69. Kennedy MC, Kent TA, Emptage M et al (1984) Evidence for the formation of a linear [3Fe-4S] cluster in partially unfolded aconitase. J Biol Chem 259:14463–14471
70. Hans M, Buckel W, Bill E (2008) Spectroscopic evidence for an all-ferrous [4Fe–4S]0 cluster in the superreduced activator of 2-hydroxyglutaryl-CoA dehydratase from Acidaminococcus fermentans. J Biol Inorg Chem 13:563–574

Chapter 16

Biochemical Approaches to Probe the Role of the Auxiliary Iron-Sulfur Cluster of Lipoyl Synthase from *Mycobacterium Tuberculosis*

Vivian Robert Jeyachandran, Jay V. Pendyala, Erin L. McCarthy, Amie K. Boal, and Squire J. Booker

Abstract

Lipoic acid is an essential sulfur-containing cofactor used by several multienzyme complexes involved in energy metabolism and the breakdown of certain amino acids. It is composed of *n*-octanoic acid with sulfur atoms appended at C6 and C8. Lipoic acid is biosynthesized de novo in its cofactor form, in which it is covalently bound in an amide linkage to a target lysyl residue on a lipoyl carrier protein (LCP). The *n*-octanoyl moiety of the cofactor is derived from type 2 fatty acid biosynthesis and is transferred to an LCP to afford an octanoyllysyl amino acid. Next, lipoyl synthase (LipA in bacteria) catalyzes the attachment of the two sulfur atoms to afford the intact cofactor. LipA is a radical S-adenosylmethionine (SAM) enzyme that contains two [4Fe-4S] clusters. One [4Fe-4S] cluster is used to facilitate a reductive cleavage of SAM to render the highly oxidizing 5′-deoxyadenosyl 5′-radical needed to abstract C6 and C8 hydrogen atoms to allow for sulfur attachment. By contrast, the second cluster is the sulfur source, necessitating its destruction during turnover. In *Escherichia coli,* this auxiliary cluster can be restored after each turnover by NfuA or IscU, which are two iron-sulfur cluster carrier proteins that are implicated in iron-sulfur cluster biogenesis. In this chapter, we describe methods for purifying and characterizing LipA and NfuA from *Mycobacterium tuberculosis,* a human pathogen for which endogenously synthesized lipoic acid is essential. These studies provide the foundation for assessing lipoic acid biosynthesis as a potential target for the design of novel antituberculosis agents.

Key words Lipoic acid, *S*-Adenosylmethionine, Crystallography, Sulfur, Radicals, NfuA, Mössbauer spectroscopy

1 Introduction

1.1 Lipoic Acid

Lipoic acid is a redox-active cofactor used in at least five multienzyme complexes, wherein it is intimately involved in the oxidative decarboxylation of pyruvate, α-ketoglutarate, branched-chain α-ketoacids, and glycine, as well as the oxidative degradation of acetoin [1–4]. All of these complexes, except for acetoin dehydrogenase, are found in organisms from all domains of life, including in

Patricia C. Dos Santos (ed.), *Fe-S Proteins: Methods and Protocols*, Methods in Molecular Biology, vol. 2353, https://doi.org/10.1007/978-1-0716-1605-5_16,

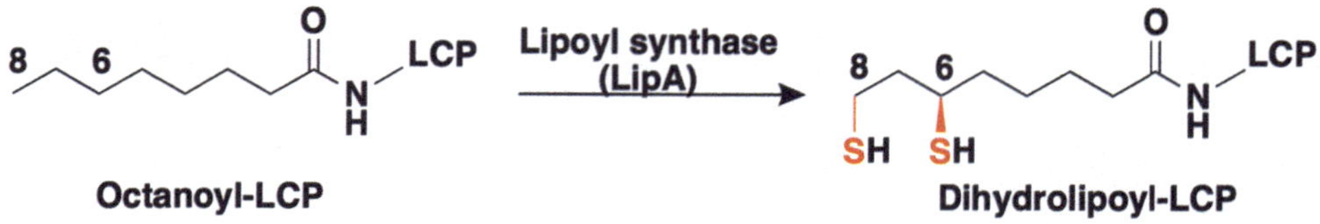

Fig. 1 The reaction catalyzed by lipoyl synthase (LipA). The inserted sulfur atoms are in red. LCP is lipoyl carrier protein

humans. Acetoin dehydrogenase is found exclusively in eubacteria [1–3]. Lipoic acid consists of an *n*-octanoic acid backbone with sulfur atoms at C6 and C8, with the stereochemical designation at C6 as *R* [5]. Very little free lipoic acid exists in cells; the majority is found in its cofactor form, in an N^6-amide linkage to a specific lysyl residue on a lipoyl carrier protein (LCP) [6, 7]. Moreover, lipoic acid is synthesized de novo in its cofactor form, rather than as free lipoic acid, in a two-step pathway that is an offshoot of type II fatty acid biosynthesis [8–10]. In the first step, catalyzed by octanoyl-transferase, an *n*-octanoyl group from octanoyl-acyl carrier protein, is transferred to the target lysyl residue of an LCP [11–13]. Next, lipoyl synthase, a radical S-adenosylmethionine (SAM) enzyme, catalyzes the attachment of sulfur atoms at C6 and C8 of the octanoyl chain to complete the biosynthetic pathway [14–16] (Fig. 1).

Many organisms, most notably *Escherichia coli* (*Ec*), also have a scavenging or salvage pathway that allows them to generate the lipoyl cofactor from free lipoic acid in a process that requires ATP. The first step of the scavenging pathway is the activation of lipoic acid by adenylation, which is then followed by displacement of adenosine monophosphate with concomitant formation of the amide linkage to give the lipoyl cofactor [17, 18]. However, the inability to synthesize the lipoyl cofactor de novo is embryonic lethal in mice and humans, and its deficiency leads to seizures, developmental delays, and death [19–21]. Lipoic acid is also an essential metabolite in *Mycobacterium tuberculosis (Mt)*. *Mt* possesses only the de novo biosynthetic pathway; it cannot scavenge lipoic acid from the environment [22, 23]. Therefore, the enzymes in the de novo pathway for lipoic acid biosynthesis are potential targets for the design of antituberculosis drugs [23].

1.2 Radical S-Adenosylmethionine Enzymes

The last step in the biosynthesis of lipoic acid, the attachment of sulfur atoms to completely unactivated aliphatic carbon centers, is catalyzed by lipoyl synthase (LipA), a radical SAM (RS) enzyme. RS enzymes compose a superfamily of structurally and mechanistically related metalloproteins that use SAM to initiate a multitude of radical-dependent transformations [24–26]. These enzymes all contain at least one [4Fe-4S] cluster, which is used to catalyze a reductive cleavage of SAM to generate *L*-methionine and a 5-′-deoxyadenosyl 5′-radical (5′-dA·). The 5′-dA· is a potent oxidant that initiates reactions typically by abstracting substrate hydrogen atoms. In 2018, the RS superfamily, designated protein family 04055 (PF04055), was reported to contain over 113,776 unique, nonredundant sequences, many of which encode enzymes with unknown functions [27]. The known chemical transformations catalyzed by RS proteins include methylation and methylthiolation of unactivated carbon and phosphorous centers, thiolation (i.e., sulfur insertion), cyclopropanation, 1,2-cross migrations, oxidations, and dehydrogenations, carbon-carbon bond formation and cleavage, oxidative decarboxylation, epimerization, and complex rearrangements, among others [19, 28–31].

A subset of RS enzymes attach sulfur atoms to unactivated carbon centers [32–35]. This subset includes lipoyl and biotin synthases, as well as the methylthiotransferases (MTTases), MiaB, MtaB, and RimO. Biotin synthase (BioB) catalyzes the insertion of a sulfur atom between carbons 6 and 9 of dethiobiotin to generate the thiophane ring of biotin [36–38]. LipA catalyzes the conversion of an octanoyllysyl-containing residue of an LCP to the lipoyl cofactor [14–16]. The MiaB subgroup of MTTases catalyzes the conversion of the hypermodified tRNA nucleoside N^6-(isopentenyl)adenosine-37 into 2-methylthio-N^6-(isopentenyl)-adenosine-37, while the MtaB family of MTTases catalyzes the conversion of N^6-(threonylcarbamoyl)adenosine-37 into 2-methylthio-N^6-(threonylcarbamoyl)adenosine-37 [39–42]. Lastly, RimO catalyzes the stereoselective attachment of a methylthio group to aspartyl 89 of protein S12 of the bacterial 30S subunit of the ribosome to afford a (*3R*)-methylthioaspartyl 89 group [43–47].

All characterized RS enzymes that catalyze sulfur attachment have at least two iron-sulfur (Fe-S clusters). One cluster, called the RS cluster, $[4Fe\text{-}4S]_{RS}$, is used to catalyze the reductive cleavage of SAM to generate the 5′-dA·, while a second, auxiliary cluster, $[4Fe\text{-}4S]_{aux}$, plays a direct role in sulfur attachment [34, 48, 49]. BioB contains a [2Fe-2S] auxiliary cluster [50–54], while LipA and the MTTases contain [4Fe-4S] auxiliary clusters [44, 47, 55–59]. A significant body of evidence indicates that the auxiliary clusters of LipA, BioB, and MiaB are degraded during catalysis to supply the attached sulfur atoms, which limits these proteins to catalyzing only a single turnover in the absence of factors to reconstitute the cluster [51, 58, 60–62].

To show that the auxiliary cluster of LipA is the source of the attached sulfur atoms during turnover, pathway intermediates have been trapped and then characterized spectroscopically and by X-ray crystallography. Electron paramagnetic resonance (EPR) and hyperfine sublevel correlation (HYSCORE) spectroscopies were used to provide evidence for a substrate-derived radical that is generated in close contact with an Fe-S cluster, presumably the $[4Fe\text{-}4S]_{aux}$ cluster of LipA [60]. Mössbauer spectroscopy was used to initially establish that the resting state of LipA contains two [4Fe-4S] clusters [56], and then to show that upon the initial sulfur attachment at C6, a $[3Fe\text{-}3S\text{-}1RS]^+$ intermediate is formed (RS indicates a C6 monothiooctanoyl ligand ligated to the cluster) [60]. Lastly, X-ray crystallography was used to allow for complete visualization of the active site in the resting state of the enzyme as well as at the C6-intermediate state, supporting the conclusions previously made by Mössbauer spectroscopy [55, 58]. Together, these approaches allowed for a detailed understanding of the structural changes that LipA undergoes during catalysis, culminating in the destruction of its $[4Fe\text{-}4S]_{aux}$ cofactor.

Recently, *Ec* NfuA, *Mt* NfuA, or IscU, all of which are Fe-S cluster carrier proteins, was shown to reconstitute the degraded auxiliary cluster in the *Ec* LipA reaction [63, 64]. This reconstitution was not rate limiting for overall catalysis, and enabled the enzyme to perform multiple turnovers. However, the pathway in other organisms for regenerating the auxiliary cluster in LipA has not been studied, and some organisms do not genetically encode NfuA. *Mt* encodes genes both for NfuA and LipA. As described above, *Mt* LipA has been crystallographically characterized with an intermediate bound, showing the formation of the $[3Fe\text{-}3S\text{-}1RS]^+$ complex, consistent with the use of the auxiliary cluster as a sulfur source. However, the impact of *Mt* NfuA on *Mt* LipA activity has not yet been demonstrated experimentally. In this chapter, we describe detailed methods for isolating and characterizing *Mt* LipA and *Mt* NfuA to enable a detailed biochemical and biophysical analysis of Fe-S cluster regeneration in this system.

2 Materials

2.1 Expression of Mt LipA and Mt NfuA

2.1.1 Cloning and Expression Vector Construction

1. *Mt lipA*: synthetic codon-optimized *Mt lipA* gene.
2. *Mt nfuA:* synthetic codon-optimized *Mt nfuA* gene.
3. pET28a: expression plasmid pET28a (*see* **Note 1**).
4. pDB1282: pAra13 plasmid containing the *Azotobacter vinelandii* isc operon.
5. Restriction enzymes: *Nde*I and *Eco*RI.
6. Competent cells: *Ec* BL21 (DE3) competent cells.

2.1.2 Overexpression of Mt LipA and Mt NfuA

1. LB: Lysogeny broth (LB).
2. M9: M9 minimal medium. Medium is made from a 20× stock of M9 salts pH 7.4, containing 136 g/L anhydrous Na_2HPO_4, 60 g/L KH_2PO_4, 10 g/L NaCl, 20 g/L NH_4Cl, 6 g/L

NaOH, 7 g/L KOH, 1 M $CaCl_2$, 1 M $MgSO_4$, and 20% D-Glucose.

3. IPTG: 1 M isopropyl-β-D-thiogalactopyranoside (IPTG).
4. Arabinose: 0.2% (*w/v*) arabinose.
5. Iron source: 500 mM $FeCl_3{\cdot}6H_2O$, or ^{57}Fe metal (98%), used exclusively for Mössbauer spectroscopy (*see* **Notes 2** and **3**).
6. Kanamycin solution: 50 mg/mL kanamycin.
7. Ampicillin solution: 100 mg/mL ampicillin.

2.2 Purification of Mt LipA and Mt NfuA

2.2.1 Affinity Chromatography

1. Special equipment: Coy vinyl anaerobic chamber with internal sonicator probe (*see* **Note 4**).
2. TALON Cobalt affinity column: 30 mL of 50% TALON Cobalt affinity resin slurry packed in a 2.5 × 10 cm chromatography column (*see* **Note 5**).
3. Lysis Buffer: 100 mM HEPES, pH 7.5, 300 mM KCl, 20 mM imidazole, 10 mM 2-mercaptoethanol, and 20% glycerol.
4. Elution Buffer: 100 mM HEPES, pH 7.5, 300 mM KCl, 250 mM imidazole, 10 mM 2-mercaptoethanol, and 20% glycerol (*see* **Note 6**).
5. DNAase: 1 mg/mL DNase.
6. Lysozyme: 100 mg/mL lysozyme.
7. PMSF: 1 mM phenylmethylsulfonyl fluoride (PMSF).

2.2.2 Chemical Reconstitution of Iron-Sulfur Clusters

1. Special equipment: Coy vinyl anaerobic chamber.
2. BSA: Bovine serum albumin standard solution.
3. Bradford reagent.
4. Iron source: 500 mM $FeCl_3{\cdot}6H_2O$ or ^{57}Fe metal (98%) to be used exclusively for Mössbauer spectroscopy.
5. Sulfur source: 500 mM $Na_2S{\cdot}9H_2O$.
6. DTT: 100 mM dithiothreitol (DTT) (*see* **Notes 8** and **9**).

2.2.3 Size Exclusion Chromatography

1. PD-10 desalting column (*see* **Note 10**).
2. Gel filtration setup: FPLC system inside a Coy vinyl anaerobic chamber fitted with a HiPrep 16/60 Sephacryl S-200 HR column.
3. S200 elution buffer: 100 mM HEPES, pH 7.5, 300 mM KCl, 5 mM DTT, and 15% glycerol (*see* **Notes 11** and **12**).

2.3 Characterization of Purified Mt LipA and Mt NfuA

2.3.1 UV-Visible Spectroscopy and Iron Analysis

1. Special equipment: UV-vis spectrophotometer.
2. ICP solution: 7% Nitric acid (*see* **Note 13**).
3. ICP-AES facility.

2.3.2 Analysis of Mt LipA Cluster Composition by Mössbauer Spectroscopy

1. 95% pure ^{57}Fe-labelled LipA protein
2. Special equipment: Mössbauer spectrometer (*see* **Note 14**).

2.3.3 Quantitative Measurement of Mt LipA Activity by LC-MS

1. Special equipment: Coy vinyl anaerobic chamber and mass spectrometer (MS).
2. Reaction buffer: 50 mM HEPES, pH 7.5, 200 mM KCl, and 8% (*v*/*v*) glycerol.
3. Protein: 1 mM *Mt* LipA.
4. Substrate: 6 mM peptide substrate (Glu-Ser-Val-[*N*6-octanoyl] Lys-Ala-Ala-Ser-Asp) (*see* **Note 15**).
5. SAH nucleosidase: 20 μM *S*-adenosylhomocysteine (SAH) nucleosidase.
6. SAM: 160 mM *S*-adenosylmethionine (SAM).
7. Reductant: 50 mM sodium dithionite (*see* **Note 16**).
8. Quench solution: 100 mM H_2SO_4, 20 μM AtsA (peptide internal standard), and 4 mM TCEP (tris(2-carboxyethyl) phosphine).
9. Peptide internal standard: 400 μM AtsA peptide (Pro-Met-Ser-Ala-Pro-Ala-Arg-Ser-Met) (*see* **Note 15**).
10. Quantification standards: 6 mM 6-thio-octanoyl peptide (Glu-Ser-Val-[N^6-6-thio-octanoyl] Lys-Ala-Ala-Ser-Asp) and 6 mM lipoyl peptide (Glu-Ser-Val-[N^6-lipoyl] Lys-Ala-Ala-Ser-Asp) (*see* **Note 15**).
11. Dilution mix: 50 mM HEPES, pH 7.5, 300 mM KCl, 10% glycerol, 50 mM H_2SO_4, 2 mM TCEP, and 20 μM AtsA internal standard.
12. MS solvent A: 0.1% formic acid in water.
13. MS solvent B: 100% acetonitrile.
14. C18 column: Zorbax Extend-C18 RRHD column (2.1 mm × 50 mm × 1.8 μm particle size).

2.3.4 Crystallization of Mt LipA

1. Special equipment: Coy vinyl anaerobic chamber.
2. Hanging drop crystal trays and cover slips.
3. Storage buffer: 10 mM HEPES, pH 7.5, 150 mM KCl, 10% (v/v) glycerol, and 5 mM DTT.

4. Protein solution: 7.3 mg/mL *Mt* LipA, 1 mM *E. coli* octanoyl-H peptide (sequence ESVK[octanoyl]AASD), and 0.9 mM SAM in storage buffer.
5. Crystallization solution: 0.1 M HEPES, pH 7.5, 12% (w/v) PEG 10000, 7.5% (v/v), and 2-methyl-2,4-pentanediol.
6. Reservoir solution: 0.5 M LiCl.
7. Cryoprotectant: Paratone-N.

2.3.5 Crystallization of Mt LipA Monothiolated Intermediate

1. Special equipment: Coy vinyl anaerobic chamber.
2. Protein: 1 mM *Mt* LipA.
3. Buffer: 50 mM HEPES, pH 7.5, 300 mM KCl, 10% (v/v) glycerol, 300 μM *Mt* octanoyl-H peptide, and 180 μM SAM.
4. 50 mM dithionite.
5. Hanging drop crystal trays, sitting drop crystal trays, and cover slips.
6. Intermediate solution: 20 mM HEPES, pH 7.5, and 200 mM NaCl.
7. Seed crystal solution: 0.1 M Bis-Tris, pH 6.5, 0.2 M Li_2SO_4, and 25% (w/v) PEG 3350.
8. Crystallization solution: 2% (v/v) Tacsimate, pH 7.0, 0.1 M imidazole, pH 7.0, 8% (w/v) PEG 3350, and 5% (v/v) isopropanol.
9. Reservoir solution: 0.5 M LiCl.
10. Cryoprotectant: 2% (v/v) Tacsimate, pH 7.0, 0.1 M imidazole, pH 7.0, 20% (w/v) PEG 3350, and 5% (v/v) isopropanol.

3 Methods

3.1 Expression of Mt LipA and Mt NfuA

The genes encoding *Mt* LipA and *Mt* NfuA were inserted into a pET28a vector to allow for their overexpression in *Ec* BL21 (DE3) via a T7 promoter [59, 63]. The resulting proteins are produced with an N-terminal hexahistidine tag, which permits their purification by immobilized metal-affinity chromatography (*see* **Note 1** regarding the placement of the affinity tag). Each of the two genes is coexpressed in the presence of a second plasmid, pDB1282, encoding the *isc* operon from *A. vinelandii*, which increases Fe-S cluster incorporation and enhances the solubility of the protein [16] (*see* **Note 4** regarding cofactor occupancy). In addition, the *Mt lipA* and *Mt nfuA* genes are typically expressed in *E. coli* cultured in M9 minimal medium supplemented with glucose, which allows for ^{57}Fe to be added during expression for characterization of the proteins by Mössbauer spectroscopy [59] (*see* **Note 2** for the source of iron).

1. Thaw 50 μL of *Ec* BL21 (DE3) competent cells on ice. After 30 min, incubate the competent cells with 10 ng each of pDB1282 and pET28a *Mt* LipA or pET28a *Mt* NfuA and flick the tube gently 4–5 times to mix the cells and plasmid. Place the mixture on ice for an additional 30 min. Heat the mixture briefly at 42 °C for 50 s. Place the mixture back on ice for 5 min. Add 500 μL of room-temperature Luria-Bertani (LB) medium to the mixture. Allow the mixture to recover in a shaking incubator at 37 °C and 250 rpm for 1–2 h. Plate the entire mixture on LB agar plates containing kanamycin (50 μg/mL) and ampicillin (100 μg/mL) and incubate the plates overnight at 37 °C. Successful transformants will arise the next day as single colonies containing both the pDB1282 and pET28a LipA or NfuA expression vectors.
2. Use a single colony to inoculate a culture containing 150 mL of LB medium supplemented with kanamycin (50 μg/mL) and ampicillin (100 μg/mL). Incubate the flask in a shaking incubator at 37 °C and 250 rpm for 14 h. Sterilize four 6 L flasks, each containing 4 L of M9 minimal medium, and prewarm in a shaking incubator overnight at 37 °C and 80 rpm.
3. On the following day, add freshly prepared kanamycin (50 μg/mL) and ampicillin (100 μg/mL) prior to inoculating each of the 6 L flasks.
4. Inoculate each 6 L flask with 30 mL of the 150 mL overnight LB culture and incubate the cultures in an incubator at 37 °C with shaking at 180 rpm.
5. Check the OD periodically. When the OD reaches ~0.3, add 0.2% (w/v) *L*-arabinose to induce the expression of genes encoded on the *isc* operon of the pDB1282 plasmid. Remove a 1 mL sample from each flask before arabinose induction (pDB1282 preinduction sample) for SDS-PAGE analysis.
6. When the OD reaches ~0.6, chill the flasks by placing them in an ice bath for 30 min. Add $FeCl_3 \cdot 6H_2O$ to a final concentration of 50 μM ($^{57}FeCl_3$ for Mössbauer sample preparation). Reduce the temperature of the shaking incubator to 18 °C. Remove a 0.5 mL sample from each flask (LipA or NfuA preinduction sample) for SDS-PAGE analysis (*see* **Note 3** for making $^{57}FeCl_3$ using ^{57}Fe metal).
7. Induce overexpression of LipA or NfuA by adding 250 μM of IPTG and incubate the culture for 18 h in a shaking incubator at 18 °C and 180 rpm.
8. On the following day, remove a 0.25 mL sample from each flask (postinduction sample) for SDS-PAGE analysis. Harvest the cells by centrifugation at 6000 × *g* for 12 min. Discard the supernatant and flash-freeze the combined cell pellet (~ 3 g/L)

in liquid nitrogen and store in a −80 °C freezer until further use.

9. Analyze the overproduction of *Mt* LipA and *Mt* NfuA by SDS-PAGE (12% polyacrylamide gel) (Fig. 2).

3.2 Purification of Mt LipA and Mt NfuA

The purification protocol is the same for both *Mt* LipA and *Mt* NfuA. *Mt* LipA and *Mt* NfuA are oxygen sensitive [59, 65–67]. All subsequent purification steps for *Mt* LipA and *Mt* NfuA must be performed under strict anaerobic conditions. Exposure to oxygen can lead to degradation of the cluster, resulting in protein aggregation and loss of enzymatic activity (*see* **Note 4** regarding cofactor occupancy).

3.2.1 Affinity Chromatography

1. Introduce the necessary lab ware, including plastic centrifuge bottles, into the Coy vinyl anaerobic chamber at least 1 day before use to let them deoxygenate. Prepare 500 mL of lysis buffer and 200 mL of elution buffer inside the anaerobic chamber using deoxygenated water (*see* **Note 6** for preparing deoxygenated water).
2. Bring the cell paste (~45 g) into the anaerobic chamber and resuspend it in 125 mL of lysis buffer containing DNase (1 mg/mL), lysozyme (100 mg/mL), and PMSF (1 mM) in a 250 mL metal sonication beaker. Allow the solution to stir for 30 min to become homogenous.
3. Place the metal beaker in an ice-water bath on a stir-plate and slowly stir. The sonicator probe should be located inside the anaerobic chamber.
4. Lyse the cells using a sonicator equipped with a 13-mm-diameter probe using a 10 s on and 10 s off pulse sequence with an amplitude of 30%. The total on-time for pulsing is 20 min. The lysate consistency should change from a thick light brown slurry to a thin dark brown translucent solution.
5. Reserve 20 μL of cell lysate sample for SDS-PAGE analysis. Pour the remaining sonicated cell lysate into a plastic centrifuge bottle. Close the cap tightly and seal the juncture between the cap and bottle with vinyl tape to avoid oxygen contamination (*see* **Note 7** regarding SDS-PAGE analysis).
6. Bring the sealed bottle out of the anaerobic chamber and centrifuge it at 23,000 × *g* for 1.5 h at 4 °C to pellet cell debris.
7. During the centrifugation step, equilibrate a column containing 30 mL of TALON cobalt resin with 100 mL of chilled lysis buffer inside the anaerobic chamber.
8. Once centrifugation of the lysate is complete, immediately transfer the centrifuge bottle into the anaerobic chamber, and carefully transfer the supernatant, without disturbing the

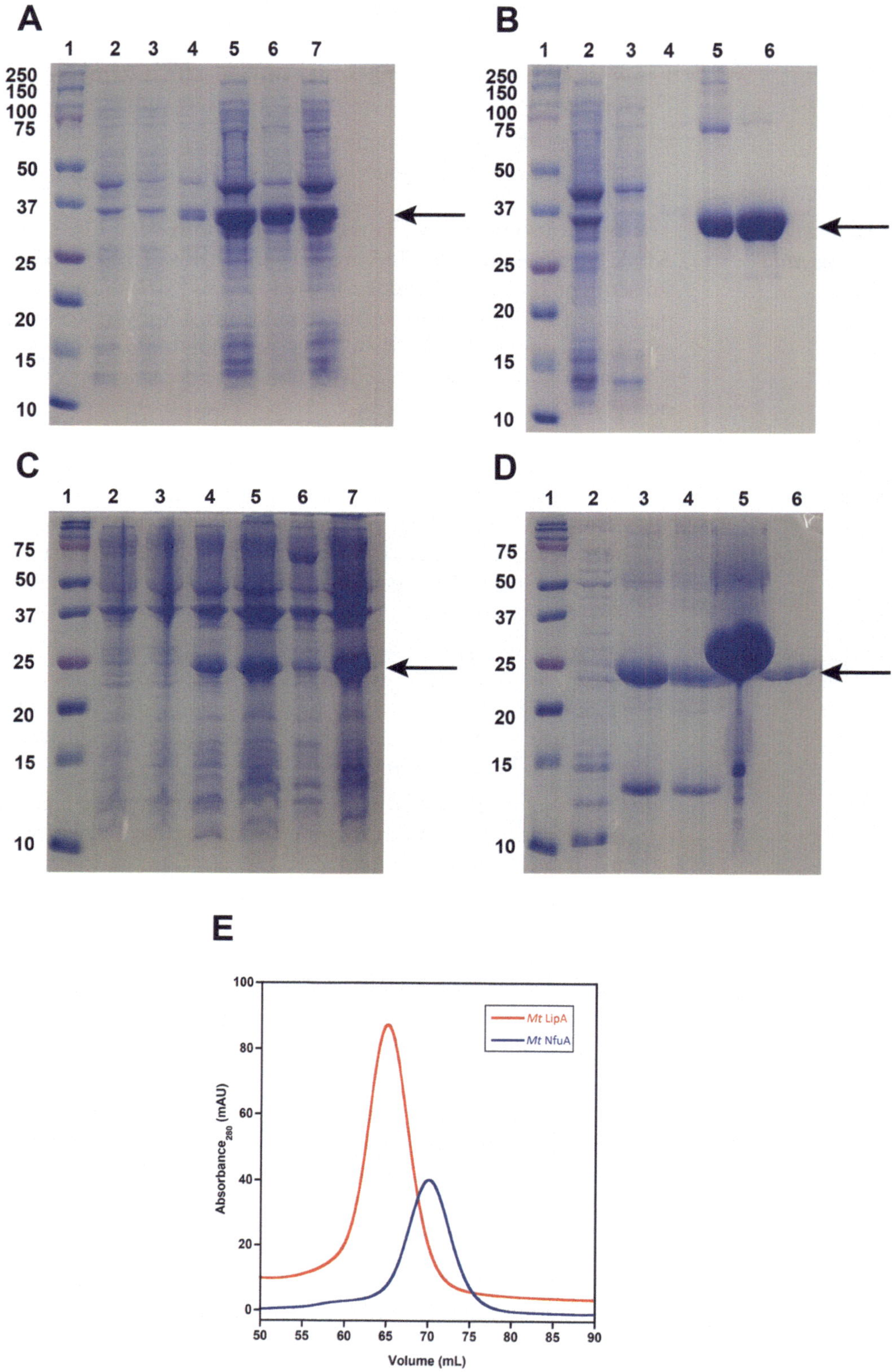

Fig. 2 SDS-PAGE analysis of (**a**) *Mt* LipA overexpression and solubility. Lanes: (1) Molecular weight marker, (2) Uninduced cells, (3) Arabinose-induced cells, (4) IPTG-induced cells, (5) Cell lysate, (6) Pellet, and

pellet, into a clean glass beaker on ice. Reserve 20 μL of supernatant sample and scrape a small amount of the pellet into an Eppendorf tube for SDS-PAGE analysis.

9. Load the supernatant onto the cobalt column and collect the flow through in a separate container. The resin should slowly turn from pink to brownish black. Reserve 20 μL of the flow-through sample for SDS-PAGE analysis.
10. Wash the column with 100 mL of chilled lysis buffer. Reserve two 20 μL of wash samples for SDS-PAGE analysis, one from the beginning of the wash and the other from the end. At this point, the resin should still retain its brownish-black color.
11. Elute the protein with 60 mL of chilled elution buffer (100 mM HEPES, pH 7.5, 300 mM KCl, 250 mM imidazole, 10 mM 2-mercaptoethanol, and 20% glycerol). The starting and ending point of the elution phase can be easily tracked by monitoring the color of the eluate. The *Mt* LipA and *Mt* NfuA Fe-S clusters both absorb in the visible region, rendering the solution a brown color. Reserve 20 μL of eluate sample for SDS-PAGE analysis (Fig. 2). Wash the resin with water and store it in 20% ethanol between uses (*see* **Note 5** for best practices regarding resin reusage and regeneration).

3.2.2 Chemical Reconstitution of Iron-Sulfur Clusters

1. Determine the protein concentration of the eluate by the Bradford method using bovine serum albumin as a standard [68]. Bradford analysis can be performed outside the anaerobic chamber. The subsequent reconstitution should be done strictly inside the anaerobic chamber.
2. Dilute the eluate to a final protein concentration of about 100 μM. Add DTT to a final concentration of 5 mM and gently invert the tube to mix. Incubate the protein on ice for 15 min.
3. Add $FeCl_3{\cdot}6H_2O$ ($^{57}FeCl_3$ for Mössbauer spectroscopy) in three equal portions to achieve a final concentration of 1 mM (10 equiv. of iron per protein) for LipA or 400 μM (4 equiv. iron per protein) for NfuA. In between each addition of the iron solution, gently invert the mixture and allow it to rest on ice for 10 min (*see* **Note 3** for making $^{57}FeCl_3$ using ^{57}Fe metal;

Fig. 2 (continued) (7) Supernatant. (**b**) *Mt* LipA purification. Lanes: (1) Molecular weight marker, (2) Flow through, (3) Wash 1, (4) Wash 2, (5) Elute, and (6) Post S200 [pure protein]. The arrow points to *Mt* LipA. (**c**) *Mt* NfuA overexpression and solubility. Lanes: (1) Molecular weight marker, (2) Uninduced cells, (3) Arabinose-induced cells, (4) IPTG-induced cells, (5) Cell lysate, (6) Pellet, and (7) Supernatant. (**d**) *Mt* NfuA purification. Lanes: (1) Molecular weight marker, (2) Flow through, (3) Wash 1, (4) Wash 2, (5) Elute, and (6) Post S200 [pure protein]. The arrow points to *Mt* NfuA. (**e**) Size exclusion chromatogram of *Mt* LipA (red) and *Mt* NfuA (blue)

see **Notes 8** and **9** for best practices regarding the amount of iron and sulfur added during chemical reconstitution).

4. Add sodium sulfide nonahydrate ($Na_2S{\cdot}9H_2O$) in three equal portions to achieve a final concentration of 1 mM (10 equiv. iron per protein) for LipA or 400 μM (4 equiv. iron per protein) for NfuA. Between each addition of the sulfide solution, gently invert the mixture and allow it to rest on ice for 20 min. Incubate the protein in an ice-filled cooler overnight inside the anaerobic chamber.
5. On the following day, filter the protein through a 0.45 μm pore-sized 25-mm-diameter syringe filter to remove any precipitate.
6. Concentrate the reconstituted protein to ~2.5 mL using ultracentrifugal filtration devices with a 10 kDa molecular weight cutoff (MWCO). When removing the tubes from the anaerobic chamber for centrifugation, make sure to close the lids tightly and to seal the juncture between the cap and the tube with vinyl tape to avoid oxygen contamination.
7. Inside the anaerobic chamber, buffer exchange the concentrated protein solution into chilled S200 elution buffer using a desalting column. Collect the exchanged protein in 3.5 mL (*see* **Note 10** regarding the importance of the desalting step).

3.2.3 Size Exclusion Chromatography

1. Equilibrate a HiPrep 16/60 Sephacryl S-200 HR column with chilled S200 elution buffer. The column should be attached to an FPLC instrument located inside an anaerobic chamber.
2. Inject 2 mL of buffer-exchanged protein per run and repeat this step until all the protein sample has been injected. During each FPLC run, the column pressure should remain under 0.2 MPa at all times. Maintain a flow rate of ~0.7 mL/min (or less if required to reduce the column pressure) and a fraction volume of 3 mL. Elute the sample with at least 1.25 column volumes of S200 elution buffer. Fractions containing protein can be identified by monitoring the absorbance of the eluate at 280 nm and 400 nm. *Mt* LipA and *Mt* NfuA should elute at 65 mL and 70 mL, respectively (Fig. 2e).
3. Collect and pool the appropriate fractions. Reserve 20 μL of the S200 combined eluate for SDS-PAGE analysis (Fig. 2).
4. Concentrate the purified protein to ~2 mL using an ultracentrifugal filtration device with a 10 kDa MWCO. When removing the tube from the anaerobic chamber for centrifugation, make sure to close the lids tightly and seal the juncture between the cap and the tube with vinyl tape to avoid oxygen contamination.

5. Determine the final concentration of the protein using the Bradford method with bovine serum albumin as a standard (*see* Subheading 3.3 regarding the Bradford correction factor) [68].
6. Record a UV-visible spectrum of the sample and prepare a sample for iron analysis by inductively coupled plasma atomic emission spectroscopy (ICP-AES) (*see* Subheading 3.3.1 for sample preparation instructions).
7. Dilute *Mt* LipA to 1 mM and *Mt NfuA* to 2 mM using the S200 elution buffer (100 mM HEPES, pH 7.5, 300 mM KCl, 5 mM dithiothreitol, and 15% (v/v) glycerol). Aliquot 100 μL of protein into 1.5 mL screw cap microcentrifuge tubes. Remove the aliquots from the anaerobic chamber and immediately flash-freeze them in liquid nitrogen. Store the protein aliquots in liquid nitrogen until further use (*see* **Note 11** for best practices regarding long-term storage).

3.3 Characterization of Purified Mt LipA and Mt NfuA

The yield and quality of purification can be monitored with a few standard techniques, which include quantitative measurements of protein, iron, and acid-labile sulfide, and spectroscopic analysis of the Fe-S clusters by UV-visible and Mössbauer spectroscopies. Protein concentration can be determined by a number of colorimetric methods, such as the method of Bradford using bovine serum albumin as a standard. However, the accuracy of the chosen method should be verified by amino acid analysis, and if necessary, amino acid analysis should be used to establish an appropriate correction factor (*see* **Note 12** regarding the importance of correction factor).

UV-visible spectroscopy is a routine technique used to analyze Fe-S proteins; however, it only provides a qualitative estimate of cluster type and occupancy. A more robust strategy to determine cluster type and occupancy is to use Mössbauer spectroscopy in concert with methods to determine the iron concentration, such as ICP-AES. Mössbauer spectroscopy provides information on the ratio of various Fe species in the protein sample, ranging from adventitiously bound Fe to various Fe-S cluster forms. This ratio can then be used to determine absolute stoichiometries of Fe-S cluster forms when accurate protein and iron concentrations have been established. Given that LipA only catalyzes one turnover in the absence of accessory proteins, activity measurements provide the proportion of active enzyme. In the final application described in this chapter, protein crystallography, protein samples must be >95% pure protein with fully occupied and homogeneous Fe-S clusters.

3.3.1 UV-Visible Spectroscopy and Iron Analysis

1. In the anaerobic chamber, dilute the protein to 20 μM in 1 mL of S200 elution buffer. Transfer the sample into a quartz cuvette and seal the top with a cap or parafilm.
2. Bring the sealed cuvette out of the glove box and immediately record a UV-visible absorption spectrum from 250 nm to 800 nm. The absorption trace should exhibit a distinct peak at 280 nm and a shoulder centered around 400 nm, the latter of which is a characteristic feature of [4Fe-4S] clusters [56]. The absorbance ratio A_{280}/A_{400} for *Mt* LipA and *Mt* NfuA should be ~2.78 and ~2.94, respectively. Poor absorbance at 400 nm directly correlates with poor [4Fe-4S] cluster occupancy. Chemical reconstitution should be performed before proceeding to activity analysis and crystallography trials (Fig. 3).
3. While UV-visible analysis provides a qualitative estimate of Fe-S cluster occupancy, a quantitative estimate of the number of iron ions bound to each protein monomer can be determined by ICP-AES. To prepare samples for ICP-AES analysis, dilute an aliquot of purified protein to 5 μM in 5 mL of 3.5% (v/v) nitric acid. Vortex the sample and let it sit for an hour at room temperature to precipitate the protein. All bound metal ions should be solubilized in the acid solution.
4. Centrifuge the sample at 10,000 × *g* for 10 min and carefully transfer the supernatant into a fresh 15 mL conical tube. Submit the sample for ICP-AES analysis to determine the amount of iron in the sample. The typical values are ~7 iron ions per *Mt LipA* monomer and ~2 iron ions per *Mt NfuA* monomer (*see* **Note 13**).

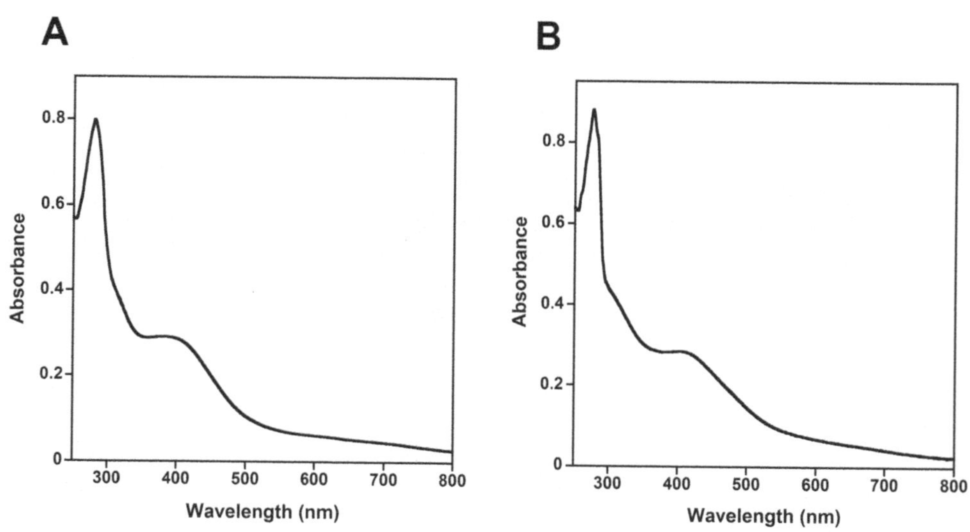

Fig. 3 UV-Visible spectra of (**a**) *Mt* LipA and (**b**) *Mt* NfuA

3.3.2 Analysis of Mt LipA Cluster Composition by Mössbauer Spectroscopy

1. Transfer frozen protein aliquots, a 350 mL Dewar flask with liquid N_2 covered in aluminum foil, and a Mössbauer cup into the anaerobic chamber (*see* **Note 14** for best practices for introducing liquid nitrogen into the anaerobic chamber).
2. Mössbauer samples should contain a total iron concentration of at least 1 mM. Dilute the protein accordingly in a final volume of 350 μL of S200 elution buffer.
3. Transfer the diluted protein into the Mössbauer cup and freeze the solution by lowering the cup into liquid N_2 using forceps.
4. Collect the Mössbauer spectra of the protein and determine the proportion of various Fe species in the sample (adventitiously bound Fe and different Fe-S clusters forms) by spectral analysis, as described previously [69] (Fig. 4). Most of the iron should be in the form of a [4Fe-4S] cluster, as noted by a single quadrupole doublet exhibiting an isomer shift $(\delta) = 0.45$ mm/s and a quadrupole splitting parameter $(\Delta E_Q) = 1.15$ mm/s.
5. Simulation of the Mössbauer spectra can be carried out using the WMOSS spectral analysis software [69].

3.3.3 Quantitative Measurement of Mt LipA Activity by LC-MS

1. Prepare and incubate all activity assay samples in an anaerobic chamber at room temperature. The final reaction contains reaction buffer, 25 μM LipA, 500 μM peptide substrate (Glu--Ser-Val-[*N*6-octanoyl]Lys-Ala-Ala-Ser-Asp), 0.5 μM SAH nucleosidase, and 1 mM SAM (*see* **Note 15**). Add the reagents in the order mentioned.
2. Prepare 500 μL of a quench solution containing 100 mM H_2SO_4, 20 μM AtsA (peptide internal standard, Pro-Met-Ser-Ala-Pro-Ala-Arg-Ser-Met), and 4 mM TCEP. Aliquot 20 μL into six microcentrifuge tubes.
3. Initiate the reaction by adding 2 mM sodium dithionite and quench the reaction at different times ($t = 0$, 10, 30, 60, 120, and 180 min) by adding 20 μL of the reaction mix into 20 μL of quench solution. Assays should be performed in triplicate. These samples can be prepared in parallel by staggering the initiation times by a convenient interval (e.g., 30 s) (*see* **Note 16** regarding sodium dithionite preparation).
4. After the assay is completed, remove the reaction tubes from the anaerobic chamber and centrifuge for at least 30 min at 14,000 × *g* to remove precipitated protein.
5. Transfer 35 μL of the supernatant from each time point sample into an LC-MS sample vial. Make sure that all particulate matter is removed from the samples.
6. Prepare all samples that compose the standard curve in the anaerobic chamber at room temperature. Prepare 60 μL of

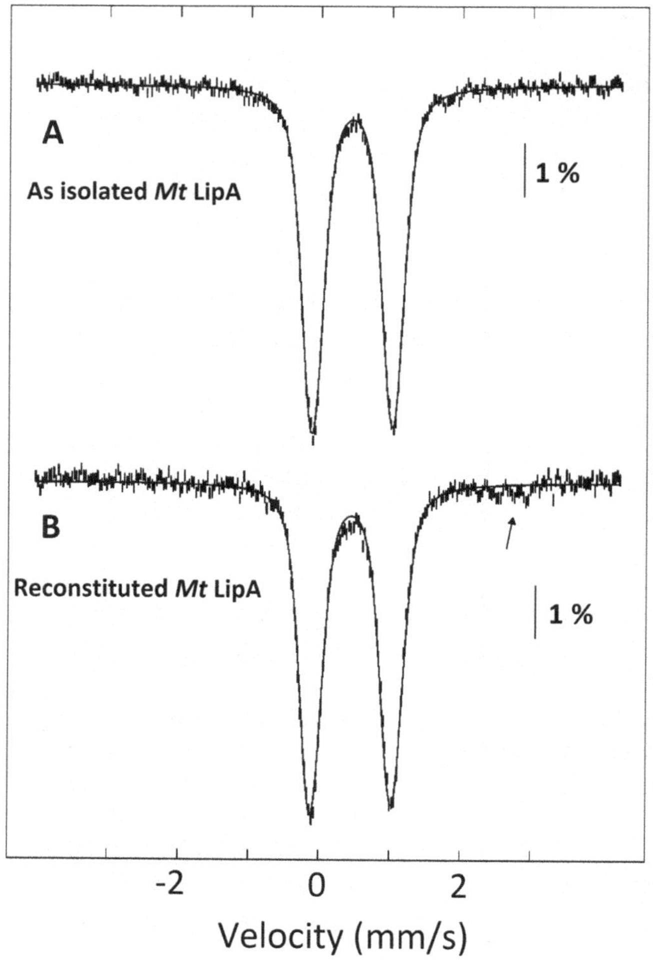

Fig. 4 Mössbauer spectra (4.2 K/53 mT) of (**a**) as-isolated *Mt* LipA and (**b**) reconstituted *Mt* LipA. The spectra can be simulated by a quadrupole doublet with identical parameters ($\delta = 0.45$ mm/s and $\Delta E_Q = 1.15$ mm/s), accounting for 94 and 92% of the total intensity, respectively (Solid lines **a** and **b**). The position of the high energy line emanating from a small amount of high-spin Fe(II) in the sample of reconstituted *Mt* LipA is indicated by the arrow (Figure adapted from Ref. 59)

500 μM of peptide standard in dilution mix containing 500 μM of octanoyl peptide, 500 μM 6-thio-octanoyl peptide, and 500 μM lipoyl peptide (*see* **Note 15** for making stock peptides).

7. Prepare 360 μL of a dilution mix.
8. Aliquot 30 μL of the dilution mix into 11 microcentrifuge tubes.
9. Dilute the standard by a factor of 2 by mixing 30 μL of the 500 μM standard with 30 μL of dilution mix. Continue the serial dilution by a factor of 2 to generate a set of standards for

quantification of substrates and products (500, 250, 125, 62.5, 31.2, 16, 8, 4, 2, 1, 0.5, and 0.25 μM). Remove the standards from the anaerobic chamber and centrifuge them for 30 min at 14,000 × *g*, and transfer them into LC-MS sample vials.

10. Analyze the quenched assay mixtures by electrospray ionization MS in positive mode (ESI^+) with a nitrogen gas temperature of 350 °C and a flow rate of 5.0 L/min, a nebulizer pressure of 45 psi, and a capillary voltage of 4000 V. Separate the assay mixture on a C18 column (Table 1) equilibrated in 92% (v/v) solvent A (0.1% formic acid) and 8% solvent B (acetonitrile). Maintain a flow rate of 0.3 mL/min throughout the chromatographic procedure. Allow the column to equilibrate for 1 min prior to the next sample injection.
11. Use the Agilent MassHunter Quantitative Analysis (QQQ) software, or any other similar software package, to measure the amount of 6-thio-octanoyl peptide intermediate and lipoyl product formed by comparing the sample's integrated intensities to the known peptide standard intensities (Fig. 5) (Table 2).

3.3.4 Crystallization of Mt LipA

1. Perform all procedures in an anaerobic chamber. Prepare necessary buffers inside the chamber with deoxygenated water. Transfer sitting drop trays and greased hanging drop trays into the anaerobic chamber at least 2 weeks before to make the trays anaerobic.
2. Dilute *Mt* LipA to 7.3 mg/mL in storage buffer containing 1 mM *E. coli* octanoyl-H peptide (sequence ESVK[octanoyl] AASD) and 0.9 mM SAM.
3. Incubate the protein at room temperature overnight and centrifuge at 2500 × *g* for 5 min before crystallization.
4. Perform crystallization trials using the hanging drop vapor diffusion method at room temperature by combining 1 μL protein solution with 1 μL crystallization solution and incubating over a reservoir of 500 μL 0.5 M LiCl.
5. Plate-shaped crystals should appear within a day. Crystals will grow to full size within 3–4 days.
6. Soak the crystals in a cryoprotectant (Paratone-N), harvest on rayon loops, and flash-freeze the crystals by direct plunge into liquid nitrogen (*see* **Note 14** for best practices in bringing liquid nitrogen into the anaerobic chamber).
7. Collect X-ray diffraction datasets at a home or synchrotron X-ray source (Fig. 6).

Table 1
Gradient for LC-MS/MS analysis of reaction intermediate and product

Time (min)	0.1% Formic acid (%)	100% Acetonitrile (%)	Flow (mL/min)
0.00	92.00	8.00	0.300
0.50	75.00	25.00	0.300
2.00	73.0	27.00	0.300
2.50	92.0	8.00	0.300

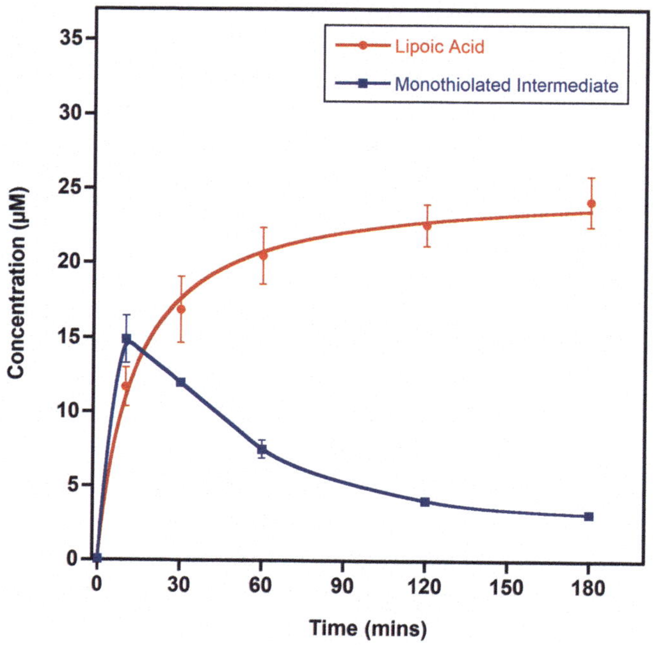

Fig. 5 *Mt* LipA activity assay showing the accumulation and subsequent conversion of 6-thio-octanoyl-peptide intermediate into lipoyl peptide as time progresses. [Concentration of *Mt* LipA: 25μM. Error bars represent the mean ± SD (standard deviation) of three replicates]

Table 2
Parameters for LC-MS/MS analysis of reaction intermediate and product

Compound	Parent ion	Fragmentator voltage	Product ion 1[a]	Product ion 2[a]
Octanoyl peptide	932.5	188	712.4 (29)	210.1 (53)
6-thio-octanoyl peptide	964.5	220	744.4 (30)	242.1 (46)
Lipoyl peptide	996.5	208	776.3 (30)	274.1 (46)
AtsA peptide (IS)	474.4	112	201.1 (18)	153.0 (26)

[a]Respective collision energies are in parentheses

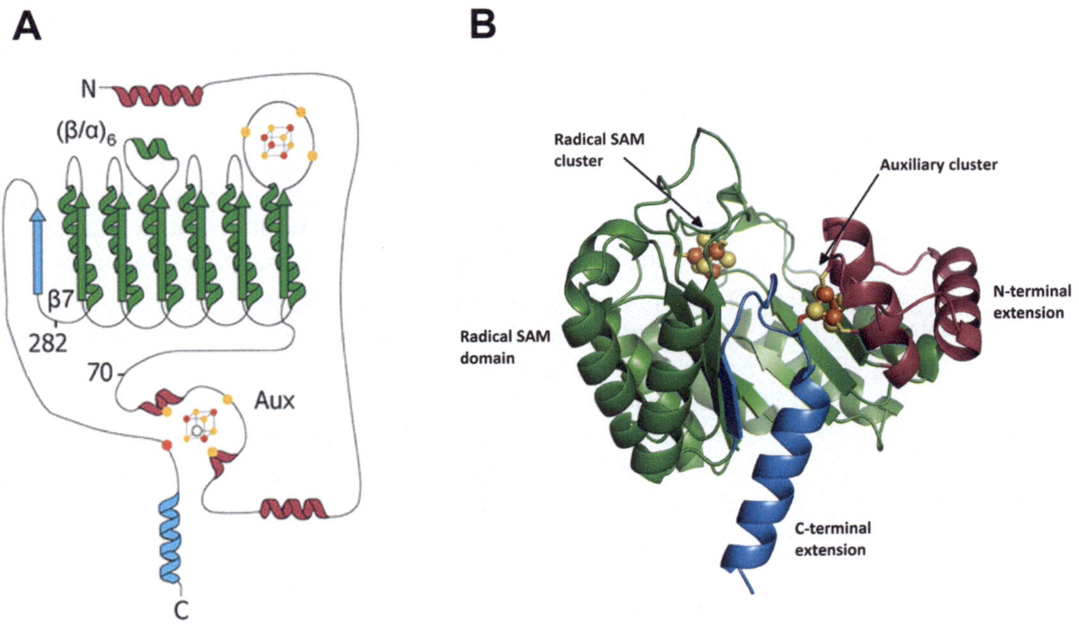

Fig. 6 (**a**) Topology diagrams of *Mt* LipA. (**b**) *Mt* LipA in ribbon representation colored by domain (PDB ID: 5EXJ). Fe-S clusters are represented in ball-and-stick form (Figure adapted from Ref. 58)

3.3.5 Crystallization of Monothiolated Peptide Intermediate Complex with Mt LipA

1. Perform all procedures in an anaerobic chamber.
2. Add 200 μM *Mt* LipA to a buffer containing 50 mM HEPES, pH 7.5, 300 mM KCl, 10% (v/v) glycerol, 300 μM *Mt* octanoyl-H peptide, and 180 μM SAM to a total volume of 1 mL.
3. Initiate the reaction with the addition of 2 mM sodium dithionite and incubate at room temperature for 45 min.
4. Apply the reaction mixture to a centrifuge tube with a 10 kDa cutoff membrane and spin for 10 min at 7000 × *g* to reduce the volume by a factor of 4 to approximately 250 μL.
5. Determine the concentration by Bradford analysis and flash-freeze the cross-linked enzyme in liquid nitrogen until further use.
6. Separately, monitor reaction progress to confirm cross-link formation by quenching aliquots of the reaction at designated times in 100 mM sulfuric acid containing 4 mM TCEP and 20 μM AtsA peptide standard for LC-MS analysis. Prepare a standard curve for analysis by LC-MS and estimate the quantities of lipoyl product and 6-thio-octanoyl peptide intermediate.
7. Dilute LipA intermediate to 10 mg/mL in intermediate solution. Incubate the protein overnight and centrifuge at 14,000 × *g* for 5 min prior to crystallization. Retain the supernatant for the next step.

8. Perform crystallization trials using the sitting drop vapor diffusion method at 20 °C by combining protein solution with seed crystal solution in a 1:1 ratio over a reservoir of seed crystal solution.
9. Needle-like crystals will appear within 2 months. Use these crystals to streak-seed fresh hanging drop trials at room temperature using the same seed crystal solution. Make seed stock using the crystals obtained from these hanging drop trials.
10. Perform crystallization trials using the sitting drop vapor diffusion method at 20 °C by combining protein solution with crystallization solution [2% (v/v) Tacsimate, pH 7.0, 0.1 M imidazole, pH 7.0, 8% (w/v) PEG 3350, and 5% (v/v) isopropanol] and seed stock [LipA intermediate seed crystals in 0.1 M Bis-Tris, pH 6.5, 0.2 M Li_2SO_4, and 25% (w/v) PEG 3350] in a 4:4:1 ratio over a reservoir of 0.5 M LiCl.
11. Rod-shaped crystals should appear within 4 days. After 9 days, soak the crystals in a cryoprotectant solution [2% (v/v) Tacsimate, pH 7.0, 0.1 M imidazole, pH 7.0, 20% (w/v) PEG 3350, and 5% (v/v) isopropanol], harvest on rayon loops, and flash-freeze by direct plunge into liquid nitrogen.
12. Collect X-ray diffraction datasets at a home or synchrotron X-ray source (Figs. 7 and 8).

4 Notes

1. N-terminal hexahistidine tags were preferred over C-terminal hexahistidine tags because *Mt* LipA coordinates its [4Fe-4S] auxiliary cluster with one ligand that is near the C-terminus of the protein, while a dimer of *Mt* NfuA binds one [4Fe-4S] cluster using two cysteines from each monomer's C-terminal Nfu-like domain.
2. Iron(III) chloride hexahydrate ($FeCl_3 \cdot 6H_2O$) or $^{57}FeCl_3$ is the source of the added iron, which is introduced into the cultures just before induction of *lipA* or *nfuA* gene expression ($OD_{600} \sim 0.6$). Adding iron to the M9 minimal medium just before induction leads to iron starvation in *E. coli* cells during preinduction, which signals the cells to release siderophores. These siderophores facilitate the transport of iron into the cell to make it available for incorporation into Fe-S proteins [70].
3. To prepare $^{57}FeCl_3$, dissolve one equivalent ^{57}Fe metal in four molar equivalents of 6 M HCl in a loosely capped Falcon centrifuge tube inside a chemical fume hood. Stir and heat the mixture overnight at 70 °C inside a chemical fume hood. The concentration of the resulting $^{57}FeCl_3$ can be determined by iron analysis using the methods of Beinert [71]. $^{57}FeCl_3$ can

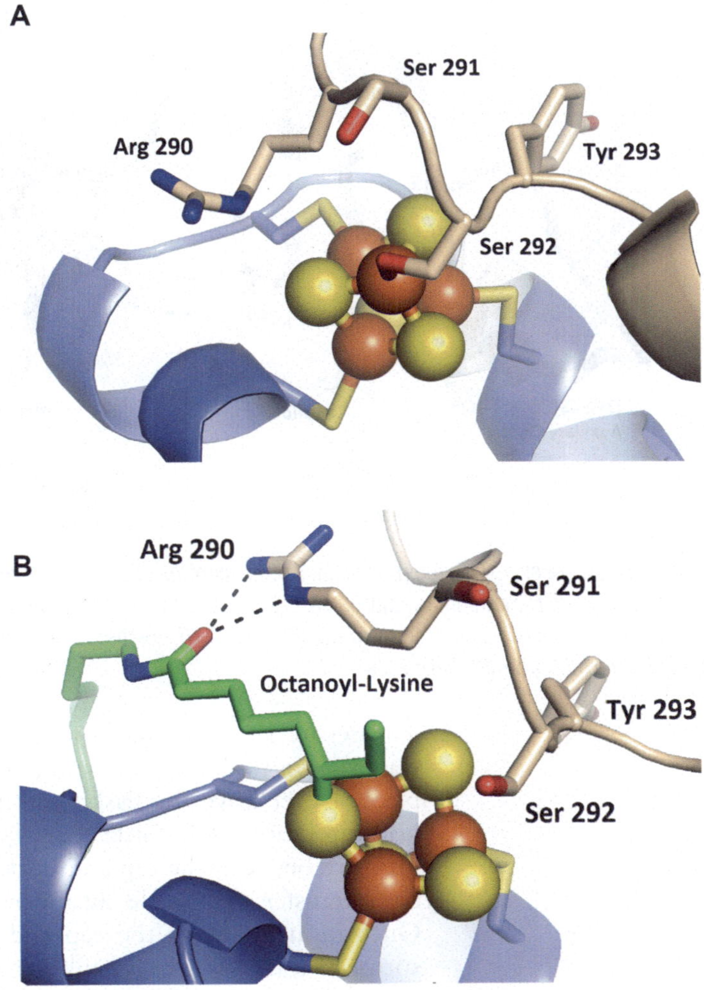

Fig. 7 (**a**) Structure of *Mt* LipA auxiliary cluster in the absence of substrate showing the conserved S292 coordinating the fourth iron and R290 blocking the substrate binding site (PDB ID: 5EXJ). (**b**) Structure of *Mt* LipA auxiliary cluster after reaction with 1 equivalent of SAM and octanoyl-H-peptide substrate showing the loss of the fourth iron, which was coordinated by S292, and R290 no longer blocking the substrate binding site but guiding substrate positioning (PDB ID: 5EXK)

be directly added to M9 minimal medium, and excess $^{57}FeCl_3$ can be stored in liquid nitrogen for future use.

4. Iron-sulfur cluster occupancy can vary with each protein preparation. To ensure full occupancy, a chemical reconstitution step is recommended for *Mt* LipA and *Mt* NfuA immediately after affinity chromatography.
5. It is highly recommended to regenerate the resin after three uses according to the manufacturer's instructions.

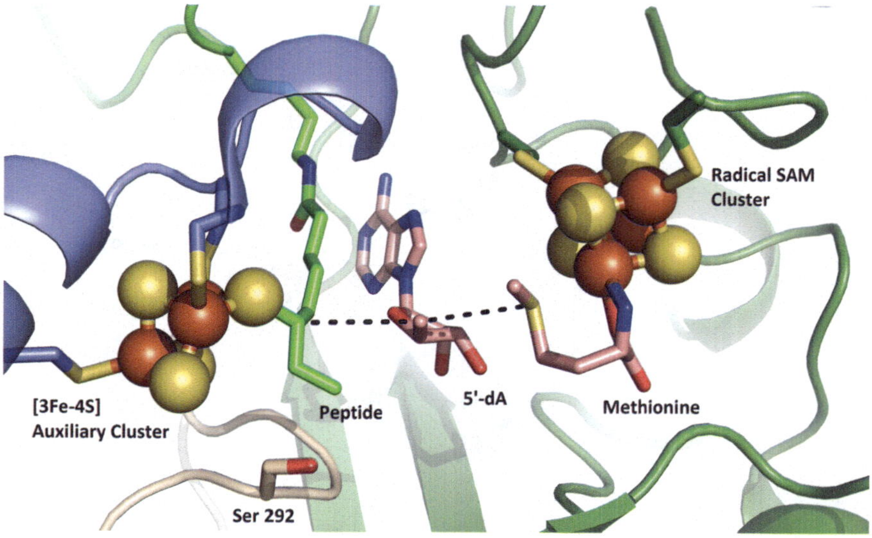

Fig. 8 *Mt* LipA active site after reaction with 1 equivalent of SAM and octanoyl-H-peptide substrate showing (1) SAM cleavage to *L*-methionine and 5′-dA·, (2) abstraction of the peptide substrate's C6 hydrogen atom, and (3) back attack on the cluster to form the [3Fe-3S-1 (*R*)-(6-mercaptooctanoyl) peptide], with (4) concomitant loss of a ferrous iron (PDB ID:5EXK)

6. To prepare large volumes of deoxygenated water, boil ~6 L water down to 4 L on a stirring hot plate in a well-ventilated place. Never leave the flask unattended. Transfer the boiling hot water into a 4 L bottle. Cap the bottle tightly and seal the juncture between the cap and bottle with vinyl tape. Immediately transfer the bottle into Coy vinyl anaerobic chamber. Once inside, remove the cap and stir water overnight on a stir-plate to allow it to equilibrate inside the anaerobic chamber. Wear goggles, heat-resistant gloves, and a lab coat while handling the boiling water.
7. SDS-PAGE analysis will give a qualitative estimate about the induction and solubility of the target protein. It should always be performed for each purification so that future induction and purification can be optimized to get a higher protein yield.
8. The acidic $^{57}FeCl_3$ solution should be neutralized with a saturated solution of $NaHCO_3$, prepared anaerobically prior to use in iron-sulfur cluster reconstitution. The concentration of the final $^{57}FeCl_3$ stock can be determined by ICP-AES or by iron analysis using the methods of Beinert [71].
9. The amount of iron and sulfur added should be in ten-fold excess of the protein concentration for LipA and in four-fold excess of the protein concentration for NfuA. These values were selected because LipA has two [4Fe-4S] clusters per protein monomer. NfuA has one [4Fe-4S] cluster per dimer, with the cluster predicted to bridge the dimer interface. In each case,

we add two additional equivalents of iron to ensure full reconstitution, but larger excesses of iron and sulfur should be avoided during chemical reconstitution because they can lead to protein denaturation and aggregation. These phenomena result in poor yields of active enzyme or protein.

10. The desalting step is crucial to avoid precipitation of protein over the S200 column due to buffer difference between the sample and the column. This step will prolong the lifespan of the column by avoiding overpressure issues due to precipitation.
11. Repeated freeze–thaw cycles of purified LipA and NfuA will lead to aggregation and loss of activity. Store the protein in small working aliquots of 100 μL.
12. The amino acid composition can affect the concentration determined by Bradford analysis. Therefore, a correction factor should be obtained through amino acid analysis. This step will be crucial if the correction factor is greater than 1.25 or less than 0.75, which can significantly affect the stoichiometry of the protein in activity assays. A correction factor is even more important for proteins that undergo only one turnover, given that the amount of product produced will be equal to the concentration of active enzyme in the reaction. The correction factors for *Mt* LipA and *Mt* NfuA are 0.893 and 0.847, respectively [59, 63].
13. This method simply estimates the amount of iron in solution. It does not provide information about the structure of the cofactor or its coordination environment. For a more precise characterization of cofactors, Mössbauer spectroscopy should be used.
14. The vacuum cycle should be performed manually by gradually increasing the vacuum every cycle. The first vacuum cycle should not exceed 3 psi to avoid ejection of liquid N_2 from the Dewar flask. In the subsequent cycles, increase the vacuum by 1 psi. Repeat this process until 10 psi is achieved. Perform three vacuum/gas purge cycles at 10 psi to remove enough oxygen from the antechamber to avoid contamination of the glove box environment.
15. The peptides used for quantification (octanoyl, 6-thio-octanoyl, lipoyl, and AtsA) are >98% pure and can be purchased from ProImmune Custom Peptides in powder form. The stock concentrations are made by dissolving the appropriate amount by weight in 20 mM anaerobic sodium hydroxide.
16. Sodium dithionite is a sufficient reducing agent for *Mt* LipA activity assays, but solutions cannot be stored. They must be prepared fresh for every set of assays.

References

1. Kruger N, Oppermann FB, Lorenzl H et al (1994) Biochemical and molecular characterization of the *Clostridium magnum* acetoin dehydrogenase enzyme system. J Bacteriol 176:3614–3630
2. Oppermann FB, Steinbuchel A (1994) Identification and molecular characterization of the *aco* genes encoding the *Pelobacter carbinolicus* acetoin dehydrogenase enzyme system. J Bacteriol 176:469–485
3. Priefer H, Hein S, Kruger N et al (1991) Identification and molecular characterization of the *Alcaligenes eutrophus* H16 *aco* operon genes involved in acetoin catabolism. J Bacteriol 173:4056–4071
4. Reed L (1974) Multienzyme complexes. Acc Chem Res 7:40–46
5. Mislow K, Meluch WC (1956) The stereochemistry of alpha-lipoic acid. J Am Chem Soc 78:5920
6. Herbert AA, Guest JR (1975) Lipoic acid content of *Escherichia coli* and other microorganisms. Arch Biochem Biophys 106:259–266
7. Reed LJ, Koike M, Levitch ME et al (1958) Studies on the nature and reactions of protein-bound lipoic acid. J Biol Chem 232:143–158
8. Cronan JE (2016) Assembly of lipoic acid on its cognate enzymes: an extraordinary and essential biosynthetic pathway. Microbiol Mol Biol Rev 80:429–450
9. Cronan JE, Zhao X, Jiang Y (2005) Function, attachment and synthesis of lipoic acid in Escherichia coli. Adv Microb Physiol 50:103–146
10. Billgren ES, Cicchillo RM, Nesbitt NM et al (2010) Lipoic acid biosynthesis and enzymology. In: Mander L, Liu H-W (eds) Comprehensive natural products II chemistry and biology. Elsevier, Oxford, pp 181–212
11. Jordan SW, Cronan JE Jr (2003) The Escherichia coli *lipB* gene encodes lipoyl(octanoyl)-acyl carrier protein:protein transferase. J Bacteriol 185:1582–1589
12. Zhao X, Miller JR, Cronan JE (2005) The reation of LipB, the octanoyl-[acyl carrier protein]:protein N-octanoyltransferase of lipoic acid synthesis, proceeds through an acyl-enzyme intermediate. Biochemistry 44:16737–16746
13. Nesbitt NM, Baleanu-Gogonea C, Cicchillo RM et al (2005) Expression, purification, and physical characterization of *Escherichia coli* lipoyl(octanoyl)transferase. Protein Expr Purif 39:269–282
14. Miller JR, Busby RW, Jordan SW et al (2000) *Escherichia coli* LipA is a lipoyl synthase: in vitro biosynthesis of lipoylated pyruvate dehydrogenase complex from octanoyl-acyl carrier protein. Biochemistry 39:15166–15178
15. Zhao S, Miller JR, Jiang Y et al (2003) Assembly of the covalent linkage between lipoic acid and its cognate enzymes. Chem Biol 10:1293–1302
16. Cicchillo RM, Iwig DF, Jones AD et al (2004) Lipoyl synthase requires two equivalents of S-adenosyl-L-methionine to synthesize one equivalent of lipoic acid. Biochemistry 43:6378–6386
17. Morris TW, Reed KE, Cronan JE Jr (1995) Lipoic acid metabolism in Escherichia coli: the *lplA* and *lipB* genes define redundant pathways for ligation of lipoyl groups to apoprotein. J Bacteriol 177:1–10
18. Green DE, Morris TW, Green J et al (1995) Purification and properties of the lipoate protein ligase of Escherichia coli. Biochem J 309:853–862
19. Landgraf BJ, Mccarthy EL, Booker SJ (2016) Radical S-adenosylmethionine enzymes in human health and disease. Annu Rev Biochem 85:485–514
20. Mayr JA, Feichtinger RG, Tort F et al (2014) Lipoic acid biosynthesis defects. J Inherit Metab Dis 37:553–563
21. Yi X, Maeda N (2005) Endogenous production of lipoic acid is essential for mouse development. Mol Cell Biol 25:8387–8392
22. Ma Q, Eddine AN, Geerlof A et al (2006) The Mycobacterium tuberculosis LipB enzyme functions as a cysteine/lysine dyad acyltransferase. Proc Natl Acad Sci U S A 103:8662–8667
23. Spalding MD, Prigge ST (2010) Lipoic acid metabolism in microbial pathogens. Microbiol Mol Biol Rev 74:200–228
24. Frey PA, Booker SJ (2001) Radical mechanisms of S-adenosylmethionine-dependent enzymes. Adv Protein Chem 58:1–45
25. Frey PA, Magnusson OT (2003) S-Adenosylmethionine: a wolf in sheep's clothing, or a rich man's adenosylcobalamin? Chem Rev 103:2129–2148
26. Sofia HJ, Chen G, Hetzler BG et al (2001) Radical SAM, a novel protein superfamily linking unresolved steps in familiar biosynthetic pathways with radical mechanisms: functional characterization using new analysis and information visualization methods. Nucleic Acids Res 29:1097–1106

27. Holliday GL, Akiva E, Meng EC et al (2018) Atlas of the radical SAM superfamily: divergent evolution of function using a "plug and play" domain. Methods Enzymol 606:1–71
28. Bauerle MR, Schwalm EL, Booker SJ (2015) Mechanistic diversity of radical S-adenosylmethionine (SAM)-dependent methylation. J Biol Chem 290:3995–4002
29. Broderick JB, Duffus BR, Duschene KS et al (2014) Radical S-adenosylmethionine enzymes. Chem Rev 114:4229–4317
30. Challand MR, Driesener RC, Roach PL (2011) Radical *S*-adenosylmethionine enzymes: mechanism, control and function. Nat Prod Rep 28:1696–1721
31. Booker SJ, Grove TL (2010) Mechanistic and functional versatility of radical SAM enzymes. F1000 Biol Rep 2:52
32. Booker SJ, Cicchillo RM, Grove TL (2007) Self-sacrifice in radical S-adenosylmethionine proteins. Curr Opin Chem Biol 11:543–552
33. Atta M, Mulliez E, Arragain S et al (2010) S-adenosylmethionine-dependent radical-based modification of biological macromolecules. Curr Opin Struct Biol 20:1–9
34. Lanz ND, Booker SJ (2012) Identification and function of auxiliary iron-sulfur clusters in radical SAM enzymes. Biochim Biophys Acta 1824:1196–1212
35. Lanz ND, Booker SJ (2015) Auxiliary iron-sulfur cofactors in radical SAM enzymes. Biochim Biophys Acta 1853:1316–1334
36. Sanyal I, Cohen G, Flint DH (1994) Biotin synthase: purification, characterization as a [2Fe-2S] cluster protein, and in vitro activity of the Escherichia coli *bioB* gene product. Biochemistry 33:3625–3631
37. Marquet A, Tse Sum Bui B, Florentin D (2001) Biosynthesis of biotin and lipoic acid. Vitam Horm 61:51–101
38. Fugate CJ, Jarrett JT (2012) Biotin synthase: insights into radical-mediated carbon-sulfur bond formation. Biochim Biophys Acta 1824 (11):1213–1222
39. Arragain S, Handelman SK, Forouhar F et al (2010) Identification of eukaryotic and prokaryotic methylthiotransferase for biosynthesis of 2-methylthio-N6-threonylcarbamoyladenosine in tRNA. J Biol Chem 285:28425–28433
40. Pierrel F, Björk GR, Fontecave M et al (2002) Enzymatic modification of tRNAs: MiaB is an iron–sulfur protein. J Biol Chem 277:13367–13370
41. Pierrel F, Douki T, Fontecave M et al (2004) MiaB protein is a bifunctional radical-S-adenosylmethionine enzyme involved in thiolation and methylation of tRNA. J Biol Chem 279:47555–47653
42. Anton BP, Russell SP, Vertrees J et al (2010) Functional characterization of the YmcB and YqeV tRNA methylthiotransferases of *Bacillus subtilis*. Nucleic Acids Res 38:6195–6205
43. Anton BP, Saleh L, Benner JS et al (2008) RimO, a MiaB-like enzyme, methylthiolates the universally conserved Asp88 residue of ribosomal protein S12 in Escherichia coli. Proc Natl Acad Sci U S A 105:1826–1831
44. Lee K-H, Saleh L, Anton BP et al (2009) Characterization of RimO, a new member of the methylthiotransferase subclass of the radical SAM superfamily. Biochemistry 48:10162–10174
45. Landgraf BJ, Booker SJ (2016) Stereochemical course of the reaction catalyzed by RimO, a radical SAM methylthiotransferase. J Am Chem Soc 138:2889–2892
46. Arragain S, García-Serres R, Blondin G et al (2010) Post-translational modification of ribosomal proteins: structural and functional characterization of RimO from *Thermotoga maritima*, a radical S-adenosylmethionine methylthiotransferase. J Biol Chem 285:5792–5801
47. Forouhar F, Arragain S, Atta M et al (2013) Two Fe-S clusters catalyze sulfur insertion by radical-SAM methylthiotransferases. Nat Chem Biol 9:333–338
48. Lanz ND, Booker SJ (2014) The role of iron-sulfur clusters in the biosynthesis of the lipoyl cofactor. In: Rouault TA (ed) Iron-sulfur clusters in chemistry and biology. Walter de Gruyter GMbH, Berlin
49. Mulliez E, Duarte V, Arragain S et al (2017) On the role of additional [4Fe–4S] clusters with a free coordination site in radical-SAM enzymes. Front Chem 5:17
50. Ugulava NB, Gibney BR, Jarrett JT (2001) Biotin synthase contains two distinct iron–sulfur binding sites: chemical and spectroelectrochemical analysis of iron–sulfur cluster interconversions. Biochemistry 40:8343–8351
51. Ugulava NB, Sacanell CJ, Jarrett JT (2001) Spectroscopic changes during a single turnover of biotin synthase: destruction of a [2Fe–2S] cluster accompanies sulfur insertion. Biochemistry 40:8352–8358
52. Cosper MM, Jameson GNL, Hernández HL et al (2004) Characterization of the cofactor composition of Escherichia coli biotin synthase. Biochemistry 43:2007–2021
53. Berkovitch F, Nicolet Y, Wan JT et al (2004) Crystal structure of biotin synthase, an S-

adenosylmethionine-dependent radical enzyme. Science 303:76–79

54. Tse Sum Bui B, Benda R, Schünemann V et al (2003) Fate of the $[2Fe\text{-}2S]^{2+}$ cluster of Escherichia coli biotin synthase during reaction: a Mössbauer characterization. Biochemistry 42:8791–8798
55. Harmer JE, Hiscox MJ, Dinis PC et al (2014) Structures of lipoyl synthase reveal a compact active site for controlling sequential sulfur insertion reactions. Biochem J 464:123–133
56. Cicchillo RM, Lee K-H, Baleanu-Gogonea C et al (2004) Escherichia coli lipoyl synthase binds two distinct [4Fe–4S] clusters per polypeptide. Biochemistry 43:11770–11781
57. Hernández HL, Pierrel F, Elleingand E et al (2007) MiaB, a bifunctional radical-S-adenosylmethionine enzyme involved in the thiolation and methylation of tRNA, contains two essential [4Fe–4S] clusters. Biochemistry 46:5140–5147
58. Mclaughlin MI, Lanz ND, Goldman PJ et al (2016) Crystallographic snapshots of sulfur insertion by lipoyl synthase. Proc Natl Acad Sci U S A 113:9446–9450
59. Lanz ND, Lee K-H, Horstmann AK et al (2016) Characterization of lipoyl synthase from Mycobacterium tuberculosis. Biochemistry 55:1372–1383
60. Lanz ND, Pandelia ME, Kakar ES et al (2014) Evidence for a catalytically and kinetically competent enzyme-substrate cross-linked intermediate in catalysis by lipoyl synthase. Biochemistry 53:4557–4572
61. Zhang B, Arcinas AJ, Radle MI et al (2020) First step in catalysis of the radical S-Adenosylmethionine Methylthiotransferase MiaB yields an intermediate with a [3Fe-4S](0)-like auxiliary cluster. J Am Chem Soc 142:1911–1924
62. Fugate CJ, Stich TA, Kim EG et al (2012) 9-Mercaptodethiobiotin is generated as a ligand to the [2Fe-2S]+ cluster during the reaction catalyzed by biotin synthase from Escherichia coli. J Am Chem Soc 134:9042–9045
63. Mccarthy EL, Rankin AN, Dill ZR et al (2019) The A-type domain in Escherichia coli NfuA is required for regenerating the auxiliary [4Fe-4S] cluster in Escherichia coli lipoyl synthase. J Biol Chem 294:1609–1617
64. Mccarthy EL, Booker SJ (2017) Destruction and reformation of an iron-sulfur cluster during catalysis by lipoyl synthase. Science 358:373–377
65. Angelini S, Gerez C, Ollagnier-De Choudens S et al (2008) NfuA, a new factor required for maturing Fe/S proteins in Escherichia coli under oxidative stress and iron starvation conditions. J Biol Chem 283:14084–14091
66. Bandyopadhyay S, Naik SG, O'carroll IP et al (2008) A proposed role for the *Azotobacter vinelandii* NfuA protein as an intermediate iron-sulfur cluster carrier. J Biol Chem 283:14092–14099
67. Jin Z, Heinnickel M, Krebs C et al (2008) Biogenesis of iron-sulfur clusters in photosystem I: holo-NfuA from the cyanobacterium Synechococcus sp. PCC 7002 rapidly and efficiently transfers [4Fe-4S] clusters to apo-PsaC in vitro. J Biol Chem 283:28426–28435
68. Bradford MM (1976) A rapid and sensitive method for the quantitation of microgram quantities of protein utilizing the principle of protein-dye binding. Anal Biochem 72:248–254
69. Pandelia ME, Lanz ND, Booker SJ et al (2015) Mössbauer spectroscopy of Fe/S proteins. Biochim Biophys Acta 1853:1395–1405
70. Mccarthy EL, Booker SJ (2018) Biochemical approaches for understanding iron-sulfur cluster regeneration in Escherichia coli lipoyl synthase during catalysis. Methods Enzymol 606:217–239
71. Beinert H (1978) Micro methods for the quantitative determination of iron and copper in biological material. Methods Enzymol 54:435–445

Chapter 17

Methods to Screen for Radical SAM Enzyme Crystallization Conditions

Lydie Martin, Xavier Vernède, and Yvain Nicolet

Abstract

Radical *S*-adenosyl-L-methionine proteins most probably belong to the widest superfamily of metalloenzymes. Thanks to their ability to catalyze difficult reactions, combined with their involvement in the biosynthesis of numbers of natural products, they sound promising for various biotechnological applications. Their structural study is often limited because they are usually challenging to crystallize. This chapter presents protocols and equipment developed to quickly screen for crystallization conditions under anaerobic conditions, as exemplified by our recent study of the nitrogenase maturase NifB.

Key words Metalloprotein, Protein expression and purification, Fe-S cluster reconstitution, Anaerobic crystallization, Crystal flash cooling. FeMo coassembly, FeMo cofactor

1 Introduction

Radical *S*-adenosyl-L-methionine (SAM) proteins are enzymes, which use the reduction of a [4Fe-4S] cluster to cleave SAM into methionine and a highly reactive 5′-deoxyadenosyl radical species (5′-dA·) [1, 2]. Then, the latter activates a specific substrate, usually through hydrogen-atom abstraction. Starting with a handful of identified enzymes at the end of the 90s, they are over 113,000 unique sequences today, catalyzing more than 80 different reactions [3, 4]. These enzymes are found at key steps in the synthesis of vitamins and cofactors, but they are also involved in the modification of proteins, peptides, and nucleic acids. The catalytic core domain corresponds to a $(\beta\alpha)_6$ motif often described as a three-fourth TIM barrel-like fold [5, 6]. This core domain is usually surrounded by additional domains at its N- and C-terminus ends, conferring further functional specificity to the protein. Thanks to the use of radical-based chemistry, the catalyzed reactions often have no direct counterpart in organic synthesis, further enhancing the general interest in the study of these enzymes. Furthermore, over the last years, many radical SAM enzymes were identified as

Patricia C. Dos Santos (ed.), *Fe-S Proteins: Methods and Protocols*, Methods in Molecular Biology, vol. 2353,
https://doi.org/10.1007/978-1-0716-1605-5_17,

being involved in the posttranslational modification of ribosomally synthesized peptides (RiPPs), leading to the production of natural products with antibiotic or anticancer properties [7]. Thus, harnessing these enzymes to use them either in synthetic biology or in vitro as cocktails of catalysts to screen on peptide libraries seems on hand when seeking new drugs. However, despite a blooming of characterizations of new radical SAM enzymes over the last years, only a few of them were studied in enough detail to understand how they recognize their substrate or how the radical-based reaction is controlled by the surrounding protein matrix [2]. Addressing these questions usually requires the combination of different techniques, including in vitro functional assays, spectroscopic studies such as EPR or electrochemistry, structural characterizations, and, whenever possible, theoretical calculations. Getting all these data at hands needs first to produce enough pure and a homogenous protein sample.

Radical SAM enzymes contain at least one [4Fe-4S] cluster. However, they frequently display additional or auxiliary Fe-S clusters essential to their function. Due to the intrinsic sensitivity of these clusters to oxygen-induced damages, almost all the sample preparations must be carried out in a confined anaerobic environment. Furthermore, the correct incorporation of these Fe-S clusters into the target protein requires the well-orchestrated action of the dedicated house-keeping *i*ron-*s*ulfur *c*luster (ISC) assembly machinery [8]. However, the host ISC machinery may not be able to provide enough Fe-S clusters when overproducing proteins like radical SAM enzymes. Yet, a well-defined cluster content is crucial for X-ray crystallography analyses because crystallizing a protein usually requires a high conformational homogeneity. Sub-optimal Fe-S cluster content may lead to protein instability and higher flexibility that would, in turn, impair crystallization.

Another possible issue when seeking crystals comes from the protein's intrinsic propensity to establish intermolecular contacts that would lead to crystal-packing interactions. The latter are usually driven by surface residues that might be difficult to mutate a priori to favor crystallization. The use of orthologs from different organisms can be an attractive alternative to modify surface residues without any prior knowledge. This chapter covers some of the methods we developed to overcome these main issues to perform structural characterization of the radical SAM enzyme NifB involved in the assembly of the nitrogenase active site.

2 Materials

1. pRSF-ISC-MetK plasmid: Expression plasmid containing the *isc* operon and *metK* gene. The *Escherichia coli* K12 ISC machinery operon without *iscR* (genes *b2530–b2524*) is

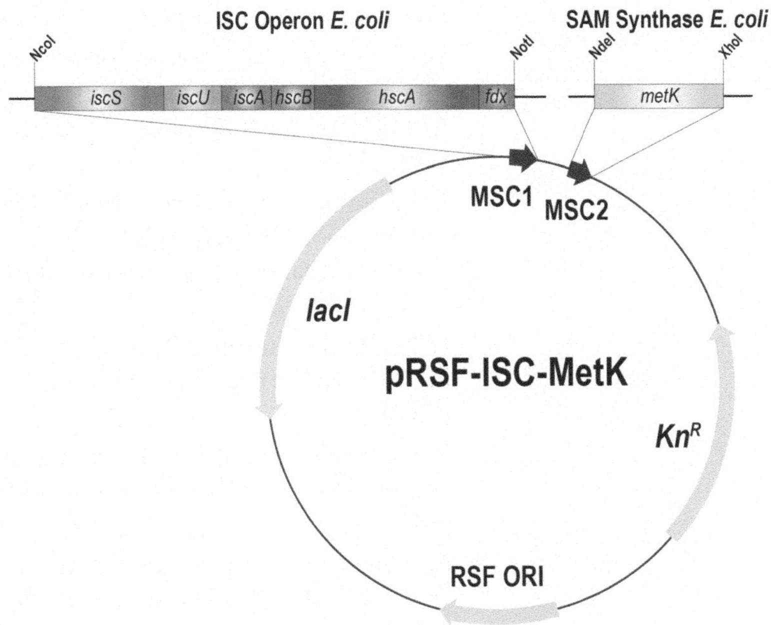

Fig. 1 Diagram of the pRSF-ISC-MetK plasmid construct. The ISC operon and metK genes are highlighted

inserted into the MCS1 region for the pRSFDuet® plasmid using *NcoI* and *NotI* restriction sites. The gene coding for *E. coli* K21 *S*-adenosyl-L-methionine synthase MetK (*b2942*) is inserted into the MCS2 region using *NdeI* and *XhoI* restriction sites (Fig. 1) [9].

2. Chemically competent expression strain: *E. coli* BL21 (DE3) strain containing the pRSF-ISC-MetK plasmid is used as a standard expression strain.
3. Terrific Broth (TB) medium: 47.6 g TB powder (Sigma-Aldrich T0918); 8 mL glycerol 100%; and deionized H_2O up to 1 L. The medium is autoclaved.
4. Fumarate stock solution: 1 M sodium fumarate dibasic in deionized H_2O. The solution is filter sterilized with a 0.22 μM filter.
5. MOPS buffer: 1 M 3-morpholino-1-propanesulfonic acid (MOPS), pH 7.4. The buffer is prepared in deionized H_2O and pH adjusted to 7.4 using drop-wise titration with 50% NaOH. The final volume is adjusted with deionized H_2O after pH calibration. The solution is filter sterilized with a 0.22 μM filter.
6. Cysteine stock solution: 0.5 M L-cysteine in deionized H_2O. The solution is filter sterilized with a 0.22 μM filter.

7. Methionine stock solution: 0.2 M L-methionine in deionized H_2O. The solution is filter sterilized with a 0.22 μM filter.
8. Vitamin stock solution: 100× MEM vitamin (SIGMA-ALDRICH M6895). Aliquoted in 10 mL and stored at −20 °C.
9. Ammonium iron(III) citrate stock solution: 1 M $C_6H_8O_7 \cdot xFe^{3+} \cdot yNH_3$ in deionized H_2O. The 1 M concentration corresponds to that of citrate. The solution is filter sterilized with a 0.22 μM filter.
10. IPTG stock solution: 1 M isopropyl-β-D-thiogalactopyranoside in deionized H_2O. The solution is filter sterilized with a 0.22 μM filter.
11. Kanamycin stock solution: 50 mg/mL kanamycin sulfate in deionized H_2O. The solution is filter sterilized with a 0.22 μM filter.
12. Ampicillin stock solution: 100 mg/mL ampicillin sodium salt in deionized H_2O. The solution is filter sterilized with a 0.22 μM filter.
13. Affinity chromatography column: 5 mL Ni-loaded HisTrap. The column can be plugged into a FPLC housed in a glove box.
14. Desalting column: HiPrep™ 26/10 desalting column. The column can be plugged into a FPLC housed in a glove box.
15. Buffer A: 100 mM 2-amino-2-(hydroxymethyl)propane-1 3-diol (Tris), pH 8 with 500 mM KCl and 5 mM β-mercaptoethanol. The solution is filtered and degassed under vacuum.
16. Buffer B: 100 mM Tris, pH 8 with 500 mM KCl, 5 mM β-mercaptoethanol, and 1 M imidazole. The solution is filtered and degassed under vacuum.
17. Ferric ion stock solution: 100 mM $FeCl_3 \cdot 6H_2O$ in deionized H_2O. The solution is filter sterilized with a 0.22 μM filter.
18. Sulfide stock solution: 100 mM $Na_2S \cdot 9H_2O$ in deionized H_2O. The solution is filter sterilized with a 0.22 μM filter.
19. DTT stock solution: 1 M DL-dithiothreitol in deionized H_2O. The solution is filter sterilized with a 0.22 μM filter.
20. Benzonase nuclease: Benzonase nuclease commercial solution. Store at −20 °C.
21. EDTA-free protease inhibitor cocktail: EDTA-free protease inhibitor cocktail tablets from Roche™. Store at −20 °C.
22. Ultrasonic liquid processor: A sonicator implemented in a glove box to break cell walls under anaerobic conditions. Using ice to keep the sample at 4 °C is not possible because

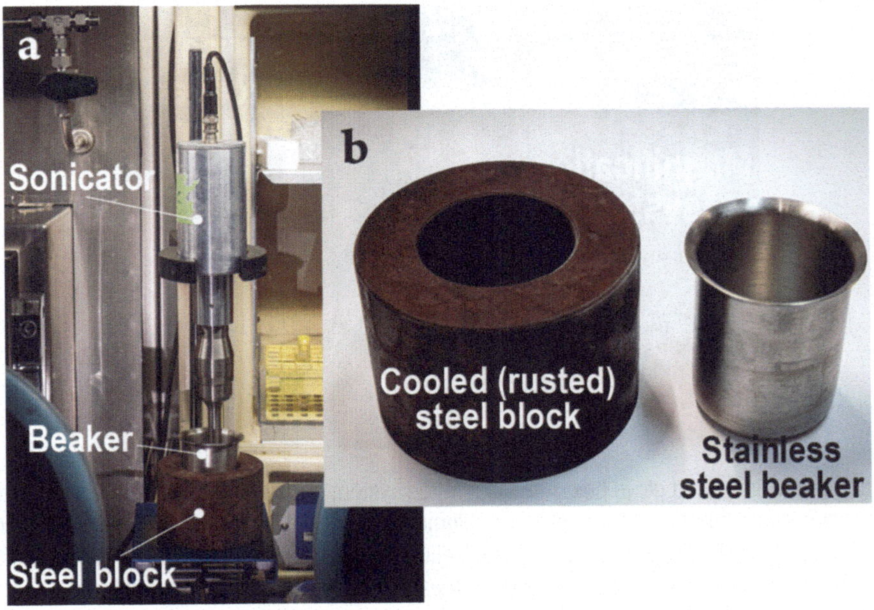

Fig. 2 Anaerobic cell-disruption setup. (**a**) A view of the Qsonica CL334 sonicator implemented in a glove box. (**b**) A homemade block of steel (left) to house a stainless-steel beaker (right)

large amounts of oxygen are trapped in it. Instead, a large block of steel was designed to house a 125 mL stainless-steel beaker during the sonication cycles. This steel block is thick enough to keep the bacterial suspension cold over the process (Fig. 2).

23. Crystallization-dedicated glove box: An anaerobic chamber containing a magnification system is required to observe, handle, and harvest crystals under anaerobic conditions. Different systems have been developed by various groups over the world. Our system corresponds to a stainless-steel glove box from JACOMEX with a Leica microscope implemented into one of the polycarbonate lateral panels (Fig. 3). A full description of this setup was reported previously in Ref. 10.

24. Anaerobic high-throughput automated crystallization setup: A stainless-steel glove box (JACOMEX) was designed to house a Crystal Gryphon robot (Art Robbins Instruments) to automatically screen for crystallization conditions under anaerobic conditions (Fig. 4a). The humidity is maintained at more than 50% thanks to a balloon heater powered by a hygrostat in order to limit the evaporation of the crystallization drops. A homemade system has been implemented to easily bring deoxygenated water in the glove box to wash the 96 syringes and the nanodispenser (Fig. 4b, c). An outside 2 L container filled with water is flushed with argon gas for 15 min prior to opening a valve that allows direct entry of the degassed water in the glove box without using any airlock (a direct evacuation is also

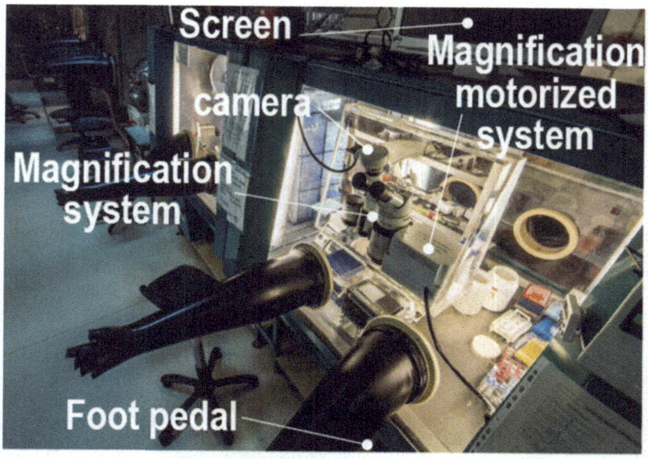

Fig. 3 Crystallization-dedicated glove box. A magnification system is integrated on the polycarbonate panel of the crystallization glove box [10]. A camera is implemented to either look at the crystallization drop on a monitor screen or to record pictures of the crystallization condition. The magnification adjustment wheel is motorized to be controlled through two foot pedals to ease the crystallization drop analysis

implemented). Our automated crystallization setup also includes a robotic arm to automatically visualize the crystallization plates at defined time lapses to monitor the crystallization process. It corresponds to an adapted six-axis G-rob robot (TX40 from Stäubli) surrounded by two shelves to host the crystallization plates (Fig. 4d). A camera with motorized magnification (same as those at synchrotron beam line) completes the setup. The whole system can automatically record pictures of the different drops and be manually controlled to further investigate a specific crystallization condition.

25. Crystallization kits: Thirteen commercially available crystallization screens stored in 96 deep-well plates are routinely used in our high-throughput crystallization setup to screen for crystallization conditions. The different screens are automatically degassed inside our glove box using the Gryphon crystallization robot. A specific program allows the 96 syringes to suck the 96 solutions and reject them into a second deep-well plate while bubbling it with the glove box atmosphere.

26. Propane: A 99.95% pure (N35) propane B5 gas cylinder from Air-Liquide equipped with a gas-reduction valve with manometer to provide a small flux of propane.

27. Crystal sample holder: 18 mm Mounted Cryoloop™—20 µm. Different diameters adapted to the crystal size from molecular dimension (Fig. 5).

28. Aluminum container: This corresponds to a home-designed 5.6-mL container made of aluminum to carry liquid and solid propane (Fig. 6). The top of the container is designed to fit a

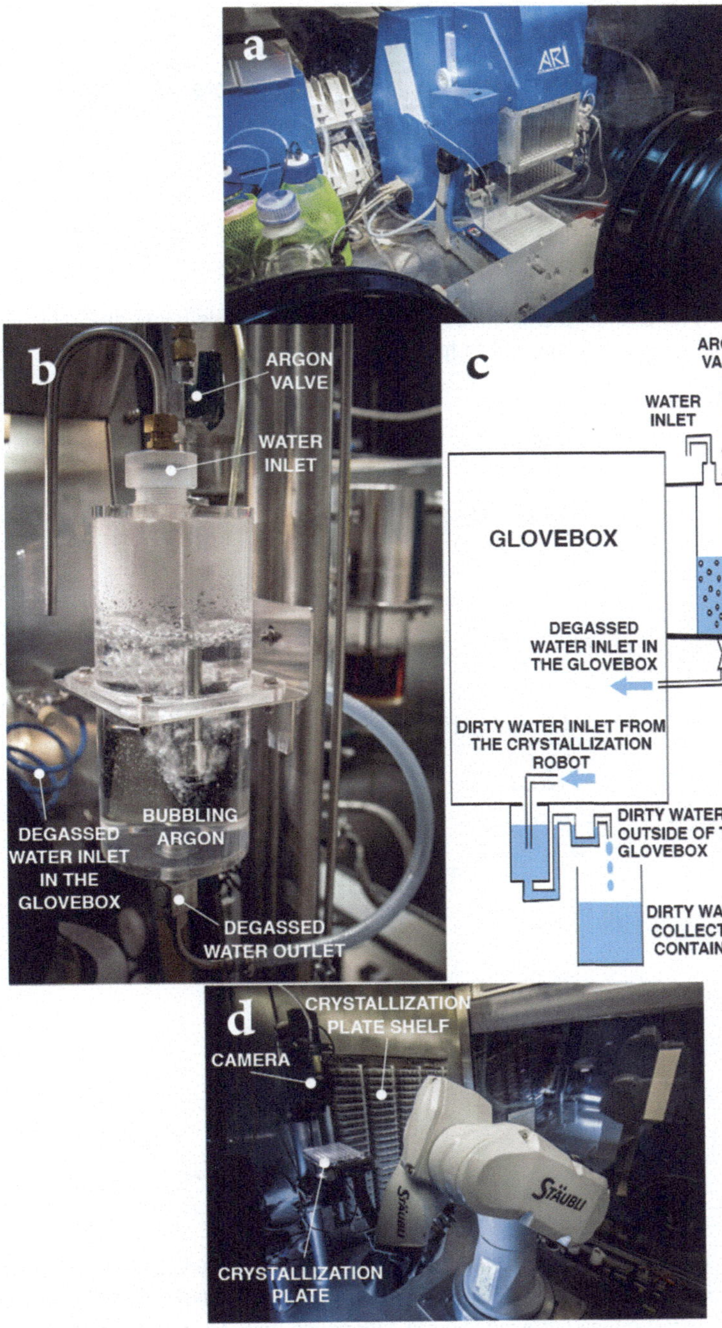

Fig. 4 Automated crystallization setup under anaerobic conditions. (**a**) A Gryphon crystallization robot in the glove box. (**b**) A homemade system to bring degassed water in the glove box. (**c**) A scheme of the homemade system to bring degassed water in the glove box. (**d**) A six-axis G-rob robot to automatically visualize the crystallization plates under anaerobic conditions

Fig. 5 Small aluminum container to freeze crystals under anaerobic conditions. The upper hole is set to house a crystal sample holder

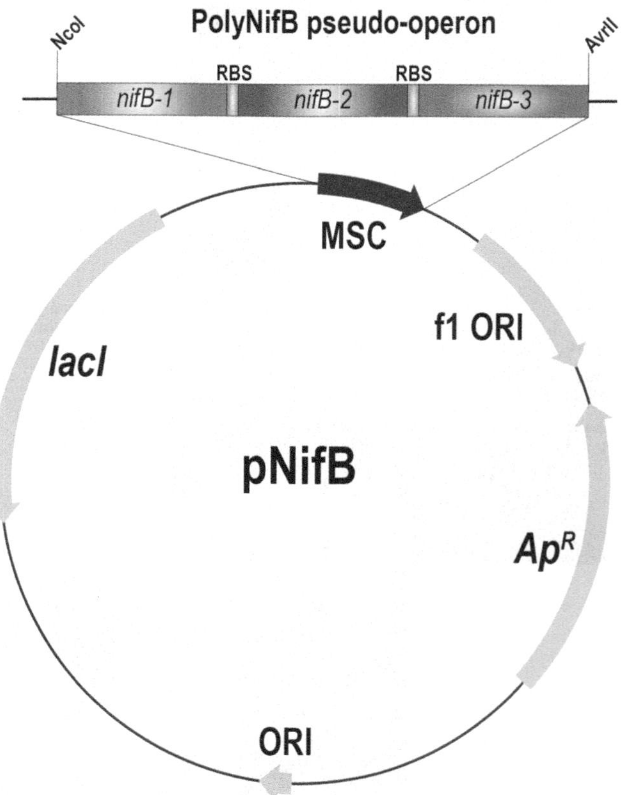

Fig. 6 Diagram of plasmid constructs used for coexpression of Radical SAM NifB enzymes. The poly-NifB pseudooperon is highlighted with *NifB-1*, *NifB-2*, and *NifB-3* genes corresponding to *NifB* genes from *Methanothrix thermoacetophila* (ABK14214.1), *Methanolobus vulcani* (WP_091709887.1), and *Methanothermobacter tenebrarum* (WP_112093130.1), respectively. RBS stands for *r*ibosome *b*inding *s*ite

standard crystal sample holder. The use of a magnetic wand to transfer crystal sample holders from the container to standard magnetic vials is not impeded due to the absence of magnetic properties for aluminum. A small pin inserted at the top makes the handling of the container easier when using flat tweezers or a plier to transfer it from and to liquid nitrogen.

3 Methods

3.1 Design of the Constructs

The feedback from the different structural genomics projects developed in the 2000s highlighted that many proteins can crystallize despite the fact that they are not purified to complete homogeneity. This has been further demonstrated by the recent crystallization of metalloproteins obtained from the fractionation of cell extracts [11, 12]. Thus, in order to increase the ability to quickly screen for different orthologs of the radical SAM protein NifB, specific constructs were designed to coexpress different NifB orthologs [13]. The method described below is applied to the structural studies of NifB. However, such an approach can be applied to any new structural project when seeking a crystal structure.

1. Select NifB ortholog sequences from a BLAST request on the NCBI server using the well-studied NifB from *Methanosarcina acetivorans* as a query sequence [14] (*see* **Note 1**).
2. Among the proposed sequences, select those with above 80% identities and that are from thermophilic organisms and pool in a subset. Finally, perform a careful look for the presence of the previously described key residues, notably those previously proposed to bind the three [4Fe-4S] clusters, to avoid selecting false-positive sequences.
3. Select six NifB orthologs and their corresponding DNA sequences. Using the *E. coli* codon usage, generate codon-optimized sequences. This can be done using various websites (free) and is usually proposed when ordering DNA sequences. Add polyhistidine tags to the 5′ end of the coding DNA sequences (*see* **Note 2**).
4. Divide the sequences into two groups of three. Build pseudooperons by adding a ribosome binding site (RBS) sequences between each His-tagged NifB (Fig. 6).
5. Ligate the resulting DNA fragments into a pETDuet® vector between the NcoI and AvrII restriction sites, to generate the pNifB vector series (Fig. 6). Please note that using such restriction sites leads to a unique multisite cloning cassette.

3.2 NifB Ortholog Coexpression

1. Transform each pNifB plasmid into an *E. coli* BL21 (DE3) pRSF-ISC-MetK strain.
2. Incubate transformed cells overnight in 100 mL of TB medium supplemented with 100 μL of ampicillin stock solution and 100 μL of kanamycin stock solution.
3. Inoculate the 100 mL of starter culture into 4 L of TB medium supplemented with 1 mL of ampicillin stock solution and 1 mL of kanamycin stock solution and incubate aerobically at 37 °C, while shaking at 200 rpm up to an OD_{600} of about 0.6 to 0.8.

4. Supplement the culture with 10 mL of MEM Vitamin stock solution and 50 mL of MOPS buffer pH 7.4 stock solution per L of culture. The vitamin solution provides pyridoxal phosphate to the cysteine desulfurase to help inserting the Fe-S clusters. The MOPS buffer prevents medium acidification upon anaerobic cell growth overnight and thus avoids a premature culture death.
5. Transfer the culture in 2 L sealed glass bottles into a glove box with an N_2 atmosphere (less than 5 ppm O_2) under constant stirring at 18 °C using a magnetic bar.
6. Open the bottles after 30 min to equilibrate the medium with the atmosphere of the glove box and to remove the remaining oxygen traces while continuing the stirring. The 30 min timelapse allows the cells to consume the remaining oxygen from the medium before switching to anaerobic metabolism.
7. Supplement the medium with 25 mL of fumarate stock solution, 750 μL of L-methionine stock solution, and 500 μL of L-cysteine stock solution, and incubate for 10 min. The addition of fumarate should favor anaerobic respiration toward fermentation.
8. Enrich the culture with 250 μL ammonium iron(III) citrate stock solution.
9. Induce protein expression when the OD_{600} reaches 1 by adding 1 mL of IPTG stock solution, still under constant agitation at 18 °C.
10. Incubate the culture overnight at 18 °C and under mild agitation, still using a magnetic bar, before transferring it into gas-tight 800 mL centrifuge bottles prior to harvesting by centrifugation at 7900 × *g* for 20 min.
11. After centrifugation, reenter bottles into the glove box to harvest the cell paste. The resulting cell paste can be frozen and stored in liquid nitrogen until further purification.

3.3 NifB Protein Copurification

In order to minimize the number of steps, the purification is limited to an affinity chromatography (here Ni column), followed by a chemical [4Fe-4S] cluster reconstitution and a desalting column. All the following steps are performed under anaerobic conditions. All buffers are prepared and filtered 1 day in advance. They are degassed through vacuum under vigorous stirring before being introduced into the glove box. A second vacuum step is performed inside the anaerobic chamber, and the remaining traces of oxygen are finally removed by a 15 min glove box atmosphere bubbling.

1. Thaw the cell paste in a glove box and resuspend in buffer A (4 mL/g of cell paste) supplemented with 2 μL of benzonase nuclease and two tablets of EDTA-free protease inhibitor cocktail.

2. Disrupt cells by sonication (2-s pulses separated by 10 s, at 90% power intensity, for a total pulse time of 2 min). In order to keep the sample cold during the sonication cycle and to avoid contamination of the glove box atmosphere by oxygen trapped in ice, use a cooled steel block (*see* **Note 3**).
3. Place the lysate into 50 mL of gas-tight centrifuge tubes prior to centrifugation at 24,600 × *g* for 45 min (*see* **Note 4**).
4. Reenter the centrifuge tubes to recover the clarified lysate, subsequently filter the lysate using 0.45 μm membrane.
5. Inject the clarified lysate to a 5 mL HisTrap™ column. The use of such a column requires a purification system implemented in the glove box. Noteworthy, this step can be replaced using Ni-NTA resin in homemade gravity columns or directly in batch. However, using a purification system allows to extensively wash the column by monitoring the absorbance at 280 nm. Perform additional washing steps using buffer A supplemented with 10–50 mM imidazole to eliminate contaminant proteins that are nonspecifically bound to the resin.
6. Once the absorbance at 280 reaches a plateau around 0, elute the bound proteins applying a 200 mL linear gradient from 50 to 500 mM imidazole using buffer B.
7. Perform a buffer exchange of the as-purified mixture of NifB proteins using a 50 mL HiPrep 26/10 desalting column (GE-Healthcare™) equilibrated with buffer A. This step removes the imidazole that may impair Fe-S cluster reconstitution during the next step. The collected fractions are pooled and diluted to 1 mg/mL protein.
8. In order to proceed to Fe-S clusters chemical reconstitution, add DTT (5 mM final concentration) under mild stirring. Thirty minuts later, add 14 equivalents of $FeCl_3$. Incubate for 30 more minutes. Slowly and drop-wise, add 14 equivalents of Na_2S (*see* **Notes 5** and **6**). Incubate the mixture under mild stirring overnight. Remove the precipitated protein and iron sulfides by filtration using a 0.22 μm cutoff membrane plugged onto a syringe.
9. Concentrate the NifB protein mixture by ultrafiltration (10 kDa MWCO) to about 30 mg/mL. The solution can be stored in liquid nitrogen (*see* **Note 7**).

3.4 Automated Crystallization Condition Screening

Because each protein sample actually corresponds to a mixture of up to three proteins, depending on the relative expression level and solubility of each selected NifB, crystallization trials are performed using various protein concentrations ranging from 15 to 30 mg/mL.

1. Dilute the different NifB protein mixtures to 15, 20, and 30 mg/mL using Buffer A to obtain about 600 μL samples per protein concentration (*see* **Note 8**).
2. Filter each by centrifugation (Stirred Cells Amicon Merck Millipore) to remove precipitates and large aggregates (*see* **Note 9**).
3. Run the crystallization robot. In order to screen for crystals, each sample is automatically mixed and incubated with 1248 commercial crystallization conditions using a Gryphon robot (Art Robbins Instruments™) set up in a glove box. Crystallization drops correspond to the mixture of 200–400 nL protein and 200–400 nL precipitant solutions. Seal each 96-well crystallization plate with an adhesive film (Viewseal sealer clear) to limit evaporation and drying prior to storage by an automated robot arm (NatX-ray™) in a dedicated space in the glove box. At given time lapses, the automated arm takes pictures of each crystallization drop to record its evolution over time.
4. Periodically, manually visualize the pictures to seek for putative hits (Fig. 7).

3.5 Manual Crystallization Optimization

Depending on whether the automatically obtained crystals are suitable for X-ray diffraction, one may need to further optimize the crystallization condition. These are standard crystallization techniques, and the most common strategy corresponding to screening around the previously identified condition, varying the protein and the precipitating agent concentration, the pH of the

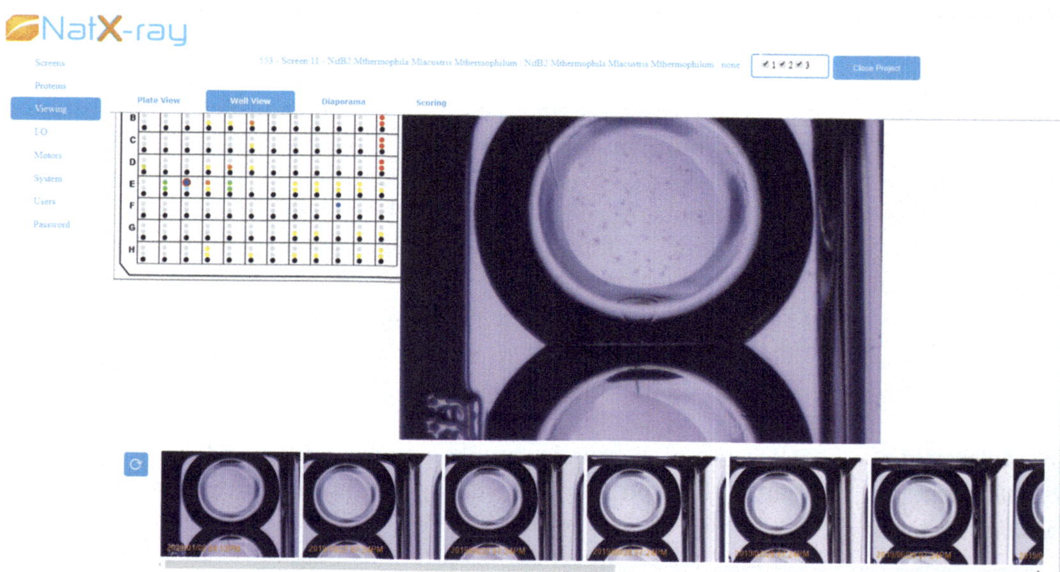

Fig. 7 Screenshot of the web interface controlling the automated crystallization plate visualization system. Small crystals of the NifB protein from *M. thermoacetophila* can be seen in the magnified drop

buffer, and adding additives to help crystal growth. Many other optimization approaches can be performed and are very well described in all macromolecular crystallography textbooks.

3.6 Crystal Harvesting Under Anaerobic Condition

Nowadays, most diffraction experiments are performed using high-energy synchrotron beam lines and require crystal samples to be frozen and stored at low temperature (typically 100 K) to avoid fast propagation of radiation-induced damages. Furthermore, Fe-S clusters can be destroyed upon oxygen exposure, and such damage may also alter the diffraction quality of the metalloprotein crystal. Thus, it is important to harvest and store crystals under anaerobic conditions. Freezing crystals under anaerobic conditions is a good way to prevent both oxygen and radiation-induced damages. It also prevents changes in the oxidation states or at least some conformational changes related to these changes.

The best strategy to freeze crystals under anaerobic conditions is to proceed directly in the glove box. However, bringing liquid nitrogen inside the anaerobic chamber through airlocks has some experimental limitations. Indeed, during the successive vacuum/refilling cycles, a significant amount of liquid nitrogen will evaporate. Therefore, a larger starting volume is required to have enough liquid nitrogen inside. Consequently, a larger airlock must be used instead of the more convenient smaller one, significantly increasing the process duration. Having such a significant amount of liquid nitrogen inside the glove box then presents potential issues. Evaporation of liquid nitrogen leads to a 700-fold volume increase leading to a potential safety risk. Indeed, any accidental spilling may result in a blow-up of the glove box, leading to polycarbonate panel break, glove tearing, or other severe damage. In order to avoid such risk, a second user may help, in case of such spilling, to quickly activate the vacuum pump while keeping open the airlock inner door, hence, reducing the pressure. In addition, nitrogen's liquefaction temperature is 77 K, whereas oxygen liquefies at 90 K. Thus, a significant amount of liquid oxygen accumulates over time in the liquid nitrogen-containing Dewar before being brought inside. This oxygen is then slowly released while the user is harvesting crystals, often leading to a significant increase in the oxygen level that may subsequently affect all the other crystallization experiments stored in the glove box. The use of liquid ethane or propane instead of liquid nitrogen is a good alternative to circumvent such problems (*see* **Note 10**). When considering propane (ethane is very similar), the gas/liquid transition occurs at 231 K, while the liquid/solid transition happens at 83 K. It is possible to liquefy and solidify a small volume of propane (about 3 mL) in a dedicated container and to transfer it using a small airlock. Propane melts in the glove box prior to crystal flash-cooling takes place.

1. Visually select the best crystals using a magnification system incorporated into the glove box. Prior to flash-cooling, they must be treated with a suitable cryoprotecting solution to avoid crystalline ice formation upon freezing. The composition of such a solution depends on the crystallization condition one would obtain.
2. Cool specific aluminum containers in a Dewar bowl filled with liquid nitrogen (*see* **Note 10**).
3. Slowly fill the containers with propane gas using a low flow to afford rapid liquefaction without letting propane gas escape. Propane readily liquefies on contact with the cooled container side (*see* **Note 10**).
4. When propane is completely solidified, introduce the container into the glove box through the airlock.
5. Wait 2 min to allow the solid propane to melt.
6. Soak the chosen crystal into the corresponding cryoprotecting solution and fish it out to quickly plunge it into liquid propane.
7. Take the container out of the glove box and transfer the crystal, frozen in a loop, into liquid nitrogen for storage. The next steps are identical to any standard crystallography technique to bring a sample to an in-house X-ray source or a synchrotron beam line.

4 Notes

1. When selecting orthologs, pick sequences not too distant to the best-characterized protein. When possible, select those from thermophilic organisms with the shortest variable loops. Finally, favor orthologs from organisms that more frequently led to crystal structures. A quick check to the calculated pI would also help avoid proteins with an isoelectronic point in the pH range of the commonly used purification buffers (i.e., 7–8.5).
2. When adding a polyhistidine tag, it is possible to choose the number of histidine residues to insert (usually 6). At this stage, it is also possible to replace the poly-His with another affinity tag depending on one's purpose. It is also possible to insert cleavage sites for subsequent removal of the affinity tag.
3. Store the steel block at −20 °C overnight to keep it cold enough over the sonication cycles.
4. When clarifying the lysate, ultracentrifugation can conveniently replace regular centrifugation.
5. When chemically reconstituting Fe-S clusters, use a protein concentration around 1 mg/mL. A higher concentration

usually leads to significant protein aggregation and precipitation in addition to lower Fe-S cluster incorporation yields. A more diluted sample is not necessary as it would subsequently require a longer time to concentrate without any significant gain in Fe-S incorporation yields. Alternatively, it would require more buffer exchange or size exclusion chromatography runs.

6. When reconstituting Fe-S clusters, sulfide ions should be added stepwise. Either add a low-concentration stock dropwise under mild agitation or use cysteine desulfurase NifS to release sulfide from L-cysteine.
7. Store NifB in liquid nitrogen or keep it at room temperature in the glove box prior to running crystallization trials to avoid aggregation upon freeze–thaw cycles.
8. Prior to crystallization, SAM and/or substrate can be mixed to the protein sample. Be careful as it is soluble enough to avoid clogging of the nanodispenser.
9. Centrifuge the protein before crystallization.
10. When preparing solid propane, use a Dewar bowl and fill it with liquid nitrogen to half the height of the aluminum containers to keep them cold enough to favor rapid solidification while avoiding spilling of liquid nitrogen into the container. Mixing liquid propane and liquid nitrogen usually impairs correct solidification.

Acknowledgments

The work presented in this chapter was supported by the French National Research Agency in the framework of the Investissements d'Avenir program (ANR-15-IDEX-02), through the funding of the "Origin of Life" project of the Univ. Grenoble-Alpes. Equipment was acquired through FRISBI (ANR-10-INBS-05-02). We also thank the DRF from the Commissariat à l'Energie Atomic for financial support to acquire and develop the automated robotic arm. We thank Juan C. Fontecilla-Camps, former head of our group, who has fostered the development of our anaerobic pipeline "from gene to structure".

References

1. Broderick JB, Duffus BR, Duschene KS, Shepard EM (2014) Radical *S*-adenosylmethionine enzymes. Chem Rev 114:4229–4317. https://doi.org/10.1021/cr4004709
2. Nicolet Y (2020) Structure-function relationships of radical SAM enzymes. Nat Catal 3:337–350. https://doi.org/10.1038/s41929-020-0448-7
3. Sofia HJ, Chen G, Hetzler BG et al (2001) Radical SAM, a novel protein superfamily linking unresolved steps in familiar biosynthetic pathways with radical mechanisms: functional

characterization using new analysis and information visualization methods. Nucleic Acids Res 29:1097–1106

4. Akiva E, Brown S, Almonacid DE et al (2014) The structure-function linkage database. Nucleic Acids Res 42:D521–D530. https://doi.org/10.1093/nar/gkt1130
5. Nicolet Y, Drennan CL (2004) AdoMet radical proteins--from structure to evolution--alignment of divergent protein sequences reveals strong secondary structure element conservation. Nucleic Acids Res 32:4015–4025. https://doi.org/10.1093/nar/gkh728
6. Vey JL, Drennan CL (2011) Structural insights into radical generation by the radical SAM superfamily. Chem Rev 111:2487–2506. https://doi.org/10.1021/cr9002616
7. Benjdia A, Balty C, Berteau O (2017) Radical SAM enzymes in the biosynthesis of Ribosomally synthesized and post-translationally modified peptides (RiPPs). Front Chem 5:87. https://doi.org/10.3389/fchem.2017.00087
8. Roche B, Aussel L, Ezraty B et al (2013) Iron/sulfur proteins biogenesis in prokaryotes: formation, regulation and diversity. Biochim Biophys Acta 1827:455–469. https://doi.org/10.1016/j.bbabio.2012.12.010
9. Nicolet Y, Zeppieri L, Amara P, Fontecilla-Camps JC (2014) Crystal structure of tryptophan lyase (NosL): evidence for radical formation at the amino group of tryptophan. Angew Chem Int Ed Engl 53:11840–11844. https://doi.org/10.1002/anie.201407320
10. Vernede X, Fontecilla-Camps JC (1999) A method to stabilize reduced and or gas-treated protein crystals by flash-cooling under a controlled atmosphere. J Appl Crystallogr 32:505–509. https://doi.org/10.1107/S0021889899002678
11. Wagner T, Koch J, Ermler U, Shima S (2017) Methanogenic heterodisulfide reductase (HdrABC-MvhAGD) uses two noncubane [4Fe-4S] clusters for reduction. Science 357:699–703. https://doi.org/10.1126/science.aan0425
12. Wagner T, Ermler U, Shima S (2016) The methanogenic CO_2 reducing-and-fixing enzyme is bifunctional and contains 46 [4Fe-4S] clusters. Science 354:114–117. https://doi.org/10.1126/science.aaf9284
13. Fajardo AS, Legrand P, Payá-Tormo LA et al (2020) Structural insights into the mechanism of the radical SAM carbide synthase NifB, a key Nitrogenase cofactor maturating enzyme. J Am Chem Soc 142(25):11006–11012. https://doi.org/10.1021/jacs.0c02243
14. Fay AW, Wiig JA, Lee CC, Hu Y (2015) Identification and characterization of functional homologs of nitrogenase cofactor biosynthesis protein NifB from methanogens. Proc Natl Acad Sci U S A 112:14829–14833. https://doi.org/10.1073/pnas.1510409112

INDEX

Patricia C. Dos Santos (ed.), *Fe-S Proteins: Methods and Protocols*, Methods in Molecular Biology, vol. 2353, https://doi.org/10.1007/978-1-0716-1605-5, © Springer Science+Business Media, LLC, part of Springer Nature 2021

The manufacturer's authorised representative in the EU is Springer Nature Customer Service Centre GmbH, Europaplatz 3, 69115 Heidelberg, Germany. If you have any concerns regarding our products, please contact ProductSafety@springernature.com

Printed and bound by CPI Group (UK) Ltd, Croydon, CR0 4YY
23/09/2026
02221542-0020